Recent Titles in This Series

116 **G. M. Fel'dman,** Arithmetic of probability distributions, and characterization problems on abelian groups, 1993

115 **Nikolai V. Ivanov,** Subgroups of Teichmüller modular groups, 1992

114 **Seizô Itô,** Diffusion equations, 1992

113 **Michail Zhitomirskii,** Typical singularities of differential 1-forms and Pfaffian equations, 1992

112 **S. A. Lomov,** Introduction to the general theory of singular perturbations, 1992

111 **Simon Gindikin,** Tube domains and the Cauchy problem, 1992

110 **B. V. Shabat,** Introduction to complex analysis Part II. Functions of several variables, 1992

109 **Isao Miyadera,** Nonlinear semigroups, 1992

108 **Takeo Yokonuma,** Tensor spaces and exterior algebra, 1992

107 **B. M. Makarov, M. G. Goluzina, A. A. Lodkin, and A. N. Podkorytov,** Selected problems in real analysis, 1992

106 **G.-C. Wen,** Conformal mappings and boundary value problems, 1992

105 **D. R. Yafaev,** Mathematical scattering theory: General theory, 1992

104 **R. L. Dobrushin, R. Kotecký, and S. Shlosman,** Wulff construction: A global shape from local interaction, 1992

103 **A. K. Tsikh,** Multidimensional residues and their applications, 1992

102 **A. M. Il'in,** Matching of asymptotic expansions of solutions of boundary value problems, 1992

101 **Zhang Zhi-fen, Ding Tong-ren, Huang Wen-zao, and Dong Zhen-xi,** Qualitative theory of differential equations, 1992

100 **V. L. Popov,** Groups, generators, syzygies, and orbits in invariant theory, 1992

99 **Norio Shimakura,** Partial differential operators of elliptic type, 1992

98 **V. A. Vassiliev,** Complements of discriminants of smooth maps: Topology and applications, 1992

97 **Itiro Tamura,** Topology of foliations: An introduction, 1992

96 **A. I. Markushevich,** Introduction to the classical theory of Abelian functions, 1992

95 **Guangchang Dong,** Nonlinear partial differential equations of second order, 1991

94 **Yu. S. Il'yashenko,** Finiteness theorems for limit cycles, 1991

93 **A. T. Fomenko and A. A. Tuzhilin,** Elements of the geometry and topology of minimal surfaces in three-dimensional space, 1991

92 **E. M. Nikishin and V. N. Sorokin,** Rational approximations and orthogonality, 1991

91 **Mamoru Mimura and Hirosi Toda,** Topology of Lie groups, I and II, 1991

90 **S. L. Sobolev,** Some applications of functional analysis in mathematical physics, third edition, 1991

89 **Valerii V. Kozlov and Dmitrii V. Treshchëv,** Billiards: A genetic introduction to the dynamics of systems with impacts, 1991

88 **A. G. Khovanskii,** Fewnomials, 1991

87 **Aleksandr Robertovich Kemer,** Ideals of identities of associative algebras, 1991

86 **V. M. Kadets and M. I. Kadets,** Rearrangements of series in Banach spaces, 1991

85 **Mikio Ise and Masaru Takeuchi,** Lie groups I, II, 1991

84 **Đào Trọng Thi and A. T. Fomenko,** Minimal surfaces, stratified multivarifolds, and the Plateau problem, 1991

83 **N. I. Portenko,** Generalized diffusion processes, 1990

82 **Yasutaka Sibuya,** Linear differential equations in the complex domain: Problems of analytic continuation, 1990

(Continued in the back of this publication)

Translations of
MATHEMATICAL MONOGRAPHS

Volume 116

Arithmetic of Probability Distributions, and Characterization Problems on Abelian Groups

G. M. Fel'dman

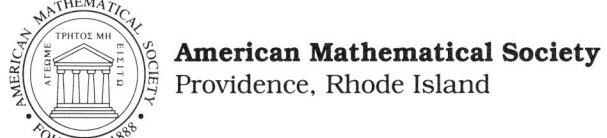

American Mathematical Society
Providence, Rhode Island

Г. М. Фельдман

АРИФМЕТИКА ВЕРОЯТНОСТНЫХ РАСПРЕДЕЛЕНИЙ И ХАРАКТЕРИЗАЦИОННЫЕ ЗАДАЧИ НА АБЕЛЕВЫХ ГРУППАХ

Translated from the Russian by Yu. Lyubarskiĭ
Translation edited by Simeon Ivanov

1991 *Mathematics Subject Classification.* Primary 60B15, 60E05; Secondary 22B05, 43A25.

ABSTRACT. The main problem of the arithmetic of probability distributions is the study of all possible representations of a given random variable as the sum of independent random variables. The book contains material concerning the applicability of that theory to random variables with values in a locally compact abelian group. The results on the arithmetic of probability distributions are then used to solve characteristic problems of mathematical statistics on abelian groups.

This book is intended for mathematician-specialists, graduate and advanced undergraduate students interested in probability, mathematical statistics and functional analysis.

Bibliography: 81 titles.

Library of Congress Cataloging-in-Publication Data

Fel′dman, G. M. (Gennadĭ Mikaĭlovich)
 [Arifmetika veroĭatnostnykh raspredelenĭi i kharakterizat͡sionnye zadachi na abelevykh gruppakh. English]
 Arithmetic of probability distributions, and characterization problems on Abelian groups/G. M. Feldman; translated from the Russian by Yu. Lyiubarskii; translated edited by Simeon Ivanov.
 p. cm.—(Translations of mathematical monographs; v. 116)
 Includes bibliographical references and index.
 ISBN 0-8218-4593-4
 1. Abelian groups. 2. Distribution (Probability theory). I. Ivanov, Simeon. II. Title. III. Series.
QA180.F4513 1993   92-45025
512′.2—dc20   CIP

Information on Copying and Reprinting can be found at the back of this volume.

This publication was typeset using $\mathcal{A}_{\mathcal{M}}\mathcal{S}$-TEX,
the American Mathematical Society's TEX macro system.

10 9 8 7 6 5 4 3 2 1 96 95 94 93 92 93

Contents

Introduction   1

Chapter I. Auxiliary Results   5
    §1. Results on duality theory and on the structure of locally
        compact abelian groups   5
    §2. Results on probability theory   10
    §3. Results on function theory and on analytic properties of
        characteristic functions   18

Chapter II. Arithmetic of Distributions   23
    §4. Group analogs of the Khinchin factorization theorems   23
    §5. Gaussian distribution   33
    §6. Decomposition of a generalized Poisson distribution   57
    §7. Group analogs of Linnik's theorems   86
    §8. General theorems on distributions of class I_0 110

Chapter III. Characterization Problems   121
    §9. Bernstein's characterization of Gaussian distribution   121
    §10. Characterization of Gaussian distribution by independence of
        linear statistics   135
    §11. Characterization of Gaussian distribution by identical
        distribution of a monomial and a linear form   151

Appendix 1. Group Analogs of the Marcinkiewicz Theorem and the
        Lukacs Theorem   173

Appendix 2. On Decomposition Stability of Distributions   181

Appendix 3. Structure of Infinitely Divisible Poisson Distributions   189

Appendix 4. On Distributions with Mutually Singular Powers   195

Comments   203

References   211

Notation   217

Subject Index   221

Author Index   223

Introduction

Probability theory on algebraic structures has been intensively developed lately (see the monographs by Grenander [Gr], Parthasarathy [P], Heyer [He2], Berg and Forst [BeF], Ruzsa and Szekely [RS]. The reasons for it are the aspiration to reach the natural possible bounds of generalization of classical results and also to solve some problems of physics, communication theory, statistics, that lead to consideration of probability distributions on algebraical structures.

The monograph is devoted to the arithmetic of probability distributions and to some characterization problems of mathematical statistics on a locally compact abelian group X.

Let a random variable ξ with values in the group X be represented as a sum of independent random variables

$$\xi = \xi_1 + \xi_2.$$

The arithmetic of probability distributions studies the following problem: given the distribution μ of the random variable ξ, describe as completely as possible the distributions μ_i of the random variables ξ_i. This problem is equivalent to that of describing all the divisors of a given distribution μ in the convolution semigroup $\mathscr{M}^1(X)$ of probability distributions on X. When passing to characteristic functions of distributions one may reduce this problem to the problem of describing possible decompositions of the characteristic function $\hat{\mu}(y)$ into factors that also are characteristic functions:

$$\hat{\mu}(y) = \hat{\mu}_1(y)\hat{\mu}_2(y). \tag{1}$$

In the classical setting, i.e., when $X = \mathbb{R}^n$, if the distribution μ decreases rapidly at infinity, its characteristic function $\hat{\mu}(y)$ admits analytic continuation to the entire n-dimensional complex space $\mathbb{C}^n$. In this case the functions $\hat{\mu}_j(y)$ also appear to be continuable to $\mathbb{C}^n$ and relation (1) is fulfilled everywhere in $\mathbb{C}^n$. This gives one the opportunity to apply methods of the theory of functions of complex variables and, in particular, method of the theory of entire functions. It is precisely in this way that the well-known Cramér theorem on decomposition of Gaussian distributions in $\mathbb{R}^n$ was proved. Presently, the arithmetic of probability distributions on $\mathbb{R}^n$ is a profound and far-developed theory, constructed mainly by Cramér, Lévy, A.Ya. Khinchin, D.A. Raĭkov, Yu.V. Linnik, I.V. Ostrovskiĭ. Several subtle investigations in this subject were recently carried out by G.P. Chistyakov. A complete exposition of this theory can be found in the monograph by Yu.V. Linnik and I.V. Ostrovskiĭ [LO], and also in the survey articles by I.V. Ostrovskiĭ [O4], [O2] and G.P. Chistyakov [Ch1].

The main and fundamental difficulty in the study of probability distributions on a group X is the absence of a natural complex structure on the group of characters, the domain of definition of the characteristic functions, and thus the impossibility of directly applying the methods of functions of complex variables.

The first results on the arithmetic of distributions on groups were obtained back in 1938. It was Lévy [Lévy1] who studied the decomposition of a Poisson distribution group $\mathbb{T}$ of rotations of the circle and Marcinkiewicz who discovered the absence of an analog for the group $\mathbb{T}$ Cramér's theorem on the decomposition of a Gaussian distribution. A significant role in the creation of the arithmetic of probabilities on groups was played by the work of Parthasarathy, Rao and Varadhan [PRV1], who found group analogs of the Khinchin theorem and the Lévy-Khinchin formula for the representation of the characteristic function of an infinitely divisible distribution. The article [Z1] of V. M. Zolotarev, where the multiplication theory of independent random variables was constructed, also exerted stimulating influence.

An exposition of the current state of the arithmetic of distributions on groups is presented in Chapter II, which contains the group analogs of the principial decomposition theorems in the classical setting. In this connection it turned out that some properties of the groups may be characterized by means of the arithmetic properties of probability distributions on them. A typical example is the group analog of the Cramér theorem: for any Gaussian distribution on the group X to have only Gaussian divisors it is necessary and sufficient that no subgroup of the group X be topologically isomorphic to the rotation group $\mathbb{T}$.

Characterization theorems of probability theory and mathematical statistics are theorems that describe the distributions of random variables via properties of some functions of these variables. An example of a characterization theorem is the following result due to S. N. Bernstein. Let ξ_1, ξ_2 be independent random variables; if the random variables $\xi_1 + \xi_2, \xi_1 - \xi_2$ also are independent, then the ξ_i are Gaussian. From the multitude of characterization problems (see the monograph by A. M. Kagan, Yu.V. Linnik, Rao [KLR]), we consider in this book only group analogs of characterization problems of Gaussian distribution in terms of the independence or identical distribution of linear statistics. The corresponding results for the case $X = \mathbb{R}$ have been obtained by S. N. Bernstein, V. P. Skitovich and Darmois, Yu. V. Linnik. The first group results for characterization problems of this type were obtained by A. L. Rukhin [Ruh1], [Ruh3], Heyer and Rall [HeR], who described the groups to which the Bernstein characterization theorem may be extended.

The current state of characterization problems is presented in Chapter 3. As in the arithmetic of distributions, it turns out that some properties of the groups themselves are fully determined by the possibility of extending to them corresponding characterization theorems. The group analog of the Bernstein theorem gives an example here. Let ξ_1 and ξ_2 be identically distributed independent random variables with values in a group X. Then the independence of the random variables $\xi_1 + \xi_2$ and $\xi_1 - \xi_2$ implies that ξ_j are Gaussian if and only if the only compact Corwin subgroup of X is the subgroup $K = \{0\}$.

In the study of characterization problems we use in an essential way the techniques developed in Chapter II for solving decomposition problems. The necessary auxiliary results are presented in Chapter I.

The author wishes to express his sincere gratitude to I. V. Ostrovskiĭ and A. I. Il'inskiĭ for their valuable advice and remarks during the preparation of this book.

CHAPTER I

Auxiliary Results

§1. Results on duality theory and on the structure of locally compact abelian groups

In this section, following mainly [HR1], we present the necessary results on duality theory and on the structure of locally compact abelian groups. As a rule we shall use the additive notation for the group operation.

1.1. Let us first introduce terminology and notation. Let X be a locally compact abelian group (in what follows, simply group). Unless otherwise stated we assume all groups to be abelian and locally compact.

The group X is called periodic if each of its elements has a finite order. If the only element of X having a finite order is the zero, then X is called a torsion-free group. By $X_1 + X_2$ we denote the direct sum of the groups X_1, X_2 and by X^n the direct sum of n copies of the group X. If $\{X_i\}_{i \in I}$ is a family of compact groups, where I is an arbitrary set of indices, $\mathsf{P}_{i \in I} X_i$ denotes the direct product (the complete direct sum) of the groups X_i, that is, the topological group coinciding (as a set) with the Cartesian product of X_i with coordinatewise operation and endowed with the Tikhonov topology(*). Obviously, $\mathsf{P}_{i \in I} X_i$ is compact. In the case when $X_i = X$ for all $i \in I$ and the cardinality of I is equal to $\mathfrak{n}$ we write $X^{\mathfrak{n}}$ instead of $\mathsf{P}_{i \in I} X_i$. In particular, we write X^∞ for the case $I = \{1, 2, \ldots\}$.

Let A_1, A_2 be subsets of the group X. Denote by $A_1 + A_2$ the arithmetic sum of A_1 and A_2:

$$A_1 + A_2 = \{x \in X : x = x_1 + x_2, \ x_1 \in A_1, \ x_2 \in A_2\}.$$

If A is a subset of X, then the direct sum of n copies of A is denoted by $(n)A$.

Let $n \in \mathbb{Z}$. Define a continuous homomorphism $f_n : X \to X$ by the relation $f_n(x) = nx$. Set $X^{(n)} = f_n(X) = \operatorname{Im} f_n$.

Let G be a closed subgroup of X. An element of the factor group X/G sometimes is denoted by $[x]$ where $x \in X$ is an arbitrary element of the conjugacy class $x + G$.

(*)*Editors note.* Product topology.

Let us enumerate the most important groups. They play a significant role in the study of the structure of an arbitrary locally compact abelian group.

1. $\mathbb{R}$—the additive real group (with the usual topology). Sometimes it will be regarded as a subset of $\mathbb{C}$, the complex plane.

2. $\mathbb{Q}$—the additive rational group (with the discrete topology).

3. $\mathbb{Z}$—the additive group of all integers (with the discrete topology).

4. $\mathbb{T} = \{z \in \mathbb{C}: |z| = 1\}$—the group of rotations of the unit circle (one-dimensional torus) with the usual topology.

5. $\mathbb{Z}(n)$—the multiplicative group of nth roots of unity (with the discrete topology).

6. $\mathbb{Z}(p^\infty)$—the multiplicative group of all p^nth roots of unity, where n runs through all nonnegative integers and p is a fixed prime number (with the discrete topology).

7. $\Delta_\mathbf{a}$—the additive group of all $\mathbf{a}$-adic integers (see [HR1, §10]). Let us describe this group in more detail. Let $\mathbf{a} = (a_0, a_1, \ldots, a_n, \ldots)$ be any fixed sequence of integers greater than 1. As a topological space, $\Delta_\mathbf{a}$ is the Cartesian product $\prod_{n=0}^\infty \{0, 1, \ldots, a_n - 1\}$ endowed with the Tikhonov topology. To define the sum $\mathbf{z} = \mathbf{x} + \mathbf{y}$ of $\mathbf{x} = (x_0, x_1, x_2, \ldots)$ and $\mathbf{y} = (y_0, y_1, y_2, \ldots) \in \Delta_\mathbf{a}$ let us construct inductively sequences $\{t_k\}_0^\infty$ and $\{z_k\}_0^\infty$ as follows. Define the integers t_0 and z_0 such that $x_0 + y_0 = t_0 a_0 + z_0$, where $z_0 \in \{0, 1, \ldots, a_0 - 1\}$ and $t_0 \in \mathbb{Z}$. Having constructed $z_0, z_1, \ldots, z_k$ and $t_0, t_1, \ldots, t_k$, we obtain z_{k+1} and t_{k+1} from the relation $x_{k+1} + y_{k+1} + t_k = t_{k+1} a_{k+1} + z_{k+1}$, where $z_{k+1} \in \{0, 1, \ldots, a_{k+1} - 1\}$ and $t_{k+1} \in \mathbb{Z}$. The sequence $\mathbf{z} = (z_0, z_1, z_2, \ldots) \in \Delta_\mathbf{a}$ is thus obtained by induction. Then $\Delta_\mathbf{a}$ is a compact totally disconnected group.

The important special case is $\mathbf{a} = (p, p, p, \ldots)$, where p is a prime number. The corresponding group is called the group of p-adic integers and is denoted by Δ_p.

8. $\Sigma_\mathbf{a}$—the $\mathbf{a}$-adic solenoid (see [HR1, §10]). Consider the group $\mathbb{R} + \Delta_\mathbf{a}$ and denote by F its subgroup $F = \{(n, n\mathbf{u})\}_{n=-\infty}^\infty$, $\mathbf{u}(1, 0, 0, \ldots)$. The factor group $\Sigma_\mathbf{a} = (\mathbb{R} + \Delta_\mathbf{a})/F$ is called the $\mathbf{a}$-adic solenoid. The group $\Sigma_\mathbf{a}$ is compact and connected.

1.2. A character of the group X is a continuous homomorphism from X into $\mathbb{T}$. Denote the set of all characters of the group X by $X^* = Y$. This set endowed with the natural abelian group structure and the topology of uniform convergence on compact sets is also a locally compact abelian group. We denote by (x, y) the value of a character $y \in Y$ on an element $x \in X$ and by τ the mapping $\tau: X \to Y^*$ defined by the formula $(y, \tau(x)) = (x, y)$.

1.3. THE PONTRYAGIN DUALITY THEOREM [HR1, §24]. *The mapping τ is a topological isomorphism of the groups X and Y^*.*

1.4. A topological isomorphism of groups will be denoted by the symbol "$\approx$". Returning to the examples described in §1.1, we note that

(1) $\mathbb{R}^* \approx \mathbb{R}$, $(t, s) = \exp\{ist\}$, where $t \in \mathbb{R}$, $s \in \mathbb{R}^*$.

(2) $\mathbb{Z}^* \approx \mathbb{T}$, $(n, \zeta) = \zeta^n$, where $n \in \mathbb{Z}$, $\zeta \in \mathbb{T}$.

(3) $(\mathbb{Z}(n))^* \approx \mathbb{Z}(n)$, $(\xi, n) = \exp\{2\pi i kl/n\}$, where $\xi = \exp\{2\pi i k/n\} \in \mathbb{Z}(n)$, $\eta = \exp\{2\pi i l/n\} \in \mathbb{Z}(n)$.

(4) $(\mathbb{Z}(p^\infty))^* \approx \Delta_p$, $(\xi, \eta) = \exp\{(2\pi i k/p^n)(x_0 + x_1 p + \cdots + x_{n-1} p^{n-1})\}$, where $\xi = \exp\{2\pi i k/p^n\} \in \mathbb{Z}(p^\infty)$ and $\eta = (x_0, x_1, x_2, \ldots) \in \Delta_p$ (see [HR1, §25]).

(5) $\mathbb{Q}^* \approx \Sigma_{\mathbf{a}}$, where $\mathbf{a} = (2, 3, 4, \ldots)$. Let $\mathbb{Q} = \{m/n! : n = 1, 2, 3, \ldots, m \in \mathbb{Z}\}$.

Then $(r, \eta) = \exp\{(2\pi i m/n!)(t - (x_0 + 2!x_1 + \cdots + (n-1)!x_{n-2}))\}$, where $r = m/n! \in \mathbb{Q}$, $\eta = [(t, \mathbf{x})] \in \Sigma_{\mathbf{a}}$ $(t, \mathbf{x}) \in \mathbb{R} + \Delta_{\mathbf{a}}$ and $\mathbf{x} = (x_0, x_1, x_2, \ldots)$ (see [HR1, §25]).

1.5. Let G be a closed subgroup of a group X. Denote by $A(Y, G)$ the annihilator of G:

$$A(Y, G) = \{y \in Y : (x, y) = 1 \; \forall x \in G\}.$$

Then the following equality

$$\text{(i)} \quad G = A(X, A(Y, G))$$

holds (see [HR1, §24]).

1.6. THEOREM [HR1, §24]. *Let G be a closed subgroup of a group X. The group G^* is topologically isomorphic to the factor group $Y/A(Y, G)$. Each character of the group G has the form $x \to (x, y)$ for some $y \in Y$. Two characters $y_1, y_2 \in Y$ define the same character of the group G if and only if $y_1 - y_2 \in A(Y, G)$. The group $(X/G)^*$ is topologically isomorphic to the group $A(Y, G)$.*

According to the duality theorem 1.3, any algebraic or topological property of the group X may be described in terms of algebraic or topological properties of its group of characters. In what follows we need some duality properties of the groups X and Y. For the sake of convenience of reference, we formulate them as a series of theorems.

1.7. THEOREM [HR1, §23]. *A group X is compact if and only if its dual group Y is discrete. A subgroup $K \subset X$ is compact if and only if its annihilator $A(Y, K)$ is an open subgroup of Y.*

1.8. Let x be an element of X. Denote by M_x the smallest closed subgroup of X containing this element. We say that the element $x \in X$ is compact if M_x is compact. We denote by X_0 the set of all compact elements of X and by C_X the connected component of zero in X.

1.9. THEOREM [HR1, §24]. *The sets* X_0 *and* C_X *are closed subgroups of the group* X, *and the following equalities hold*:

$$Y_0 = A(Y, C_X), \qquad C_X = A(X, Y_0).$$

1.10. COROLLARY. *In order that every element of the group* X *be compact it is necessary and sufficient that the group* Y *be totally disconnected. In order that the group* X *be connected it is necessary and sufficient that the group* Y *contain no nonzero compact subgroup. The group* X *is both compact and connected if and only if* Y *is a discrete torsion-free group.*

1.11. THEOREM [HR1, §24]. *A compact group* X *satisfies the second axiom of countability if and only if the group* Y *is countable.*

1.12. A set $A \subset X$ is called independent if for any $x_1, \ldots, x_n \in A$ and $k_1, \ldots, k_n \in \mathbb{Z}$ the equality $k_1 x_1 + \cdots + k_n x_n = 0$ implies $k_1 = \cdots = k_n = 0$.

Let G be a discrete torsion-free group. By the rank of the group G we mean the cardinality of a maximal independent system of elements of G. We denote by $r(G)$ the rank of G and by $\dim X$ the dimension of a connected group X.

1.13. THEOREM [HR1, §24]. *Let* X *be a connected compact group. Then*

$$\dim X = r(Y).$$

Below we shall use the following results on the structure of a locally compact abelian group X.

1.14. THEOREM [HR1, §24]. *Every group* X *is topologically isomorphic to the group*

(i) $\mathbb{R}^n + G$,

where $n \geq 0$ *and the group* G *contains a compact open subgroup* K.

1.15. THEOREM [HR1, §9]. *Every connected group* X *is topologically isomorphic to the group*

(i) $\mathbb{R}^n + K$,

where $n \geq 0$ *and* K *is a connected compact group.*

1.16. THEOREM [HR1, §25]. *Every compact torsion-free group* K *is topologically isomorphic to the group*

(i) $(\Sigma_\mathbf{a})^\mathfrak{n} + \mathsf{P}_{p \in \mathscr{P}} \Delta_p^{\mathfrak{n}_p}$,

where $\mathbf{a} = 2, 3, 4, \ldots,$ $\mathscr{P}$ *is the set of all prime numbers, and* $\mathfrak{n}$ *and* $\mathfrak{n}_p$ *are some cardinal numbers.*

1.17. REMARK. Let $\mathbf{a} = (2, 3, 4, \ldots)$. Then, for any prime p, the group $\Sigma_\mathbf{a}$ contains a subgroup topologically isomorphism to Δ_p (see [HR1, §25]).

1.18. THEOREM [HR1, §25]. *Every compact periodic group K is topologically isomorphic to the group*

$$\text{(i)} \quad \mathsf{P}_{i \in I} \mathbb{Z}(b_i),$$

where I is some set of indices and only a finite number of integers b_i are distinct.

1.19. THEOREM [HR1, §24]. *Let X be a compact group and let U be any neighborhood of zero in X. Then there exists a closed subgroup $G \subset X$ such that $G \subset U$ and the factor group X/G is topologically isomorphic to the group $\mathbb{T}^n + F$, where $n \geq 0$ and F is a finite group.*

1.20. Let groups X_1 and X_2 be given, let Y_1 and Y_2 be their groups of characters, and let $p \colon X_1 \to X_2$ be a continuous homomorphism. Define a mapping $\tilde{p} \colon Y_2 \to Y_1$ by the formula $\tilde{p}(y_2) = y_2 \circ p$ (that is, $(p(x_1), y_2) = (x_1, \tilde{p}(y_2))$ for all $x_1 \in X_1$, $y_2 \in Y_2$). This mapping $\tilde{p}$, the adjoint to the homomorphism p, is a continuous homomorphism.

Let us enumerate some properties of adjoint homomorphisms (see [HR1, §24]):

(a) The relation $p = \tilde{\tilde{p}}$ holds.

(b) $\operatorname{Ker} \tilde{p} = A(Y_2, \overline{p(X_1)})$. In particular, the homomorphism $\tilde{p}$ is a monomorphism if and only if the group $p(X_1)$ is dense in X_2.

(c) Let us define the homomorphism $f_n \colon X \to X$ by the formula $f_n(x) = nx$.

Then the adjoint homomorphism $\tilde{f}_n \colon Y \to Y$ has the form $\tilde{f}_n(y) = ny$.

1.21. REMARK. Let X be a connected compact group that satisfies the second axiom of countability and is not topologically isomorphic to $\mathbb{T}$. Then there exists a continuous monomorphism $\psi \colon \mathbb{R} \to X$ such that the subgroup $\psi(\mathbb{R})$ is dense in X. The range of a continuous homomorphism of $\mathbb{R}$ into X is called one-parameter subgroup of X.

We also need some results on topological direct summands of X and on properties of closed subgroups of $\mathbb{R}^n$.

1.22. PROPOSITION [HR1, §25]. *Let the group X contain a subgroup G that is topologically isomorphic to $\mathbb{T}^{\mathfrak{n}}$, where $\mathfrak{n}$ is a cardinal number. Then G is a topological direct summand of X.*

1.23. THEOREM [B1, Chapter VII, §1]. *Let G be a closed subgroup of $\mathbb{R}^n$. Then*

(a) *the group G is topologically isomorphic to the group $\mathbb{R}^p + \mathbb{Z}^q$, where $p + q \leq n$;*

(b) *for a group G to be the direct sum of a given closed subgroup $G_1 \subset G$ and some other closed subgroup $G_2 \subset G$ it is necessary and sufficient that G_1 be the intersection of G and some subspace of $\mathbb{R}^n$.*

1.24. THEOREM [HR1, Appendix A]. *Let X be the discrete group generated by elements $\{x_1, x_2, \ldots, x_k\}$. Then*

$$X \approx \mathrm{P}_{j=1}^{k} X_j,$$

where either $X_j = \mathbb{Z}$ or $X_j = \mathbb{Z}(n_j)$ for $j = 1, 2, \ldots, k$ and $n_j \geq 1$.

§2. Results on probability theory

In this section we present results on probability distributions on groups we shall need. The proofs, as a rule, are omitted (one can find them in monographs [Gr], [He2], [Be], [P]).

2.1. Let X be a topological abelian group. Throughout this book by a measure on the group X we mean a nonnegative countably additive function defined on the σ-algebra $\mathscr{B}(X)$ of all Borel sets (i.e., the smallest σ-algebra that contains all open subsets of X). A measure μ is called finite if $\mu(X) < \infty$. Denote the set of all finite measures on X by $\mathscr{M}_+(X)$. A measure $\mu \in M_+(X)$ is called a probability distribution if $\mu(X) = 1$. In what follows we shall write simply "distribution" instead of "probability distribution". We shall denote the set of all distributions on the group X by $\mathscr{M}^1(X)$.

By a random variable with values in the group X we mean a function on a probability space $(\Omega, \mathfrak{A}, \mathbf{P})$ with values in X such that the preimages of open sets are measurable. Every random variable ξ with values in X generates a distribution μ_ξ on X by the relation $\mu_\xi(E) = \mathbf{P}\{\xi \in E\}$ for $E \in \mathscr{B}(X)$.

Let the group X satisfy the second axiom of countability. Then every measure $\mu \in \mathscr{M}_+(X)$ is regular, i.e., $\mu(E) = \sup_{A \subset E} \mu(A)$, where A is a closed set. There exists the smallest closed set A such that $\mu(A) = \mu(X)$. This set is called the support $\sigma(\mu)$ of the measure μ. To each pair of measures $\mu, \nu \in \mathscr{M}_+(X)$ one can associate the measure $\mu * \nu \in \mathscr{M}_+(X)$ defined by the formula

$$(\mu * \nu)(E) = \int_X \mu(E - x) \, d\nu(x), \qquad E \in \mathscr{B}(X).$$

This measure is called the convolution of μ and ν. The sets $\mathscr{M}_+(X)$ and $\mathscr{M}^1(X)$ are abelian semigroups with respect to convolution. The convolution of n copies of a measure $\mu \in \mathscr{M}_+(X)$ is denoted by μ^{*n}. We also have $\sigma(\mu * \nu) = \overline{\sigma(\mu) + \sigma(\nu)}$.

Let ξ and η be independent random variables with values in X and with the distributions μ_ξ and μ_η. Then their sum $\xi + \eta$ also is a random variable and has the distribution $\mu_{\xi+\eta} = \mu_\xi * \mu_\eta$.

For any measure μ we define a function $\overline{\mu}$ by the relation $\overline{\mu}(E) = \mu(-E)$, where $E \in \mathscr{B}(X)$ and $-E$ is the set of the elements that are opposite to the elements of E. The function $\overline{\mu}$ is also a measure. The measure μ is called symmetric if $\mu = \overline{\mu}$.

Let us endow $\mathscr{M}^1(X)$ with the weak topology, defined as follows: a sequence of distributions $\{\mu_n\}$ converges to a distribution μ $(\mu_n \Rightarrow \mu)$ if $\int_X f \, d\mu_n \to \int_X f \, d\mu$ for every bounded continuous function f. The weak topology in $\mathscr{M}_+(X)$ is defined in the same way.

We denote by E_x the degenerate distribution concentrated at the point $x \in X$, i.e., such that $\sigma(E_x) = \{x\}$. The set of all degenerate distributions on the group X is denoted by $D(X)$. By a shift of a distribution μ we mean a convolution of the form $\mu * E_x$.

A distribution α is called a divisor of a distribution μ if $\mu = \alpha * \beta$ for some distribution β. (Terms "factor" and "component" are also used.) A nondegenerate distribution μ is called indecomposable if it has only degenerate distributions or shifts μ as divisors.

A set of distributions $N \subset \mathscr{M}^1(X)$ is called shift-compact if every sequence of distributions $\{\mu_n\}$, $\mu_n \in N$, contains a subsequence $\{\mu_{n_m}\}$ such that for some sequence $\{x_m\}$, $x_m \in X$, the sequence $\mu_{n_m} * E_{x_m}$ converges.

The following theorem is a convenient tool when studying distributions on groups. It was proved by Parthasarathy, Ranga Rao, and Varadhan [PRV2] (see the proof in [P, §III.2], [He2, §1.2]).

2.2. THEOREM. *Let N be a relatively compact subset of $\mathscr{M}^1(X)$. Then the set $F(N)$ of all divisors of distributions from N is shift-compact.*

2.3. COROLLARY. *Let $\mu \in \mathscr{M}^1(X)$, and let $\{\nu_n\}$ be a sequence of divisors of μ such that ν_n is a divisor of ν_{n+1}. Then, for every n, one can choose a shift ν_n' of the distribution ν_n in such a way that the sequence $\{\nu_n'\}$ converges.*

2.4. We shall say that a measure μ is concentrated on a set $A \in \mathscr{B}(X)$, if $\mu(E) = 0$ for any $E \in \mathscr{B}(X)$ such that $A \cap E = \varnothing$. (Here the set A does not have to be closed and, in general, is not uniquely determined).

2.5. PROPOSITION. *Let G be a Borel subgroup of X, $\mu \in \mathscr{M}^1(G)$, and $\mu = \mu_1 * \mu_2$ for some $\mu_i \in \mathscr{M}^1(X)$. Then one can choose shifts μ_i' of the distributions μ_i such that $\mu = \mu_1' * \mu_2'$ and $\mu_i' \in \mathscr{M}^1(G)$.*

PROOF. The equality

$$1 = \mu(G) = \int_X \mu_1(G - x) \, d\mu_2(x)$$

implies $\mu_2\{x \in X : \mu_1(G - x) = 1\} = 1$. Hence we have $\mu_1(G - x_0) = 1$ for some $x_0 \in X$, i.e., the distribution μ_1 is concentrated on the set $G - x_0$ and the distribution $\mu_1' = \mu_1 * E_{x_0}$ is concentrated on the set G. Let $\mu_2' = \mu_2 * E_{-x_0}$. Then $\mu = \mu_1' * \mu_2'$ and

$$1 = \mu(G) = \int_X \mu_2'(G - x) \, d\mu_1'(x) = \int_G \mu_2'(G - x) \, d\mu_1'(x).$$

It follows that $\mu_1\{x \in G : \mu_2'(G-x) = 1\} = 1$. Therefore, $\mu_2'(G-x_1) = 1$ for some $x_1 \in G$. Since G is a group, we have $\mu_2'(G) = 1$, i.e., $\mu_2' \in \mathcal{M}^1(G)$. $\square$

2.6. PROPOSITION. *Let X_1 and X_2 be topological groups and p an algebraic isomorphism $p \colon X_1 \to X_2$ such that the images and preimages of Borel sets under the mapping p also are Borel sets. Then p generates a semigroup isomorphism $p \colon \mathcal{M}_+(X_1) \to \mathcal{M}_+(X_2)$ by the formula*

$$\text{(i)}\quad p(\mu)(E) = \mu(p^{-1}(E)),$$

where $\mu \in \mathcal{M}_+(X_1)$, $E \in \mathcal{B}(X_2)$. Relation (i) also establishes an isomorphism between the semigroups $\mathcal{M}^1(X_1)$ and $\mathcal{M}^1(X_2)$.

The proof can be reduced to a series of standard verifications and is omitted.

In the remaining part of this section we assume X to be a locally compact abelian group.

2.7. By the characteristic function of a measure $\mu \in \mathcal{M}_+(X)$ we mean the function $\hat{\mu}(y)$ defined on the group Y by the equality

$$\text{(i)}\quad \hat{\mu}(y) = \int_X (x, y)\, d\mu(x).$$

A function $f(y)$ on the group Y is called a characteristic function if $f(y) = \hat{\mu}(y)$ for some $\mu \in \mathcal{M}_+(X)$.

2.8. BOCHNER-KHINCHIN THEOREM [HR2, §33]. *A function $f(y)$ on the group Y is characteristic if and only if $f(y)$ is continuous and for any integer n the inequality*

$$\text{(i)}\quad \sum_{i,j=1}^{n} f(y_i - y_j)\xi_i\bar{\xi}_j \geq 0$$

holds for all $y_1, \ldots, y_n \in Y$ and $\xi_1, \ldots, \xi_n \in \mathbb{C}$.

A function $f(y)$ defined on the group Y and satisfying inequality (i) is called a positive definite function. The Bochner-Khinchin theorem allows one to make no distinction between characteristic functions and continuous positive definite functions.

2.9. REMARK [HR2, §32]. Let X be an arbitrary group, G a subgroup of X, and $f_0(x)$ a positive definite function on G. Define a function $f(x)$ on X by the equality

$$f(x) = \begin{cases} f_0(x), & x \in G, \\ 0, & x \notin G. \end{cases}$$

The function $f(x)$ is positive definite on X.

Indeed, let E be a finite subset of X. We need to prove that

$$\sum_{x \in E} \sum_{y \in E} f(x-y)\xi_x\bar{\xi}_y \geq 0 \tag{1}$$

for any family $\{\xi_x, \ x \in E\}$ of complex numbers. The set E intersects only finitely many conjugacy classes of the group X with respect to the subgroup G, say the classes $z_1 + G$, $z_2 + G$, ..., $z_m + G$. Denote $E_k = E \cap (z_k + G)$, $k = 1, 2, \ldots, m$. Clearly the left-hand side of (1) equals

$$\sum_{k=1}^{m} \sum_{x \in E_k} \sum_{y \in E_k} f(x - y) \xi_x \overline{\xi}_y. \tag{2}$$

For each k we have

$$\sum_{x \in E_k} \sum_{y \in E_k} f(x - y) \xi_x \overline{\xi}_y = \sum_{x \in E_k} \sum_{y \in E_k} f_0((x - z_k) - (y - z_k))$$

$$= \sum_{u \in E_k - z_k} \sum_{v \in E_k - z_k} f_0(u - v) \xi_{u+z_k} \overline{\xi}_{v+z_k} \geq 0. \tag{3}$$

Combining (3) and (2), we obtain (1).

2.10. Let a distribution $\mu \in \mathcal{M}^1(X)$ be given. The characteristic function $\hat{\mu}(y)$ has the following properties (see [Gr, §3.3]):

(a) $|\hat{\mu}(y)| \leq \hat{\mu}(0) = 1$;

(b) $\hat{\mu}(y)$ uniquely defines μ;

(c) $\hat{\mu}(y)$ is uniformly continuous with respect to y; moreover, $|\hat{\mu}(y_1) - \hat{\mu}(y_2)|^2 \leq 2(1 - \operatorname{Re} \hat{\mu}(y_1 - y_2))$ for every y_1, $y_2 \in Y$;

(d) $\widehat{(\mu * \nu)}(y) = \hat{\mu}(y)\hat{\nu}(y)$, μ, $\nu \in \mathcal{M}^1(X)$;

(e) $\hat{\bar{\mu}} = \overline{\hat{\mu}(y)}$;

(f) Let $\{\mu_n\}$ be a sequence in $\mathcal{M}^1(X)$; then $\mu_n \Rightarrow \mu$ if and only if $\hat{\mu}_n(y) \to \hat{\mu}(y)$, uniformly on every compact subset Y;

(g) Let $\mu_n \in \mathcal{M}^1(X)$, and let $\hat{\mu}_n(y)$ converge to a limit uniformly on every compact subset of Y. Then there exists $\mu \in \mathcal{M}^1(X)$ such that

(i) $\hat{\mu}(y) = \lim_{n \to \infty} \hat{\mu}_n(y)$,

(ii) $\mu_n \Rightarrow \mu$;

(h) Let H be a closed subgroup of Y. The restriction to H of the characteristic function $\hat{\mu}(y)$ is the characteristic function of the distribution $\nu \in \mathcal{M}^1(X/A(X, H))$ defined by the relation $\nu = \tau(\mu)$, where $\tau \colon X \to X/A(X, H)$ is the natural homomorphism;

(i) Let H be a closed subgroup of Y. If $|\hat{\mu}(y)| \equiv 1$ for all $y \in H$, then there exists an $[x] \in X/A(X, H) \approx H^*$ such that $\hat{\mu}(y) = ([x], y)$ for all $y \in H$.

2.11. PROPOSITION [B2, Chapter II, §1]. *Let $p \colon X_1 \to X_2$ be a continuous group homomorphism and $p \colon \mathcal{M}_+(X_1) \to \mathcal{M}_+(X_2)$ the corresponding mapping defined by the formula (i) in 2.6. For a measure $\mu \in \mathcal{M}_+(X_1)$ the characteristic function of the measure $p(\mu)$ has the form $\hat{p}(\mu)(y_2) = \hat{\mu}(\check{p}(y_2))$, where $y_2 \in Y_2$ and $\check{p} \colon Y_2 \to Y_1$ is the homomorphism conjugate to p.*

2.12. THEOREM [HR1, Chapter IV]. *There exists a measure m_X on the group X such that*

(a) $m_X(E + x) = m_X(E)$ *for all* $E \in \mathscr{B}(X)$ *and* $x \in X$,

(b) $m_X(-E) = m_X(E)$ *for each* $E \in \mathscr{B}(X)$.

The measure m_X *is called the Haar measure. For a compact group* X *the measure* m_X *is finite and we assume* $m_X \in \mathscr{M}^1(X)$.

2.13. PROPOSITION. *Let* $\mu \in \mathscr{M}^1(X)$. *The set* $E = \{y \in Y : \hat{\mu}(y) = 1\}$ *is a subgroup of* Y, *the characteristic function* $\hat{\mu}(y)$ *is constant on every conjugacy class of the group* Y *modulo* E, *and* $\sigma(\mu) \subset A(X, E)$.

PROOF. The inequality

$$1 - \mathrm{Re}(x, y_1 + y_2) \leq 2[(1 - \mathrm{Re}(x, y_1)) + (1 - \mathrm{Re}(x, y_2))]$$

is valid for all $x \in X$, $y_1, y_2 \in Y$. This inequality implies

$$1 - \mathrm{Re}\, \hat{\mu}(y_1 + y_2) \leq 2[(1 - \mathrm{Re}\, \hat{\mu}(y_1)) + (1 - \mathrm{Re}\, \hat{\mu}(y_2))]. \tag{1}$$

Hence E is a subgroup. It follows from 2.10(c) that, for $y_1 - y_2 \in E$, we have $\hat{\mu}(y_1) = \hat{\mu}(y_2)$. Let $H = Y/E$. By Theorem 1.6, $H^* \approx A(X, E)$. As was already mentioned, one can regard the function $\hat{\mu}(y)$ as a function $f([y])$ on H. The function $f([y])$ is continuous and positive definite. By Theorem 2.8 $f([y]) = \hat{\lambda}([y])$ for some $\lambda \in \mathscr{M}^1(A(X, E))$. We have

$$\hat{\mu}(y) = \int_X (x, y)\, d\mu(x) - f([y]) - \int_{A(X, E)} (x, [y])\, d\lambda(x)$$

$$= \int_X (x, y)\, d\lambda(x) = \hat{\lambda}(y).$$

Now 2.10(b) implies $\mu = \lambda$. $\square$

2.14. A distribution μ on the group X is called idempotent if $\mu^{*2} = \mu * E_x$ for some $x \in X$. The set $I(X)$ of all idempotent distributions on X coincides with the set of all shifts on the Haar distribution m_K of compact subgroups K of the group X. Indeed, let $\mu^{*2} = \mu * E_x$. Setting $\lambda = \mu * E_{-x}$, we have $\lambda^{*2} = \lambda$. Therefore either $\hat{\lambda}(y) = 0$ or $\hat{\lambda}(y) = 1$. Consider the set $E = \{y \in Y : \hat{\lambda}(y) = 1\}$. Combining Proposition 2.13 and Theorem 1.7, one can easily see that $\lambda = m_K$, where $K = A(X, E)$ is a compact group.

Let us also mention that if K is a compact subgroup of X, then the function $\hat{m}_K(y)$ has the form

$$\text{(i)} \quad \hat{m}_K(y) = \begin{cases} 1, & y \in A(Y, K), \\ 0, & y \notin A(Y, K). \end{cases}$$

Now let us pass to the results of Parthasarathy, Rao, and Varadhan [PRV1] on properties of infinitely divisible distributions on the group X. For a detailed account of the subject see [He2], [P].

2.15. A distribution μ on the group X is called infinitely divisible if for any natural n there exist an element $x_n \in X$ and a distribution μ_n such that $\mu = \mu_n^{*n} * E_{x_n}$.

In contrast to the classical definition, we introduce here the shift by an element x_n. This is done because of the fact that an element of the group X does not have to be divisible. Not introducing a shift E_{x_n} might result in the class of infinitely divisible distributions being unjustifiably restricted. In particular, degenerated distributions might not belong to this class.

Examples of infinitely divisible distributions are idempotent distributions as well as shifts of the distribution $e(\Phi)$ that is generated by a finite measure Φ:

$$ e(\Phi) = \exp\{-\Phi(X)\} \left(E_0 + \Phi + \frac{\Phi^{*2}}{2!} + \cdots + \frac{\Phi^{*n}}{n!} + \cdots \right). \qquad (1) $$

It should be noted that the characteristic function of the distribution $e(\Phi)$ has the form

$$ \widehat{e(\Phi)}(y) = \exp \left\{ \int_X [(x, y) - 1] \, d\Phi(x) \right\}. \qquad (2) $$

2.16. PROPOSITION. *The infinitely divisible distributions form a closed subsemigroup of $\mathscr{M}^1(X)$.*

PROOF. We have only to check that the subsemigroup of all infinitely divisible distributions is closed. Let $\{\mu_k\}$ be a sequence of infinitely divisible distributions converging to μ. For each natural n let

$$ \mu_k = \mu_{kn}^{*n} * E_{x_{kn}}. \qquad (1) $$

By Theorem 2.2 the sequence $\{\mu_{kn}\}_{k=1}^{\infty}$ contains a subsequence that after appropriate shifts converges to some distribution ν_n. It follows from the convergence $\mu_k \Rightarrow \mu$ and from (1) that there exists an element x_n such that $\mu = \nu_n^{*n} * E_{x_n}$. □

2.17. PROPOSITION. *For the characteristic function of an infinitely divisible distribution to vanish at some point it is necessary and sufficient that μ has a nondegenerate idempotent divisor.*

PROOF. The sufficiency is clear. To prove the necessity suppose that $\hat{\mu}(y_0) = 0$. By the definition of infinite divisibility for any n there exist an element x_n and a distribution μ_n such that $\mu = \mu_n^{*n} * E_{x_n}$. Then $\hat{\mu}_n(y_0) = 0$ for every n. By Theorem 2.2 the sequence $\{\mu_n\}$ is shift-compact. Let λ be any limit of shifts of the μ_n. Since, obviously, $\hat{\lambda}(y_0) = 0$, the distribution λ is nondegenerate. Every power of λ clearly is a divisor of μ. Therefore, the sequence $\{\lambda^{*n}\}$ is shift-compact and any limit of the shifts λ^{*n} is a nondegenerate idempotent divisor of μ. □

2.18. COROLLARY. *Let μ be an infinitely divisible distribution. Then the set $E = \{y \in Y : \hat{\mu}(y) \neq 0\}$ is a subgroup of Y.*

PROOF. Let H be the subgroup generated by the set E, and let $f(y)$ be the restriction to H of the function $\hat{\mu}(y)$. By 2.10(h) we have $f(y) = \hat{\nu}(y)$ for some $\nu \in \mathcal{M}^1(X/A(X, H))$. The distribution ν clearly is infinitely divisible. By construction, ν does not have nondegenerate idempotent divisors; this follows from the fact that if any distribution $\delta \in \mathcal{M}^1(X)$ has such a divisor, then the set $\{y \in Y : \hat{\delta}(y) \neq 0\}$ does not generate the whole group Y. Taking into consideration Proposition 2.17, we have $\hat{\nu}(y) \neq 0$ for $y \in H$. Since $\hat{\mu}(y) = \hat{\nu}(y)$ on H, we obtain $E = H$. $\square$

Sometimes this corollary reduces the investigation of an arbitrary infinitely divisible distribution to the case when such a distribution is free of nondegenerate idempotent divisors.

2.19. LEMMA. *There exists a function* $g(x, y)$ *defined on the product* $X \times Y$ *with the following properties*:

(a) $g(x, y)$ *is continuous with respect to* x *and* y;

(b) $\sup_{x \in X} \sup_{y \in A} |g(x, y)| < \infty$ *for all compact sets* $A \subset Y$;

(c) $g(x, y_1 + y_2) = g(x, y_1) + g(x, y_2)$, $g(-x, y) = -g(x, y)$;

(d) *For any compact set* $A \subset Y$ *one can find a neighborhood* V_A *of zero in* X *such that* $(x, y) = \exp\{ig(x, y)\}$ *for all* $x \in V_A$ *and* $y \in A$;

(e) *For any compact set* $A \subset Y$ *the function* $g(x, y)$ *tends to zero uniformly with respect to* $y \in A$ *as* x *tends to the zero of* X.

2.20. Let us give examples of the function $g(x, y)$ for some groups.

1. Let $X = Y = \mathbb{R}^n$. Then

$$g(x, y) = \sum_{i=1}^{n} \varphi_i(x_i)y_i,$$

where $x = (x_1, \ldots, x_n) \in X$, $y = (y_1, \ldots, y_n) \in Y$, and $\varphi_i(t)$, $i = 1, 2, \ldots, n$, are bounded continuous functions on the real line such that $\varphi_i(t) = t$ in some neighborhood of zero and $\varphi_i(-t) = -\varphi_i(t)$.

2. Let the group X be totally disconnected. In this case by Corollary 1.10 the group Y consists of compact elements and, hence, the only continuous homomorphism of the group Y into $\mathbb{R}$ is the trivial one. Therefore, $g(x, y) \equiv 0$ for all $x \in X$ and $y \in Y$.

2.21. THEOREM. *Let* μ *be an infinitely divisible distribution on the group* X. *Then its characteristic function* $\hat{\mu}(y)$ *admits the representation*

(i) $\quad \hat{\mu}(y) = (x_0, y)\hat{m}_K(y) \exp\left\{ \int_{X \setminus \{0\}} [(x, y) - 1 - ig(x, y)] d\Phi(x) - \varphi(y) \right\}$;

here x_0 *is some element of* X, K *is a compact subgroup of* X, $g(x, y)$ *is a function on* $X \times Y$ *that satisfies the conditions of Lemma 2.19 and does not depend on* μ, Φ *is a measure on* X *that is finite on the complement of every neighborhood of zero and such that all* $y \in Y$

$$\int_{X \setminus \{0\}} [1 - \text{Re}(x, y)] \, d\Phi(x) < \infty,$$

and $\varphi(y)$ *is a continuous nonnegative function on* Y *such that for all* $y_1, y_2 \in Y$ *one has*

$$\varphi(y_1 + y_2) + \varphi(y_1 - y_2) = 2[\varphi(y_1) + \varphi(y_2)].$$

The measure Φ in the representation (i) is called the Lévy measure of the infinitely divisible distribution μ. In the case when μ has no nondegenerate idempotent divisors, this representation was obtained by Parthasarathy, Rao, and Varadhan in [PRV1], and, in the general case, by Parthasarathy and Sazonov in [PS]. The proof may also be found in [He2], [P].

It should be noted that in the case $X = \mathbb{R}^n$ representation (i) formally differs from the classical Lévy representation of the characteristic function of an infinitely divisible distribution μ (see [Lin O, Chapter VI, §1])

$$\hat{\mu}(s) = \exp\left\{ i\langle \beta, s \rangle - \langle As, s \rangle + \int_{\mathbb{R}^n \setminus \{0\}} \left(e^{i\langle t, s \rangle} - 1 - \frac{i\langle t, s \rangle}{1 + \|t\|^2} \right) d\Phi(t) \right\},$$

where $\beta \in \mathbb{R}^n$, A is a positive semidefinite matrix, and the measure Φ is σ-finite and satisfies the condition

$$\int_{\mathbb{R}^n \setminus \{0\}} \frac{\|t\|^2}{1 + \|t\|^2} \, d\Phi(t) < \infty.$$

2.22. REMARK. The representation

$$(\text{i}) \quad \hat{\mu}(y) = (x_0, y) \exp\left\{ \int_{X \setminus \{0\}} [(x, y) - 1 - i g(x, y)] \, d\Phi(x) - \varphi(y) \right\}$$

of the characteristic function of an infinitely divisible distribution μ without nondegenerate idempotent divisors will be called, for brevity, the (x_0, Φ, φ) representation. If the characteristic function of an infinitely divisible distribution μ admits both (x_0, Φ, φ) and (x_0', Φ', φ') representations, then $\varphi = \varphi'$ and $\Phi = \Phi'$ on the complement of the subgroup of all compact elements of X.

For the representation (i) to be unique it is necessary and sufficient that either the group X does not contain nonzero compact subgroup or any nonzero compact element of X has order 2. It should be mentioned that the set of groups X for which the representation (i) is unique contains, in particular, the group $\mathbb{R} + \mathbb{Z}(2)$. Infinitely divisible distributions on this group have been studied by Zolotarev [Z] in connection with a general multiplication theory for independent random variables that he constructed.

2.23. A triangular sequence of distributions $\{\mu_{nj}\}$, $j = 1, 2, \ldots, j_n$, $n = 1, 2, 3, \ldots$, is called infinitesimal if

$$(i) \quad \lim_{n\to\infty} \sup_{1\le j\le j_n} \sup_{y\in A} |\hat{\mu}_{nj}(y) - 1| = 0$$

for any compact set $A \subset Y$.

2.24. THEOREM. *If $\{\mu_{nj}\}$ is infinitesimal, $\mu_n = \mu_{n1} * \cdots * \mu_{nj_n}$, and $\mu_n \Rightarrow \mu$, then μ is an infinitely divisible distribution.*

2.25. REMARK. Theorem 2.24 remains true after replacing the infinitesimal $\{\mu_{nj}\}$ by a shift-compact sequence $\{\mu_{nj}\}$ such that each limit of shifts of μ_{nj} is degenerate.

2.26. By a charge on a group X we mean any function λ defined on $\mathscr{B}(X)$ that admits the representation $\lambda = \mu_1 - \mu_2$, where $\mu_i \in \mathscr{M}_+(X)$. If the measures μ_1 and μ_2 in this representation are concentrated on disjoint sets, the quantity $\|\lambda\| = \mu_1(X) + \mu_2(X)$ is called the norm of the charge λ.

One can define the convolution of two charges μ and ν, the characteristic function of the charge μ and the charge $\hat{\mu}$ in the same way as for finite measures. The properties (c), (d), (e) in 2.10 remain valid both for finite measures and for charges. Let Φ be a charge on the group X. Denote by $e(\Phi)$ the charge defined by formula 2.15(1). The characteristic function of the charge $e(\Phi)$ is defined by formula 2.15(2).

§3. Results on function theory and on analytic properties of characteristic functions

In this section we present the needed results on the theory of functions of one and several complex variables and on analytic properties of characteristic functions.

3.1. By an entire function we mean a function analytic in the whole complex plane $\mathbb{C}$. In order for the series

$$f(z) = \sum_{n=0}^{\infty} a_n z^n, \qquad z \in \mathbb{C},$$

to represent an entire function it is necessary and sufficient that $\overline{\lim}_{n\to\infty} \sqrt[n]{|a_n|} = 0$.

Let $f(z)$ be an entire function. Set

$$M_f(r) = \max_{\|z\|\le r} |f(z)|.$$

By the order of the entire function $f(z)$ we mean the quantity

$$\rho = \rho(f) = \overline{\lim_{r\to\infty}} \frac{\ln\ln M_f(r)}{\ln r}.$$

The order may also be defined as the greatest lower bound of the set of numbers $a > 0$ such that the inequality

$$M_f(r) < \exp\{r^a\}$$

holds for all sufficiently large r.

By the type of an entire function $f(z)$ of order ρ we mean the quantity

$$\sigma = \sigma(f) = \varlimsup_{r \to \infty} \frac{\ln M_f(r)}{r^\rho}.$$

The type may also be defined as the greatest lower bound of the set of all numbers $b > 0$ such that the inequality

$$M_f(r) < \exp\{br^\rho\}$$

holds for all sufficiently large r.

An entire function f is called an entire function of exponential type σ, $0 \le \sigma < \infty$, if either $\rho(f) < 1$ or $\rho(f) = 1$, $\sigma(f) \le \sigma$.

The following result is a particular case of the Hadamard representation of an entire function of finite order [Mark, Chapter VII, §2].

3.2. THEOREM. *Let $f(z)$ be an entire function. If $f(z) \ne 0$ for all $z \in \mathbb{C}$, then $f(z) = \exp\{\varphi(z)\}$, where $\varphi(z)$ is an entire function. If in addition the order of $f(z)$ does not exceed ρ, then $\varphi(z)$ is a polynomial whose degree does not exceed ρ.*

3.3. THEOREM [Mark, Chapter VII, §5]. *Let $f(z)$ be a T-periodic entire function of exponential type σ. Then*

$$\text{(i)} \quad f(z) = \sum_{p=-n}^{n} d_p \exp\left\{\frac{2\pi i z p}{T}\right\}, \qquad n = \left[\frac{|T|\sigma}{2\pi}\right].$$

3.4. CARATHÉODORY INEQUALITY [Le, Chapter I, §6]. *Let $f(z)$ be an entire function. Then*

$$M_f(r) \le \frac{2r}{R-r}[A_f(r) - \operatorname{Re} f(0)] + |f(0)|, \qquad 0 \le r < R,$$

where $A_f(r) = \max_{\|z\| \le r} \operatorname{Re} f(z)$.

3.5. PHRAGMÉN-LINDELÖF THEOREM [Mark, Chapter VI, §3]. *Let A be an angle of opening π/α, $\frac{1}{2} \le \alpha < \infty$, and vertex at the origin, and let $f(z)$ be a function analytic both inside A and on its boundary. Suppose that the inequality*

$$\text{(i)} \quad |f(z)| \le M$$

is valid on the boundary of A and that the inequality

$$\text{(ii)} \quad f(z) = O(\exp\{|z|^\rho\}), \qquad |z| \to \infty, \quad \rho < \alpha,$$

is valid everywhere inside A. *Then inequality* (i) *is valid everywhere inside* A.

3.6. SCHWARZ FORMULA [Mark, Chapter VI, §1]. *Let* $f(z)$ *be a function analytic in the disk* $|z| \leq 1$, *and let* $u(z) = \operatorname{Re} f(z)$. *Then*

$$f(z) = \int_{-\pi}^{\pi} u(e^{it}) \frac{e^{it} + z}{e^{it} - z} \, dt + ic, \qquad c \in \mathbb{R}.$$

3.7. A function $f(z)$, $z = (z_1, \ldots, z_n)$, defined in a domain $A \subset \mathbb{C}^n$ is said to be analytic in A if in a neighborhood of every point $z^0 \in A$ it may be expanded into a series

$$f(z) = \sum_{\|k\|=0}^{\infty} a_k(z^0)(z - z^0)^k, \qquad z \in \mathbb{C}^n,$$

where $k = (k_1, \ldots, k_n)$, $\|k\| = k_1 + \cdots + k_n$, $z^k = z_1^{k_1} \cdots z_n^{k_n}$, $k_j \geq 0$.

HARTOGS THEOREM [Fuks, Chapter 1]. *For a function* $f(z)$ *to be analytic in a domain* $A \subset \mathbb{C}^n$ *it is necessary and sufficient that for every* $z^0 \in A$ *and for every* $j = 1, 2, \ldots, n$ *the function* $f(z_1^0, \ldots, z_{j-1}^0, z_j, z_{j+1}^0, \ldots, z_n^0)$ *be analytic with respect to* z_j *in some neighborhood of the point* z_j^0.

By an entire function we mean a function analytic in the whole space $\mathbb{C}^n$. In order for the sum of the series

$$f(z) = \sum_{\|k\|=0}^{\infty} a_k z^k$$

to be an entire function it is necessary and sufficient that for every $r = (r_1, \ldots, r_n)$, where $r_j \geq 0$, for $j = 1, 2, \ldots, n$, the inequality

$$\text{(i)} \qquad \sum_{\|k\|=0}^{\infty} |a_k| r^k < \infty$$

be fulfilled.

3.8. THEOREM. *Let* $f(z)$ *be an entire function. If* $f(z) \neq 0$ *for any* $z \in \mathbb{C}^n$, *then* $f(z) = \exp\{\varphi(z)\}$, *where* $\varphi(z)$ *is an entire function.*

Let $f(z)$ be an entire function. Set $B_r = \{z \in \mathbb{C}^n : |z_j| \leq r, \, j = 1, 2, \ldots, n\}$,

$$M_f(r) = \max_{z \in B_r} |f(z)|, \qquad A_f(r) = \max_{z \in B_r} \operatorname{Re} f(z).$$

3.9. THEOREM [Ro, Chapter I, §3]. *Let* $f(z)$ *be an entire function. If*

$$A_f(r) = O(r^p), \qquad r \to \infty,$$

then $f(z)$ *is a polynomial of degree not exceeding* p.

3.10. Let $\mu \in \mathscr{M}^1(\mathbb{R}^n)$. The characteristic function $\hat{\mu}(s)$ is said to be entire if there exists an entire function $f(s)$, $s \in \mathbb{C}^n$, such that

$$f(s) = \hat{\mu}(s) \quad \text{for } s \in \mathbb{R}^n.$$

By the uniqueness theorem for analytic functions, this function $f(s)$ is uniquely defined. We shall denote it by $\hat{\mu}(s)$ as well.

3.11. LÉVY-RAĬKOV THEOREM [LinO, Chapter II, §28]. *Let $\mu \in \mathscr{M}^1(\mathbb{R})$. If the characteristic function $\hat{\mu}(s)$ is entire, then the representation*

$$\text{(i)} \quad \hat{\mu}(s) = \int_{-\infty}^{\infty} \exp\{its\}\, d\mu(t)$$

is valid for all $s \in \mathbb{C}$ and the integral on the right-hand side of (i) *converges uniformly on every compact subset of $\mathbb{C}$.*

3.12. LÉVY THEOREM [LinO, Chapter III, §1]. *Let $\mu \in \mathscr{M}^1(\mathbb{R})$, and let the characteristic function $\hat{\mu}(s)$ be entire. If μ_1 is a divisor of μ, then the characteristic function $\hat{\mu}_1(s)$ is entire as well.*

3.13. LINNIK THEOREM [Ram, Addendum to Chapter VI]. *Let $\mu \in \mathscr{M}^1(\mathbb{R})$ be an infinitely divisible distribution whose Lévy measure Φ is concentrated on a segment $[-a, b]$, where $0 \leq a \leq b$, and let μ_1 be a divisor of μ. Then $\ln \hat{\mu}_1(s)$ is an entire function of exponential type σ, $\sigma \leq b$.*

We shall also need multidimensional analogs (proved by Ostrovskiĭ and Cuppens) of Theorems 3.11 and 3.12. These analogs are particular cases of Theorems 6.1.3 and 6.2.1 from [LinO].

3.14. THEOREM. *Let $\mu \in \mathscr{M}^1(\mathbb{R}^n)$. If the characteristic function $\hat{\mu}(s)$ is entire, then the representation*

$$\text{(i)} \quad \hat{\mu}(s) = \int_{\mathbb{R}^n} \exp\{i\langle t, s\rangle\}\, d\mu(t)$$

holds for every $s \in \mathbb{C}^n$ and the integral on the right-hand side of (i) *converges uniformly on every compact subset of $\mathbb{C}^n$.*

3.15. THEOREM. *Let $\mu \in \mathscr{M}^1(\mathbb{R}^n)$, and let the characteristic function $\hat{\mu}(s)$ be entire. If μ_1 is a divisor of μ, then the characteristic function $\hat{\mu}_1(s)$ is entire as well.*

CHAPTER II

Arithmetic of Distributions

Throughout this chapter, unless otherwise stated, we assume that X is a locally compact abelian group satisfying the second axiom of countability. In this case the group X is separable and metric.

§4. Group analogs of the Khinchin factorization theorems

The following fundamental results on distributions on the real line have been obtained by A. Ya. Khinchin (see [Lin O, Chapter III]).

THE FIRST KHINCHIN THEOREM. *Any distribution μ on the group $X = \mathbb{R}$ is either a finite or a countable convolution of indecomposable distributions and a distribution without indecomposable divisors.*

THE SECOND KHINCHIN THEOREM. *If a distribution μ on the group $X = \mathbb{R}$ has no indecomposable divisors, then μ is infinitely divisible.*

A remark is appropriate here. In contrast to the case $X = \mathbb{R}$, a distribution on an arbitrary group can have an idempotent divisor. This is the reason why the formulations of theorems on the arithmetic of distributions on the real line differ from their group analogs (compare with Theorem 2.21). Before considering the group analogs of the Khinchin theorems, let us prove the following assertion that permits one to extract the maximal idempotent divisor of a given distribution.

4.1. THEOREM. *Any distribution μ on a group X can be represented as the convolution $\mu = m_K * \lambda$, where λ is a distribution without nondegenerate idempotent divisors and m_K is the maximal idempotent divisor of μ.*

PROOF. Let $E = \{y \in Y : \hat{\mu}(y) \neq 0\}$, and let H be the subgroup generated by the set E. The subgroup H is open, and, by virtue of Theorem 1.7, its annihilator $K = A(X, H)$ is compact. The distribution m_K is just the required maximal idempotent divisor of μ. Let $\tau : X \to X/K$ be the natural homomorphism. According to 2.10(h), the restriction to H of the characteristic function $\hat{\mu}(y)$ is the characteristic function of the distribution $\nu = \tau(\mu) \in \mathscr{M}^1(X/K)$. By construction, ν has no nondegenerate idempotent divisors. As follows from a result of Mackey [Mac, p. 102], there exists a Borel subset $A \subset X$, such that τ is a one-to-one mapping of A onto X/K.

Since any one-to-one continuous mapping preserves Borel sets [Kur, §39, IV], the images and pre-images of Borel sets under the mapping τ are again Borel sets. So ν induces a distribution $\lambda \in \mathcal{M}^1(X)$. As can easily be seen, this distribution satisfies the conditions of the theorem. □

4.2. Let $\mu \in \mathcal{M}^1(X)$, and let μ have no nondegenerate idempotent divisors. Let $F(\mu)$ be the set of all divisors of μ. We define on $F(\mu)$ a function $\theta(\alpha)$ analogous to the Khinchin functional. (Recall that A. Ya. Khinchin, when studying distributions on $\mathbb{R}$, considered the functional

$$N_a[\alpha] = -\int_0^a \ln \hat{\alpha}(t)\, dt,$$

where a is some sufficiently small positive number.) Since μ has no nondegenerate idempotent divisors, a sequence $y_1, y_2, \ldots, y_n, \ldots$ of characters in Y exists such that $\hat{\mu}(y_i) \neq 0$, $i = 1, 2, 3, \ldots$, and the smallest closed subgroup generated by this sequence is the whole of Y. Choose a sequence $\{\varepsilon_n\}$, $\varepsilon_n > 0$, in such a way that

$$-\sum_{n=1}^{\infty} \varepsilon_n \ln |\hat{\mu}(y_n)| < \infty.$$

The function

$$\theta(\alpha) = -\sum_{n=1}^{\infty} \varepsilon_n \ln |\hat{\alpha}(y_n)|$$

is clearly finite at each divisor α of the distribution μ and has the following properties:

  (a) $\theta(\alpha) \geq 0$,
  (b) $\theta(\alpha) = 0 \Leftrightarrow \alpha = E_x$ for some $x \in X$,
  (c) $\theta(\alpha_1 * \alpha_2) = \theta(\alpha_1) + \theta(\alpha_2)$,
  (d) If $\alpha_n \Rightarrow \alpha$, then $\theta(\alpha_n) \to \theta(\alpha)$,
  (e) $\theta(\alpha) = \theta(\beta)$, if $\alpha = \beta * E_x$ for some $x \in X$.

Let us denote by $I_0 = I_0(X)$ the class of distributions on the group X having neither indecomposable nor nondegenerate idempotent divisors. (Note that this definition can directly be used for any topological group.)

The following result is a group analog of the second Khinchin theorem.

4.3. THEOREM. *If $\mu \in I_0$, then μ is an infinitely divisible distribution.*

PROOF. As follows from Remark 2.25, it suffices to construct a system of decompositions

$$\mu = \mu_{n1} * \mu_{n2} * \cdots * \mu_{n2^n}, \tag{1}$$

such that any limit of the shifts of μ_{nj} is degenerate. As follows from the properties of the function θ, it is sufficient to construct the decomposition (1) with $\theta(\mu_{nj}) = 2^{-n}\theta(\mu)$. If $\mu \in I_0$ and ν is a divisor of μ, then $\nu \in I_0$. Consequently, it is enough to prove that an arbitrary distribution μ from the class I_0 admits the decomposition $\mu = \mu_1 * \mu_2$, where $\theta(\mu_1) = \theta(\mu_2) = \frac{1}{2}\theta(\mu)$.

Observe now that

$$\inf_{\alpha \in F(\mu),\ \theta(\alpha) \neq 0} \theta(\alpha) = 0.$$

Indeed, let us assume the contrary. Since the set $F(\mu)$ is shift-compact, the infimum is attained on some distribution. This distribution has to be indecomposable, which is impossible.

So there exist divisors of μ that make the function θ arbitrarily small. Let μ_1 and μ_2 be such that $\mu = \mu_1 * \mu_2$ and $|\theta(\mu_1) - \theta(\mu_2)|$ is minimal. Such μ_1 and μ_2 exist because $F(\mu)$ is shift-compact. The minimum must be equal to zero. Indeed, otherwise the value $|\theta(\mu_1) - \theta(\mu_2)|$ could be decreased by transferring a divisor of μ_1 or μ_2 on which the value of θ is small enough to μ_2 or μ_1. □

It follows from Theorem 4.3 that to describe the class I_0 one has to describe conditions on the measure Φ and the function φ from representation 2.22(i), needed to ensure that $\mu \in I_0$.

Now let us prove the group analog of the first Khinchin theorem.

4.4. THEOREM. *Let $\mu \in \mathcal{M}^1(X)$. Then μ may be represented in the form*

$$\text{(i)} \quad \mu = m_K * \lambda_1 * \lambda_2,$$

where m_K is the maximal idempotent divisor of μ, λ_1 is a convolution of either finite or countable number of indecomposable distributions, and $\lambda_2 \in I_0$.

PROOF. According to Theorem 4.1, the distribution μ may be represented in the form $\mu = m_K * \lambda$, where m_K is the maximal idempotent divisor of μ and λ has no nondegenerate idempotent divisors. Therefore a function θ on the set $F(\lambda)$ may be defined so that it would have the properties 4.2(a)–(e). Let δ_1 be the supremum of the values of θ on the set of all indecomposable divisors of λ. If $\delta_1 > 0$, then λ admits a representation $\lambda = \alpha_1 * \beta_1$, where α_1 is an indecomposable distribution and $\theta(\alpha_1) \geq \delta_1/2$. Denote by δ_2 the supremum of the values of θ on the set of all indecomposable divisors of the distribution β_1. If $\delta_2 > 0$, then β_1 admits a representation $\beta_1 = \alpha_2 * \beta_2$, where α_2 is an indecomposable distribution, and $\theta(\alpha_2) \geq \delta_2/2$ and so on. If this process stops at the nth step, then $\lambda = \alpha_1 * \cdots * \alpha_n * \beta_n$ and $\delta_{n+1} = 0$, i.e., β_n does not have indecomposable divisors, that is, $\beta_n \in I_0$. Now let this process be infinite. Since the series $\sum_{n=1}^{\infty} \theta(\alpha_n)$ converges, $\theta(\alpha_n) \to 0$ as $n \to \infty$. The sequence $\nu_n = \alpha_1 * \cdots * \alpha_n$ satisfies the condition of Corollary 2.3. Therefore, it converges after making proper shifts. Shifting the distributions α_n in the appropriate way, one can see that $\alpha_1 * \cdots * \alpha_n$ converges to λ_1. In addition, β_n would converge automatically to some distribution λ_2. As can be easily seen, any indecomposable divisor α of λ_2 is a common divisor of all β_n. Therefore, $\theta(\alpha) \leq \delta_n$ for all $n = 1, 2, 3, \ldots$. But $\delta_n \leq 2\theta(\alpha_n)$, and we have $\delta_n \to 0$. Thus $\theta(\alpha) = 0$, i.e., $\lambda_2 \in I_0$. □

4.5. Let us discuss the problem of whether an indecomposable distribution on the group X exists. In the case $X \not\approx \mathbb{Z}(2)$ an indecomposable distribution on X does exist. Indeed, if $X^{(2)} \neq \{0\}$, assume that $x \in X$ is such that $x \neq 0$, $2x \neq 0$. Then the distribution $\mu = \frac{1}{2}(E_0 + E_x)$ obviously is indecomposable. If $X^{(2)} = \{0\}$, i.e., all nonzero elements of the group X are of the order 2, then, as can easily be seen, the distribution $\mu = \frac{1}{3}(E_0 + E_{x_1} + E_{x_2})$, where x_1 and x_2 are distinct nonzero elements of X, is indecomposable.

Let $X \approx \mathbb{Z}(2)$ and $x \in X$, $x \neq 0$. Then one can easily check that $\mathscr{M}^1(X) = \{m_X, e(aE_x), e(aE_x) * E_x\}$, where $a \geq 0$. Thus, all distributions on X are infinitely divisible, and hence no distribution on X is indecomposable.

To strengthen Theorem 4.3, let us prove the following statement of independent interest.

4.6. PROPOSITION. *Let* $\mu \in \mathscr{M}^1(X)$, *and let* μ *have a nondegenerate idempotent divisor. Then either* $\mu = m_K * E_x$, *where* $K \subset X$ *and* $K \approx \mathbb{Z}(2)$, *or* μ *has an indecomposable divisor.*

PROOF. Let m_K be the maximal idempotent divisor of μ (by Theorem 4.1 such a divisor exists). Assume first that $K \not\approx \mathbb{Z}(2)$. It follows from the previous section that one can choose an indecomposable distribution on K. Any such distribution is an indecomposable divisor of μ. It remains to consider the case $K \approx \mathbb{Z}(2)$. By Theorem 4.1, the decomposition $\mu = m_K * \lambda_0$ takes place, where λ_0 has no nondegenerate divisors. Assuming that $\mu \neq m_K * E_x$, let us show that a decomposition

$$\mu = m_K * \lambda, \tag{1}$$

exists, in which the distribution λ is either indecomposable itself or has an indecomposable divisor.

Denote by ζ an element of order 2 in K, and rewrite (1) in the form

$$\mu(E) = \frac{1}{2}[\lambda(E) + \lambda(E + \zeta)], \qquad E \in \mathscr{B}(X).$$

Obviously,

$$\mu(E) = \mu(E + \zeta) \tag{2}$$

for any $E \in \mathscr{B}(X)$. The following three cases are possible, depending on the structure of the support $\sigma(\mu)$ of the distribution μ.

1. The support $\sigma(\mu)$ consists of two points.

Taking (2) into account, it can easily be checked that this is possible if and only if $\mu = m_K * E_x$.

2. The support $\sigma(\mu)$ consists of four points.

Let $\sigma(\mu) = \{x_1, x_2, x_3, x_4\}$. Without loss of generality, we may suppose that $x_1 = 0$ and $x_{j+2} = x_j + \zeta$, $\mu\{x_j\} = \mu\{x_{j+2}\}$, $j = 1, 2$. If $2x_2 \neq 0$, then we define the distribution λ as follows:

$$\lambda(\{x_j\}) = \begin{cases} 2\mu(\{x_j\}), & j = 1, 2, \\ 0, & j = 3, 4. \end{cases}$$

As can easily be seen, the distribution λ satisfies (1) and is indecomposable. If $2x_2 = 0$, then $\mu \in \mathscr{M}^1(G)$, where $G \subset X$, $G \approx (\mathbb{Z}(2))^2$. In this case we define the distribution λ as follows:

$$\lambda(\{x_j\}) = \begin{cases} \mu(\{x_1\}), & j = 1, 3, \\ 2\mu(\{x\}), & j = 2, \\ 0, & j = 4. \end{cases}$$

One can check directly that λ is indecomposable and satisfies (1).

3. The support $\sigma(\mu)$ contains at least six points.

Let $\{x_j\}$, $j = 1, 2, \ldots, 6$, be six distinct points in $\sigma(\mu)$ such that $x_{j+3} = x_j + \zeta$, $j = 1, 2, 3$. If V is a neighborhood of zero in the group X, denote $V_x = V + x$, $x \in X$. Let the neighborhood V be chosen in such a way that the neighborhoods V_{x_j}, $j = 1, 2, \ldots, 6$, are pairwise disjoint. Denote $H = A(Y, K)$ and choose an element $\eta \in Y$ such that $(\zeta, \eta) = -1$. Then the decomposition of the group Y with respect to the subgroup H has the form $Y = H \cup (\eta + H)$. Define a function $f(x)$ on the group X by the relation

$$f(x) = \begin{cases} 1, & x \notin \bigcup_{j=1}^6 V_{x_j}, \\ 1 + \alpha_j, & x \in V_{x_j}, \; j = 1, 2, 3, \\ 1 - \alpha_j, & x \in V_{x_{j+3}}, \; j = 1, 2, 3, \end{cases}$$

where $-1 < \alpha_j < 1$. Set $\lambda = f\mu$ and check that this distribution satisfies (1). Turning to the characteristic functions in (1) and taking into consideration 2.14(i), we see that it is enough to check that the characteristic functions $\hat{\mu}(y)$ and $\hat{\lambda}(y)$ coincide on H. We have

$$\hat{\lambda}(y) = \int_X (x, y) \, d\lambda(x) = \int_X (x, y) f(x) \, d\mu(x)$$

$$= \hat{\mu}(y) + \sum_{j=1}^3 \alpha_j \left[\int_{V_{x_j}} (x, y) \, d\mu(x) - \int_{V_{x_{j+3}}} (x, y) \, d\mu(x) \right], \qquad y \in Y.$$

$$(3)$$

Taking into account relation (2) and that the equality $(x + \zeta, y) = (x, y)$ holds for $y \in H$, we obtain

$$\int_{V_{x_{j+3}}} (x, y) \, d\mu(x) = \int_{V_{x_j + \zeta}} (x, y) \, d\mu(x) = \int_{V_{x_j}} (x, y) \, d\mu(x),$$

$$y \in H, \qquad j = 1, 2, 3.$$

Consequently,

$$\hat{\lambda}(y) = \hat{\mu}(y), \qquad y \in H. \tag{4}$$

We note that no distribution λ satisfying (1) can have nondegenerate idempotent divisors distinct from a translation of m_K. This follows from the fact that there exists a distribution $\lambda = \lambda_0$ without such divisors, which satisfies (1). If $\lambda \neq \mu$, then m_K cannot be a divisor of λ, since otherwise we would have $\hat{\mu}(y) = \hat{m}_K(y) = \hat{\lambda}(y) = 0$ at $y \in \eta + H$, and with accounting for (4), $\hat{\mu}(y) = \hat{\lambda}(y)$ for all $y \in Y$. Because of 2.10(b), $\mu = \lambda$. Therefore, if at least one of the numbers α_j is nonzero, then $\lambda \neq \mu$, and, hence, λ has no nondegenerate idempotent divisors. Let us show that one can choose numbers α_j, $j = 1, 2, 3$, not simultaneously equal to zero, such that the equality $\hat{\lambda}(y_1) = 0$ holds for some $y_1 \in \eta + H$. Then λ must have an indecomposable divisor. Indeed, by the construction, λ has no nondegenerate idempotent divisors. If λ had no indecomposable divisors either, then, by Theorem 4.3, λ would have to be infinitely divisible, which contradicts Proposition 2.17.

Let $y_1 \in \eta + H$. Then, as can easily be seen,

$$\int_{V_{x_{j+3}}} (x, y_1)\, d\mu(x) = -\int_{V_{x_j}} (x, y_1)\, d\mu(x) \tag{5}$$

and substituting (5) into (3), we obtain

$$\hat{\lambda}(y_1) = \hat{\mu}(y_1) + 2\sum_{j=1}^{3} \alpha_j \int_{V_{x_j}} (x, y_1)\, d\mu(x) = 2\sum_{j=1}^{3} \alpha_j \int_{V_{x_j}} (x, y_1)\, d\mu(x),$$

since $\hat{\mu}(y_1) = 0$. It is clear that there exist α_j, $-1 < \alpha_j < 1$, $j = 1, 2, 3$, not all zero, such that $\hat{\lambda}(y_1) = 0$. $\square$

The two previous sections yield the following strengthening of Theorem 4.4.

4.7. COROLLARY. *Let a distribution on the group X have no indecomposable divisors. Then it is infinitely divisible.*

Consider the decomposition 4.4(i). The next problem we want to consider is the problem of determining conditions for a distribution μ to have an indecomposable divisor. We shall prove the corresponding result (Theorem 4.17) for distributions on compact groups. To do this we need a number of technical lemmas. Some of them will be used also in our study of the Poisson generalized distribution (§6).

4.8. LEMMA. *Let $\mu \in \mathcal{M}_+(X)$ and $\mu = \mu_1 * \mu_2$, $\mu_j \in \mathcal{M}_+(X)$, $j = 1, 2$. If $c = \mu_2(\{0\}) > 0$, then*

$$\text{(i)} \quad \mu_1(E) \leq \frac{\mu(E)}{c}$$

for every $E \in \mathcal{B}(X)$. In particular, the measure μ_1 is absolutely continuous with respect to μ.

PROOF. Let $E \in \mathcal{B}(X)$. Then we have

$$\mu(E) = \int_X \mu_1(E - x) \, d\mu_2(x) \geq \int_{\{0\}} \mu_1(E - x) \, d\mu_2(x) = \mu_1(E) \cdot \mu_2(\{0\})$$

whence (i) follows.   □

4.9. LEMMA. *Let the charges ν_1 and ν_2 on the group X be absolutely continuous with respect to finite measures μ_1 and μ_2, respectively. Then the charge $\nu = \nu_1 * \nu_2$ is absolutely continuous with respect to the measure $\mu = \mu_1 * \mu_2$.*

PROOF. Obviously, it suffices to prove the lemma under the assumption that ν_1 and ν_2 are measures. Let $E \in \mathcal{B}(X)$ and $\mu(E) = 0$. Then $0 = (\mu_1 * \mu_2)(E) = \int_X \mu_1(E - x) \, d\mu_2(x)$, i.e., $\mu_2(\{x : \mu_1(E - x) > 0\}) = 0$. Observe that $\{x : \mu_1(E - x) = 0\} \subset x : \nu_1(E - x) = 0\}$. Therefore $\{x : \mu_1(E - x) > 0\} \supset \{x : \nu_1(E - x) > 0\}$. It follows from $\mu_2(\{x : \mu_1(E - x) > 0\}) = 0$ that $\nu_2(\{x : \mu_1(E - x) > 0\}) = 0$. Hence $\nu_2(\{x : \nu_1(E - x) > 0\}) = 0$, and, consequently, $\int_X \nu_1(E - x) \, d\nu_2(x) = 0$, i.e., $\nu(E) = 0$.   □

4.10. LEMMA. *Let $\mu \in \mathcal{M}^1(X)$ and $\mu(\{0\}) = q > 1/2$. Then any divisor of μ has the form $e(\Phi) * E_x$, where Φ is a charge such that*

$$\text{(i)} \quad \|\Phi\| \leq \ln \frac{1}{2q - 1}.$$

If $\mu = e(P)$, where $P \in \mathcal{M}_+(X)$, then the charge Φ is absolutely continuous with respect to μ.

PROOF. Let $\mu = \mu_1 * \mu_2$. The condition of the lemma implies that for some point $x_1 \in X$ the inequality $\mu_1(\{x_1\}) \geq q$ holds. We assume that $x_1 = 0$; otherwise we go over to the shifts $\mu_1 * E_{-x_1}$ and $\mu_2 * E_{x_1}$. Setting $c = \mu_1(\{0\})$ and $\nu = \mu_1 - cE_0$, we have

$$\nu(X) = 1 - c \leq 1 - q < q \leq c. \tag{1}$$

Consider the charge Φ defined by the series

$$\Phi = \sum_{m=1}^{\infty} \frac{(-1)^{m+1}}{mc^m} \nu^{*m}. \tag{2}$$

It follows from (1) that this series is norm convergent. We have $\hat{\Phi}(y) = \ln(1 + \hat{\nu}(y)/c) = \ln \hat{\mu}_1(y) - \ln c$. Setting $y = 0$ here, we obtain $\ln c = -\hat{\Phi}(0) = -\int_X d\Phi(x) = -\Phi(X)$. So, $\ln \hat{\mu}_1(y) = \int_X [(x, y) - 1] d\Phi(x)$ and, hence, $\mu_1 = e(\Phi)$.

Rewrite (2) in the form

$$\Phi = \Phi^+ - \Phi^- = \sum_{n=1}^{\infty} \frac{\nu^{*(2n-1)}}{(2n-1)c^{2n-1}} - \sum_{n=1}^{\infty} \frac{\nu^{*2n}}{2nc^{2n}}.$$

By using (1) we obtain

$$\Phi^+(X) = \sum_{n=1}^{\infty} \frac{[\nu(X)]^{(2n-1)}}{(2n-1)c^{2n-1}} \leq \sum_{n=1}^{\infty} \frac{(1-q)^{2n-1}}{(2n-1)q^{2n-1}} = \frac{1}{2}\ln\frac{1}{2q-1}, \qquad (3)$$

$$\Phi^-(X) = \sum_{n=1}^{\infty} \frac{[\nu(X)]^{2n}}{2nc^{2n}} \leq \sum_{n=1}^{\infty} \frac{(1-q)^{2n}}{2nq^{2n}} = \frac{1}{2}\ln\frac{q^2}{2q-1}. \qquad (4)$$

Inequality (i) follows from (3) and (4).

In what follows we shall assume $\Phi^+(\{0\}) = \Phi^-(\{0\}) = 0$. This normalization does not change the distribution $e(\Phi)$, and inequality (i) is preserved. Let $\mu = e(P)$. From $\mu(\{0\}) > \frac{1}{2}$ and $\mu_1(\{0\}) > \frac{1}{2}$, one can easily see that $\mu_2(\{0\}) > 0$. By Lemma 4.8, the distribution μ_1 is absolutely continuous with respect to μ. Hence the measure ν is absolutely continuous with respect to μ. By Lemma 4.9, the measure ν^{*n} is absolutely continuous with respect to the measure μ^{*n} for all natural n. Since $\mu = e(P)$, the distributions μ and μ^{*n} are mutually absolutely continuous. Therefore the measure ν^{*n} is absolutely continuous with respect to μ. From the norm convergence of the series (2) conclude that the charge Φ is absolutely continuous with respect to μ. $\square$

4.11. LEMMA. *Let Φ_1 and Φ_2 be charges on the group X, with $\Phi_1(\{0\})$ $= \Phi_2(\{0\}) = 0$. If*

$$\text{(i)} \quad e(\Phi_1) = e(\Phi_2) * E_x,$$

then $\Phi_1(X) = \Phi_2(X)$.

PROOF. Let $\Phi = \Phi_1 - \Phi_2$. Passing to characteristic functions in (i), we obtain

$$\exp\{\hat{\Phi}(y) - \Phi(X)\} = (x, y). \qquad (1)$$

Consider the charge $\overline{\Phi}$. We have

$$\exp\{\hat{\overline{\Phi}}(y) - \Phi(X)\} = \overline{(x, y)}. \qquad (2)$$

Using (1) and (2), we find $(\hat{\Phi} + \hat{\overline{\Phi}})(y) - 2\Phi(X) = 2k_y\pi i$ for some $k_y \in \mathbb{Z}$. But the left-hand side of this equality is real. Therefore, $(\hat{\Phi} + \hat{\overline{\Phi}})(y) = 2\Phi(X)$ for all $y \in Y$. By virtue of 2.10(b), $(\Phi + \overline{\Phi}) = 2\Phi(X)E_0$. On the other hand, $(\Phi + \overline{\Phi})(\{0\}) = 0$. Hence $\Phi(X) = 0$, i.e., $\Phi_1(X) = \Phi_2(X)$. $\square$

4.12. LEMMA. *Let Φ_1 and Φ_2 be charges on the group X, and*

$$\text{(i)} \quad \Phi_1(\{0\}) = \Phi_2(\{0\}) = 0,$$

$$\text{(ii)} \quad \|\Phi_1\| + \|\Phi_2\| < 2\pi,$$

$$\text{(iii)} \quad e(\Phi_1) = e(\Phi_2).$$

Then $\Phi_1 = \Phi_2$.

PROOF. Passing to characteristic functions, we obtain from (iii) that $\exp\{\hat{\Phi}_1(y) - \Phi_1(X)\} = \exp\{\hat{\Phi}_2(y) - \Phi_2(X)\}$. By Lemma 4.11, it follows from (i) and (iii) that $\Phi_1(X) = \Phi(X)$. Taking this into account, we have $\hat{\Phi}_1(y) - \hat{\Phi}_2(y) = 2k_y\pi i$, $k_y \in \mathbb{Z}$. Since, by (ii), $|\hat{\Phi}_1(y)| + |\hat{\Phi}_2(y)| < 2\pi$, we obtain $k_y = 0$ for all $y \in Y$. Therefore, $\hat{\Phi}_1(y) \equiv \hat{\Phi}_2(y)$, and by 2.10(b) we have $\Phi_1 = \Phi_2$. □

4.13. LEMMA. *Let* $\mu \in \mathcal{M}^1(X)$ *and* $\mu(\{0\}) = q > \frac{1}{2}$. *Assume that an infinitely divisible distribution* ν, *whose characteristic function has the representation* $(x, \Phi, 0)$, *is a divisor of* μ. *Then* $\Phi \in \mathcal{M}_+(X)$ *and* $\Phi(X) \leq \ln \frac{q}{(2q-1)}$.

PROOF. Let $A \in \mathcal{B}(X)$. Set $\chi_A(x) = 1$, for $x \in A$, and $\chi_A(x) = 0$, for $x \notin A$. Consider a divisor ν_1 of μ, whose characteristic function has the representation $(0, \chi_{X\setminus U}\Phi, 0)$, where U is a neighborhood of zero in X. By Lemma 4.10, ν_1 has the form $\nu_1 = e(\Phi_1) * E_{x_1}$, where Φ_1 is some charge on X, and by Lemma 4.11, $\Phi_1(X) = (\chi_{X\setminus U}\Phi)(X) = \Phi(X\setminus U)$. According to Lemma 4.10, $\|\Phi_1\| \leq \ln \frac{1}{2q-1}$. Therefore $\Phi(X\setminus U) \leq \ln \frac{q}{2q-1}$, and since U is arbitrary, $\Phi(X) \leq \ln \frac{q}{2q-1}$. □

4.14. LEMMA. *Let* $\mu_\varepsilon = e(\Phi_\varepsilon) \in \mathcal{M}^1(X)$, *where* Φ_ε *is a charge on* X *that is not a measure and* $\|\Phi_\varepsilon\| \to 0$ *as* $\varepsilon \to 0$. *Then for sufficiently small* ε *the distribution* μ_ε *is not infinitely divisible.*

PROOF. Suppose that is not so. Since $\hat{\mu}_\varepsilon(y) \neq 0$ for all $y \in Y$ and the function φ in the representation 2.22(i) is uniquely determined (see Remark 2.22), the characteristic function $\hat{\mu}_\varepsilon(y)$ has the representation $(x_\varepsilon, P_\varepsilon, 0)$, i.e.,

$$\hat{\mu}_\varepsilon(y) = (x_\varepsilon, y) \exp\left\{ \int_{X\setminus\{0\}} [(x, y) - 1 - ig(x, y)] \, dP_\varepsilon(x) \right\}. \tag{1}$$

It follows from the conditions of the lemma that $\mu(\{0\}) = 1 + o(\varepsilon)$ as $\varepsilon \to 0$. Take ε small enough to ensure that

$$\mu_\varepsilon(\{0\}) = q > \tfrac{2}{3}. \tag{2}$$

Then by Lemma 4.13 $P_\varepsilon \in \mathcal{M}_+(X)$ and

$$P_\varepsilon(X) \leq \ln \frac{q}{2q - 1} < \ln 2. \tag{3}$$

So the measure P_ε is finite. The properties of the function $g(x, y)$ from Lemma 2.19 via the duality Theorem 1.3 imply that

$$\exp\left\{ -i \int_{X\setminus\{0\}} g(x, y) \, dP_\varepsilon(x) \right\} = (x'_\varepsilon, y)$$

for some $x'_\varepsilon \in X$. Therefore it follows from (1) that

$$\hat{\mu}_\varepsilon(y) = (x''_\varepsilon, y) \exp\left\{ \int_{X\setminus\{0\}} [(x, y) - 1] \, dP_\varepsilon(x) \right\}. \tag{4}$$

Relations (3) and (4) imply $\mu_\varepsilon(\{x_\varepsilon''\}) > \frac{1}{2}$. Therefore, if $x_\varepsilon'' \neq 0$, then $\mu_\varepsilon(\{0\}) < \frac{1}{2}$, in contradiction to (2). So, $x_\varepsilon'' = 0$ in (4), i.e., $\mu_\varepsilon = e(P_\varepsilon)$. Since one can assume $\|\Phi_\varepsilon\| + \|P_\varepsilon\| < 2\pi$, Lemma 4.12 yields $\Phi_\varepsilon = P_\varepsilon$, in contradiction to the condition of the lemma. $\square$

4.15. COROLLARY. *Let* $\mu \in \mathscr{M}^1(X)$, *and let the characteristic function* $\hat{\mu}(y)$ *admit the representation*

(i) $\quad \hat{\mu}(y) - \exp\left\{(k - k_1)\int_{A_1} [(x, y) - 1]\, dm_X(x)\right.$

$$\left. -k_1 \int_{A_2} [(x, y) - 1]\, dm_X(x)\right\},$$

where $A_1, A_2 \in \mathscr{B}(X)$, $A_1 \cap A_2 = \varnothing$, $0 \notin A_1 \cup A_2$, $m_X(A_1) < \infty$, $0 < m_X(A_2) < \infty$, $0 < k_1 < k$. *Then for sufficiently small* k *the distribution* μ *is not infinitely divisible.*

4.16. REMARK. *Let* G *be a compact group. If* $\mu \in \mathscr{M}^1(G)$ *and the inequality*

(i) $\quad \mu(E) \geq am_G(E), \qquad 0 < a < 1,$

holds for every $E \in \mathscr{B}(X)$, *then* μ *may be decomposed in the form*

(ii) $\quad \mu = \dfrac{\mu - am_G}{1 - a} * [(1 - a)E_0 + am_G].$

For the proof it suffices to remove the parentheses on the right-hand side of (ii) and to observe that for any $\nu \in \mathscr{M}^1(G)$ the equality $\nu * m_G = m_G$ is valid.

4.17. THEOREM. *Let* G *be a compact group,* $G \not\approx \mathbb{Z}(2)$, *and let* $\mu \in \mathscr{M}^1(G)$ *satisfy condition* 4.16(i). *Then* μ *has an indecomposable divisor.*

PROOF. Let us use expansion 4.16(ii) and note that

$$e(km_G) = e^{-k} \sum_{n=0}^\infty \frac{(km_G)^{*n}}{n!} = e^{-k}E_0 + (1 - e^{-k})m_G, \qquad k > 0,$$

i.e.,

$$e(km_G) = (1 - a)E_0 + am_G, \qquad a = 1 - e^{-k}. \tag{1}$$

Therefore, it is sufficient to prove that the distribution $e(km_G)$ has an indecomposable divisor.

Choose nonzero elements x_1, x_2, and x_3 in G such that $x_3 = x_1 + x_2$ (x_1 and x_2 being not necessarily different). This can be done since $G \not\approx \mathbb{Z}(2)$. Let U_{x_1} and U_{x_3} be neighborhoods of the elements x_1 and x_3 such that $\overline{U}_{x_3} = x_2 + \overline{U}_{x_1}$, $\overline{U}_{x_1} \cap \overline{U}_{x_3} = \varnothing$, and $0, x_2 \notin \overline{U}_{x_3}$. Then the open set $V = G \backslash \overline{U}_{x_3}$ satisfies the condition $(2)V = G$. Take $\nu = \chi_V m_G$. Then for any $E \in \mathscr{B}(X)$ the inequality $\nu^{*2}(E) \geq a_1 m_G(E)$, $0 < a_1 < 1$, obviously holds, and therefore $e(k\chi_V m_G)(E) \geq a_2 m_G(E)$, $0 < a_2 < 1$. Thus the distribution $e(k\chi_V m_G)$ satisfies condition 4.16(i). By using 4.16(ii) and (1) this yields

$$e(k\chi_V m_G) = e(k_1 m_G) * \mu_1, \qquad (2)$$

where $\mu_1 \in \mathscr{M}^1(G)$, and one may assume $k_1 < k$. As may be seen from (2), the characteristic function $\hat{\mu}_1(y)$ of the distribution μ_1 admits the representation 4.15(i), where $A_1 = V\backslash\{0\}$, $A_2 = G\backslash V$. Since k may be supposed to be arbitrarily small and the sets A_1 and A_2 satisfy the conditions of Corollary 4.15, the distribution μ_1 is not infinitely divisible. This distribution obviously does not have nondegenerate idempotent divisors. By Theorem 4.3, μ_1 has an indecomposable divisor. Hence the distribution $e(k\chi_V m_G)$ has an indecomposable divisor, and thus $e(km_G)$ as well. $\square$

Theorem 4.17 has a number of useful consequences. They will be presented in §§5 and 6.

Returning to the decomposition 4.4(i), let us note that when studying the arithmetic of distributions the attention is mainly focused on the description of the class I_0 (see §§5–8). In this connection the following simple but useful assertion is significant.

4.18. PROPOSITION. *Let X_1 and X_2 be topological groups and $p: X_1 \to X_2$ a continuous monomorphism. A distribution $\mu \in \mathscr{M}^1(X_1)$ belongs to the class $I_0(X_1)$ if and only if the distribution $p(\mu) \in \mathscr{M}^1(X_2)$ belongs to the class $I_0(X_2)$.*

PROOF. Since a one-to-one and continuous image of a Borel set is a Borel set [Kur, §39, IV], p is an algebraic isomorphism of the topological groups X_1 and $p(X_1)$, which preserves Borel sets. If $\nu \in \mathscr{M}^1(X_1)$ is an indecomposable distribution, then Propositions 2.5 and 2.6 imply that $p(\nu)$ is also an indecomposable distribution. The converse follows directly from Proposition 2.6. If $\lambda \in \mathscr{M}^1(X_1)$ is an idempotent distribution, then, by the Proposition 2.6, $p(\lambda)$ is an idempotent distribution as well. Furthermore, if $p(\lambda)$ is an idempotent distribution, i.e., $(p(\lambda))^{*2} = (p(\lambda)) * E_{x_2}$, $x_2 \in X_2$, then obviously $x_2 = p(x_1)$, $x_1 \in X_1$. Proposition 2.6 implies that $\lambda^{*2} = \lambda * E_{x_1}$, i.e,. λ is an idempotent distribution.

Now Proposition 4.18 follows from what is said above and Proposition 2.5. $\square$

4.19. REMARK. It follows from its proof that Proposition 4.18 remains valid if p is an algebraic isomorphism of the groups X_1 and X_2 that preserves Borel sets.

§5. Gaussian distribution

In this section we define Gaussian distribution on a group X and study its properties. In particular, we prove that an arbitrary symmetric Gaussian distribution can be "linearized" in the sense that it is a continuous homomorphic image of a Gaussian distribution on a linear space. We completely describe those groups X to which the Cramér theorem on decomposition of the Gaussian distribution on $\mathbb{R}^n$ can be extended. We also give a complete description of all groups X with the following property: a distribution on X transformed into a Gaussian distribution on $\mathbb{T}$ by every character, is itself Gaussian.

5.1. DEFINITION. A distribution μ on a group X is called Gaussian if its characteristic function may be represented in the form

$$\text{(i)} \quad \hat{\mu}(y) = (x, y) \exp\{-\varphi(y)\},$$

where $\varphi(y)$ is a nonnegative continuous function on Y satisfying the equation

$$\text{(ii)} \quad \varphi(y_1 + y_2) + \varphi(y_1 - y_2) = 2[\varphi(y_1) + \varphi(y_2)]$$

for any $y_1, y_2 \in Y$.

It can easily be seen that, for the groups $X = \mathbb{R}^n$ and $X = \mathbb{T}^n$, this definition coincides with the classical one. (Recall that a Gaussian distribution on the group $\mathbb{T}^n$ is an image of a Gaussian distribution on $\mathbb{R}^n$ under the natural homomorphism $\mathbb{R}^n \to \mathbb{T}^n$. In particular, a nondegenerate Gaussian distribution μ on the group $\mathbb{T}$ has the density

$$\frac{1}{2\pi} \frac{d\mu}{dx} = \frac{1}{\sqrt{2\pi}\sigma} \sum_{n=-\infty}^{\infty} \exp\left\{-\frac{1}{2\sigma^2}(x - a + 2\pi n)^2\right\}$$

with respect to m_T, and its characteristic function is

$$\hat{\mu}(n) = \exp\left\{-\frac{1}{2}\sigma^2 n^2 + ina\right\}, \quad n \in \mathbb{Z}.)$$

We denote by $\Gamma(X)$ the set of all Gaussian distributions on the group X. A Gaussian distribution is called symmetric if $x = 0$ in (i). We denote the set of symmetric Gaussian distributions on the group X by $\Gamma^s(X)$. (Obviously, if $\mu \in \Gamma^s(X)$, then $\mu = \bar{\mu}$. Furthermore, if $\mu = \bar{\mu}$, then one can easily see that $2x = 0$ in (i). Indeed, if $\mu = \bar{\mu}$, then the function $\hat{\mu}(y)$ is real; therefore, (x, y) is a real function for all $y \in Y$. Hence $(x, y) = \pm 1$ and $(2x, y) = (x, y)^2 = 1$ for all $y \in Y$, i.e., $2x = 0$).

5.2. REMARK. Let $\psi(y_1, y_2)$, $y_1, y_2 \in Y$, be a real continuous function having the following properties:

(a) $\psi(y_1, y_2) = \psi(y_2, y_1)$,
(b) $\psi(y_1 + y_2, y_3) = \psi(y_1, y_3) + \psi(y_2, y_3)$,
(c) $\psi(y, y) \geq 0$.

As can easily be seen the function $\varphi(y) = \psi(y, y)$ satisfies 5.1(ii). Conversely, any nonnegative function φ satisfying equation 5.1(ii) can be obtained in this way. The corresponding function ψ is determined by the formula

$$\psi(y_1, y_2) = \frac{1}{2}[\varphi(y_1 + y_2) - \varphi(y_1) - \varphi(y_2)].$$

5.3. REMARK. Let $\varphi(y)$, $y \in Y$, be a continuous function satisfying equation 5.1(ii). Then $\varphi(y_2) = 0$ for any $y_2 \in Y_0$ and $\varphi(y_1 + y_2) = \varphi(y_1)$ for any $y_1 \in Y$, $y_2 \in Y_0$.

Indeed, let Y_1 be a compact subgroup of Y. Integrating equality 5.1(ii) over Y_1 with respect to the measure $dm_{Y_1}(y_1)$, we obtain $\varphi(y_2) = 0$ for $y_2 \in Y_2$.

Let $y_1 \in Y$, $y_2 \in Y_0$, and let M_{y_2} be the closed subgroup generated by y_2. Consider the function $P(l) = \varphi(y_1 + ly_2)$ on the group $\mathbb{Z}$. It follows from equation 5.1(ii) that the function $P(l)$ satisfies the equation $P(l+3) - 3P(l+2) + 3P(l+1) - P(l) = 0$. The latter implies (see [Ge, Chapter V, §3]) that $P(l) = a_0 l^2 + a_1 l + a_2$, where the a_j depend on y_1 and y_2. Since the function $\varphi(y)$ is continuous on the compact set $y_1 + M_{y_2}$, it is bounded on it. Therefore $a_0 = a_1 = 0$ and hence $\varphi(y_1 + y_2) = \varphi(y_1)$.

To study characterization problems on groups in Chapter III, we shall need the following property of the function $\varphi(y)$.

5.4. PROPOSITION. *Let H be an open subgroup of Y and $\varphi_0(h)$ a continuous nonnegative function on H satisfying equation 5.1(ii). Then there exists a continuous nonnegative function $\varphi(y)$ on Y satisfying equation 5.1(ii) and such that its restriction to H coincides with $\varphi_0(h)$.*

PROOF. Let $y_1 \notin H$. The standard reasoning that we are omitting here shows it is sufficient to extend the function $\varphi_0(h)$ (with its properties preserved) from the subgroup H to the open subgroup $H_1 = \{y \in Y : y = ny_1 + h, \ n \in \mathbb{Z}, \ h \in H\}$. Two cases are possible:

1. $ny_1 \cap H = \varnothing$ for arbitrary $n \in \mathbb{Z}$, $n \neq 0$. Put $\varphi(ny_1 + h) = \varphi_0(h)$ for $n \in \mathbb{Z}$, $h \in H$.

2. $ny_1 \in H$ for some $n \in \mathbb{Z}$, $n \neq 0$. Let n_0 be the least natural number such that $n_0 y_1 \in H$. Then $n_0 y \in H$ for any $y \in H_1$. Set $\varphi(y) = (1/n_0^2)\varphi_0(n_0 y)$, $y \in H_1$.

As can easily be seen, the function $\varphi(y)$ thus defined yields the desired extension. $\square$

Now let us study the structure of the support of a Gaussian distribution. Apparently, it suffices to restrict ourselves to the case of a symmetric Gaussian distribution.

5.5. PROPOSITION. *Let $\mu \in \Gamma^s(X)$. Then $\sigma(\mu) = G$, where G is a connected subgroup of the group X.*

PROOF. Consider $E = \{y \in Y : \mu(y) = 1\}$ and put $G = A(X, E)$. As follows from Proposition 2.13, $\sigma(\mu) \subset G$, i.e., $\mu \in \Gamma^s(G)$. Denote $G^* = H$. The function $\hat{\mu}(h)$, $h \in H$, equals 1 at $h = 0$ only. From Remark 5.3 we obtain that $H_0 = \{0\}$ and, according to Corollary 1.10, the group G is connected.

Suppose that $\mu(U) = 0$ for some open set $U \subset G$, and let U_0 be an open subset of U such that $U_0 + V \subset U$ for some neighborhood V of zero in G. The group G is connected. Therefore in view of Theorems 1.15 and 1.19 the neighborhood V contains a subgroup K such that the group G/K is topologically isomorphic to $\mathbb{R}^n + \mathbb{T}^m$ for some $n, m \geq 0$. Denote this isomorphism by τ, and put $p = \tau \circ \pi$, where $\pi: G \to G/K$ is the natural homomorphism. Then $p(\mu) \in \Gamma^s(\mathbb{R}^n + \mathbb{T}^m)$, $p(\mu)(p(U_0)) = 0$, and $p(\hat\mu)(\zeta) = 1$ only at $\zeta = 0$, $\zeta \in \mathbb{R}^n + \mathbb{Z}^m$, but this is impossible since the set $p(U_0)$ is open. The contradiction we have come to shows that $\mu(U) \neq 0$. $\square$

Let us prove now that the "linearization" is possible for an arbitrary symmetric Gaussian distribution. First consider the case of finite-dimensional groups.

5.6. PROPOSITION. *Let X be a connected group of finite dimension l. Then there exists a continuous homomorphism $p: \mathbb{R}^l \to X$ having the property*: *for any symmetric Gaussian distribution μ on X there exists a symmetric Gaussian distribution ν on $\mathbb{R}^l$ such that $\mu = p(\nu)$.*

PROOF. Introduce the following notation. If $t = (t_1, \dots, t_n) \in \mathbb{R}^n$ and $s = (s_1, \dots, s_n) \in \mathbb{R}^n$, then put $\langle t, s \rangle = \sum_{j=1}^n t_j s_j$.

By Theorem 1.15, $X \approx \mathbb{R}^n + K$, where $n \geq 0$, K being a connected compact group. Let $D = K^*$. First we prove the proposition assuming that $X = K$. According to Corollary 1.10 and Theorem 1.13, in this case D is a discrete torsion-free group of rank l. Let $d_1, \dots, d_l$ be a maximal independent system of elements of D. Then for each $d \in D$ there exist integers $k, k_1, \dots, k_l$, such that

$$kd = k_1 d_1 + \cdots + k_l d_l. \tag{1}$$

The independence of $\{d_j\}$ implies that the set $\{k_j/k\}$ is uniquely determined by d. Since D is a torsion-free group, the mapping

$$f(d) = \left(\frac{k_1}{k}, \dots, \frac{k_l}{k} \right)$$

is a monomorphism $f: D \to \mathbb{R}^l$.

Let the function $\varphi(d)$ on the group D satisfy equation 5.1(ii). In view of Remark 5.2, equality (1) yields

$$k^2 \varphi(d) = \varphi(kd) = \psi(kd, kd) = \psi\left(\sum_{j=1}^l k_j d_j, \sum_{j=1}^l k_j d_j \right) = \sum_{i,j=1}^l \alpha_{ij} k_i k_j,$$

where $\alpha_{ij} = \psi(d_i, d_j)$, $1 \leq i, j \leq l$. Consequently,

$$\varphi(d) = \sum_{i,j=1}^l \alpha_{ij} \left(\frac{k_i}{k} \right) \left(\frac{k_j}{k} \right) = \langle Af(d), f(d) \rangle, \qquad A = (\alpha_{ij})_{i,j=1}^l. \tag{2}$$

If $\varphi(d) \geq 0$ for $d \in D$, then the symmetric matrix A is positive semidefinite, i.e., the quadratic form $\langle As, s \rangle$ is nonnegative for any $s = (s_1, \dots, s_l) \in \mathbb{R}^l$.

Let $\mu \in \Gamma^s(K)$ and the function $\varphi(d)$ in equation 5.1(ii) correspond to the characteristic function $\hat{\mu}(d)$. As follows from formula (2),

$$\hat{\mu}(d) = \exp\{-\langle Af(d), f(d)\rangle\}. \tag{3}$$

Denote by ν a Gaussian distribution in $\mathbb{R}^l$ with the characteristic function

$$\hat{\nu}(s) = \exp\{-\langle As, s\rangle\}. \tag{4}$$

Let p, $p\colon \mathbb{R}^l \to K$, be the adjoint homomorphism to f (§1.20). It follows from §1.20, Theorem 2.11 and formula (4) that

$$\hat{p}(\nu)(d) = \exp\{-\langle Af(d), f(d)\rangle\}. \tag{5}$$

Comparing (3) and (5), we obtain from 2.10(b) that $\mu = p(\nu)$.

In the general case $X \approx \mathbb{R}^n + K$ one may assume $Y = \mathbb{R}^n + D$ and the proof is similar. A continuous monomorphism $f\colon Y \to \mathbb{R}^l$ is determined in the following way: $f((s; d)) = f((s_1, \ldots, s_n; d)) = (s_1, \ldots, s_n, k_1/k, \ldots, k_m/k)$, $l = n+m$, where $s = (s_1, \ldots, s_n) \in \mathbb{R}^n$, $d \in D$, and $k, k_1, \ldots, k_m$ are chosen as in the case $X = K$. One can easily see that the matrix $A = (\alpha_{ij})_{i,j=1}^l$ has the form: $a_{ij} = \psi(e_i, e_j)$, $1 \le i, j \le n$; $\alpha_{i,j+n} = \psi(e_i, d_j)$, $1 \le i \le n$, $1 \le j \le m$; $\alpha_{i+n,j+n} = \psi(d_i, d_j)$, $1 \le i, j \le m$, where $e_1, \ldots, e_n$ is the standard basis of $\mathbb{R}^n$. $\square$

In order to formulate an analog of Proposition 5.6 for infinite-dimensional groups, we need a definition of a Gaussian distribution on an infinite-dimensional space.

5.7. Denote by $\mathbb{R}^\infty$ the space of all real number sequences endowed with the Tikhonov topology of a Cartesian product. Convergence of elements $t^{(k)} \to t$ in this topology means coordinate-wise convergence. The space $\mathbb{R}^\infty$ may be regarded as the projective limit of the directed set of spaces $\mathbb{R}^n$. The topological group $\mathbb{R}^\infty$ is not locally compact, but it satisfies the second axiom of countability.

Denote by $\mathbb{R}_0^\infty$ the space of all finitary sequences of real numbers (*) with the topology of the inductive limit of spaces $\mathbb{R}^n$. Converges of the elements $s^{(k)} \to s$ in this topology means that all $s^{(k)}$ lie in the same $\mathbb{R}^n$ and converge there. The topological group $\mathbb{R}_0^\infty$ also is not locally compact. Let $t = (t_1, \ldots, t_n, \ldots,) \in \mathbb{R}^\infty$, $s = (s_1, \ldots, s_n, 0, \ldots) \in \mathbb{R}_0^\infty$. Denote by $\langle t, s \rangle = \sum_{j=1}^\infty t_j s_j$. Fixing $s \in \mathbb{R}_0^\infty$, let us consider the function

$$(t, s) = \exp\{i\langle t, s\rangle\}, \qquad t \in \mathbb{R}^\infty \tag{1}$$

on the group $\mathbb{R}^\infty$.

This function is a character of the group $\mathbb{R}^\infty$ since it is a continuous homomorphism of the group $\mathbb{R}^\infty$ into $\mathbb{T}$. As can easily be seen, any character of the group $\mathbb{R}^\infty$ possesses the form (1).

(*) *Editor's note.* I.e., infinite sequences with finitely many nonzero terms.

Let $\mathscr{B}(\mathbb{R}^\infty)$ be the σ-algebra of Borel sets in $\mathbb{R}^\infty$, that is, the minimal σ-algebra containing all sets of the form $U_j^n + \mathbb{R}^{\infty/J}$, where $J = (j_1, \ldots, j_n)$, U_j^n is an open set in $\mathbb{R}_j^n = \{t \in \mathbb{R}^\infty : t_j = 0 \text{ for } j \notin J\}$, and $\mathbb{R}^{\infty/J} = \{t \in \mathbb{R}^\infty : t_j = 0 \text{ for } j \in J\}$.

From the definitions of the topology in $\mathbb{R}^\infty$ and of the σ-algebra $\mathscr{B}(\mathbb{R}^\infty)$ it follows that each continuous function on $\mathbb{R}^\infty$ is measurable. For an arbitrary distribution μ on $\mathbb{R}^\infty$ we define its characteristic function

$$\hat{\mu}(s) = \int_{\mathbb{R}^\infty} (t, s) \, d\mu(t), \qquad s \in \mathbb{R}_0^\infty.$$

This function has the following properties: a) $\hat{\mu}$ is continuous; b) $\hat{\mu}$ is positive definite on each subspace $\mathbb{R}_j^n \subset \mathbb{R}_0^\infty$; c) $\hat{\mu}(0) = 1$. It follows directly from the Kolmogorov theorem and Theorem 2.8 that each function $g(s)$ on $\mathbb{R}_0^\infty$ satisfying the conditions a)–c) uniquely corresponds to some distribution μ_g on $\mathbb{R}^\infty$ such that $\hat{\mu}_g(s) = g(s)$.

Evidently, a characteristic function has the properties 2.10(a)–(e).

We now define a Gaussian distribution on the space $\mathbb{R}^\infty$.

5.8. Let $A = (\alpha_{ij})_{i,j=1}^\infty$ be a symmetric positive semidefinite matrix, i.e., such that the quadratic form $\langle As, s \rangle = \sum_{i,j=1}^\infty \alpha_{ij} s_i s_j$ is nonnegative for any $s = (s_1, \ldots, s_n, 0, \ldots) \in \mathbb{R}_0^\infty$.

DEFINITION. A distribution μ on $\mathbb{R}^\infty$ is called Gaussian, if its characteristic function has the form

$$\hat{\mu}(s) = (t, s) \exp\{-\langle As, s \rangle\},$$

where $t \in \mathbb{R}^\infty$ and $A = (\alpha_{ij})_{i,j=1}^\infty$ is a symmetric positive semidefinite matrix. The notion of a symmetric Gaussian distribution is introduced in the same way as in Definition 5.1, as well as the notation $\Gamma(\mathbb{R}^\infty)$ and $\Gamma^s(\mathbb{R}^\infty)$.

Let us prove now an analog of Proposition 5.6 for infinite-dimensional groups.

5.9. PROPOSITION. *Let X be a connected infinite-dimensional group. Then there exists a continuous homomorphism $p \colon \mathbb{R}^\infty \to X$ having the following property: for any symmetric Gaussian distribution μ on X there exists a symmetric Gaussian distribution ν on $\mathbb{R}^\infty$ such that $\mu = p(\nu)$.*

PROOF. By Theorem 1.15, $X \approx \mathbb{R}^n + K$, where $n \geq 0$ and K is a connected compact group. We restrict ourselves to the case $X = K$. The changes needed to be made for the general case are the same as in Proposition 5.6. Put $D = K^*$. According to Corollary 1.10 and Theorems 1.11 and 1.13, D is a countable torsion-free discrete group of infinite rank. Let $d_1, \ldots, d_l, \ldots$ be a maximal independent system of elements in D. Then for each $d \in D$ there exist integers $k, k_1, \ldots, k_l$ such that the equality $kd = k_1 d_1 + \cdots + k_l d_l$ holds. In a similar way as in Proposition 5.6, we consider a monomorphism

$f\colon D \to \mathbb{R}_0^\infty$ defined by the formula

$$f(d) = \left(\frac{k_1}{k}, \dots, \frac{k_l}{k}, 0, \dots \right).$$

Let $p\colon \mathbb{R}^\infty \to K$ be the continuous adjoint homomorphism to f, i.e., $(p(t), d) = (t, f(d))$ for any $t \in \mathbb{R}^\infty$, $d \in D$ (clearly, p is continuous).

Note that whatever a distribution α on $\mathbb{R}^\infty$, the characteristic function of the distribution $p(\alpha)$ has the form

$$\widehat{p(\alpha)}(d) = \hat{\alpha}(f(d)). \tag{1}$$

(This assertion does not follow from Proposition 2.11 since the group $\mathbb{R}^\infty$ is not locally compact. It can be proven independently.)

Just as in the proof of Proposition 5.6, we see that any nonnegative function $\varphi(d)$ on D satisfying equation 5.1(ii) can be represented in the form

$$\varphi(d) = \langle Af(d), f(d) \rangle, \tag{2}$$

where $A = (\alpha_{ij})_{i,j=1}^\infty$ is a symmetric positive semidefinite matrix $(\alpha_{ij} = \psi(d_i, d_j))$.

Let $\mu \in \Gamma^s(K)$ and let the function $\varphi(d)$ be connected with the characteristic function $\hat{\mu}(d)$ as in 5.1(i). Then (2) implies that

$$\hat{\mu}(d) = \exp\{-\langle Af(d), f(d) \rangle\}. \tag{3}$$

Denote by ν a Gaussian distribution on $\mathbb{R}^\infty$ having a characteristic function

$$\hat{\nu}(s) = \exp\{-\langle As, s \rangle\}. \tag{4}$$

It follows from (1) and (4) that

$$\widehat{p(\nu)}(d) = \exp\{-\langle Af(d), f(d) \rangle\}. \tag{5}$$

Comparing (3) and (5), and using 2.10(b) we conclude that $\mu = p(\nu)$. □

Combining Propositions 5.6 and 5.9, we obtain the following assertion.

5.10. THEOREM. *Let X be a connected group of finite dimension l (or of infinite dimension). Then there exists a continuous homomorphism p from the group $\mathbb{R}^l$ (respectively $\mathbb{R}^\infty$) into X with the following property: for any $\mu \in \Gamma^s(X)$ there exists $\nu \in \Gamma^s(\mathbb{R}^l)$ ($\nu \in \Gamma^s(\mathbb{R}^\infty)$) such that $\mu = p(\nu)$.*

5.11. REMARK. Let $\varphi(y)$ be a continuous nonnegative function on Y that satisfies equation 5.1(ii). Then there exists a distribution $\mu \in \mathscr{M}^1(X)$ whose characteristic function has the form $\hat{\mu}(y) = \exp\{-\varphi(y)\}$. The distribution μ is infinitely divisible. Actually, according to Remark 5.3, the function $\varphi(y)$ defines a continuous nonnegative function $\tilde{\varphi}([y])$ that satisfies equation 5.1(ii) on the group Y/Y_0 according to the formula $\tilde{\varphi}([y]) = \varphi(y)$. As follows from Theorems 1.6 and 1.9, $C_X^* \approx Y/Y_0$. The arguments in the proofs of Propositions 5.6 and 5.9 demonstrate that $\exp\{-\tilde{\varphi}([y])\} = \hat{\mu}([y])$ for some distribution $\mu \in \mathscr{M}^1(C_X)$. Here the characteristic function of the distribution μ, regarded as a distribution on X, has the form $\hat{\mu}(y) = \exp\{-\varphi(y)\}$. The fact that μ is an infinitely divisible distribution is obvious.

5.12. REMARK. As we have proved in Proposition 5.5, if $\mu \in \Gamma^s(X)$, then $\sigma(\mu) = G$, where G is a connected subgroup of the group X. It turns out that for any connected group G there exists a distribution $\mu \in \Gamma^s(G)$ such that $\sigma(\mu) = G$. By Theorem 1.15, it suffices to prove this assertion for a connected compact group K. Put $D = K^*$. Let $\mu \in \Gamma^s(K)$, and let the characteristic function $\hat{\mu}(d)$, $d \in D$, have the form

$$\hat{\mu}(d) = \exp\{-\langle Af(d), f(d) \rangle\}.$$

Here the homomorphism f is either as in Proposition 5.6 if $\dim K < \infty$, or as in Proposition 5.9 if $\dim K = \infty$. Then $\hat{\mu}(d) = 1$ only at $d = 0$, and it follows from Proposition 5.5 that $\sigma(\mu) = K$.

5.13. Let X be a nondiscrete group and $\mu \in \mathcal{M}_+(X)$. Expand the measure μ into the sum

$$\text{(i)} \quad \mu = \mu_{\text{ac}} + \mu_{\text{s}} + \mu_{\text{d}},$$

where μ_{ac} is a measure absolutely continuous with respect to m_X, μ_{s} is a measure singular with respect to m_X, and μ_{d} is a discrete measure. We call such an expansion the structure of the measure μ. Let us study the structure of a Gaussian distribution.

Let X be a connected group and $\mu \in \Gamma(X)$. Accounting for Proposition 5.5, when studying the structure of the measure μ we assume $\mu \in \Gamma^s(X)$ and $\sigma(\mu) = X$.

First note that $\mu_{\text{d}} = 0$. Indeed, by Theorem 1.19, a subgroup G of X can be found such that the group X/G is topologically isomorphic to $\mathbb{T}$. Denote this isomorphism by τ and set $p = \tau \circ \pi$, where $\pi: X \to X/G$ is the natural homomorphism. Then $p(\mu) \in \Gamma(\mathbb{T})$. If $\mu(\{x_0\}) > 0$, then $p(\mu)(\{p(x_0)\}) = \mu(\{p^{-1}(p(x_0))\}) \geq \mu(\{x_0\}) > 0$, but the latter is impossible.

Let $\mathbb{T}^\infty$ be the infinite-dimensional torus. Denote by $\mathbb{Z}_0^\infty$ the additive discrete group of all finitary sequences of integers. Then $(\mathbb{T}^\infty)^* \approx \mathbb{Z}_0^\infty$.

Further analysis of the structure of a Gaussian distribution will be carried out separately for groups that are and that are not locally connected. Recall that a connected group X is locally connected if and only if X is topologically isomorphic either to the group $\mathbb{R}^n + \mathbb{T}^m$ or to the group $\mathbb{R}^n + \mathbb{T}^\infty$ (see [HR1, §24]).

5.14. PROPOSITION. *Let X be a connected, not locally connected group and $\mu \in \Gamma^s(X)$. Then $\mu = \mu_{\text{s}}$.*

PROOF. Let L denote either the group $\mathbb{R}^l$, l being a nonnegative integer in the case $\dim X = l$ or the group $\mathbb{R}^\infty$ in the case $\dim X = \infty$. As follows from Theorem 5.10, there exists a continuous homomorphism $p: L \to X$ such that $\mu = p(\nu)$, $\nu \in \Gamma(L)$. Therefore the distribution μ is concentrated on the subgroup $p(L)$. This subgroup cannot coincide with X. Indeed, if $p(L) = X$, then the group X is the union of its one-parameter subgroups. Since X satisfies the second axiom of countability, we have either $X \approx \mathbb{R}^n + \mathbb{T}^m$ or $X \approx \mathbb{R}^n + \mathbb{T}^\infty$ [B2, Chapter 2], in contradiction to the hypothesis.

It should be noted now that if $A \in \mathscr{B}(X)$ and $m_X(A) > 0$, then the difference set $A - A = \{x \in X : x = u - v, \ u, v \in A\}$ contains a neighborhood of zero of the group X. Therefore, if A is a subgroup of X, then $A - A = A$, and, hence, $A = X$ due to connectedness of X. Since we have already proved that $p(L) \neq X$, it follows that $m_X(p(L)) = 0$, and, hence, $\mu = \mu_s$. $\square$

5.15. Consider now the case of a connected locally connected group X. If $\dim X < \infty$, then $X \approx \mathbb{R}^n \to \mathbb{T}^m$. The structure of Gaussian distributions on such groups is well known. If $\mu \in \Gamma^s(X)$, then either $\mu = \mu_{ac}$ or $\mu = \mu_s$. (This fact can easily be seen from Proposition 5.6. Indeed, set $G = \sigma(\nu)$. Then, if $p(G) = X$, then $\mu = \mu_{ac}$. On the other hand, if $p(G) \neq X$, then $\mu = \mu_s$.)

Now let $\dim X = \infty$. Then $X \approx \mathbb{R}^n \to \mathbb{T}^\infty$. We need the Kakutani theorem [Ka], which is stated as follows.

THEOREM. *Let $\{\mu_k\}$ and $\{\nu_k\}$ be two sequences of distributions on a space Ω. Assume that for all $k = 1, 2, 3, \ldots$ the distributions μ_k and ν_k are mutually absolutely continuous. Then the direct products of the distributions*

$$\mu = \bigotimes_{k=1}^\infty \mu_k, \qquad \nu = \bigotimes_{k=1}^\infty \nu_k$$

are mutually absolutely continuous if the product

$$\prod_{k=1}^\infty \int_\Omega \sqrt{\frac{d\nu_k}{d\mu_k}} \, d\mu_k$$

converges ($d\nu_k/d\mu_k$ is the Radon-Nikodým derivative), and mutually singular if this product diverges.

To simplify notation, we suppose that $X = \mathbb{T}^\infty$. If the matrix A corresponding to the distribution $\mu \in \Gamma^s(\mathbb{T}^\infty)$ (see Proposition 5.9) is diagonal $A = \operatorname{diag}\{\frac{1}{2}\sigma_1^2, \ldots, \frac{1}{2}\sigma_k^2, \ldots\}$, then $\mu = \bigotimes_{k=1}^\infty \mu_k$, where $\mu_k \in \Gamma^s(\mathbb{T})$ and $\hat{\mu}_k(n) = \exp\{-\frac{1}{2}\sigma_k^2 n^2\}$. Inasmuch as $m_{\mathbb{T}^\infty} = \bigotimes_{k=1}^\infty m_{\mathbb{T}_k}$, $\mathbb{T}_k = \mathbb{T}$, it immediately follows from the Kakutani theorem that either $\mu = \mu_{ac}$ or $\mu = \mu_s$. Either case may occur here. If the matrix A is arbitrary, the structure of the corresponding Gaussian distribution remains unclear.

It is well known [Ga] that any two Gaussian distributions in a linear space are either mutually absolutely continuous or mutually singular. Theorem 5.10 enables us to prove this alternative for finite-dimensional groups and for some class of infinite-dimensional groups.

5.16. PROPOSITION. *Let X be a connected group of finite dimension l. Then any two Gaussian distributions on X are either mutually absolutely continuous or mutually singular.*

PROOF. By Theorem 1.15, $X \approx \mathbb{R}^n + K$, where $n \geq 0$ and K is a connected compact group. To avoid complicated notation, we restrict ourselves to considering the case $X = K$, $\dim K = l$. Put $D = K^*$. Let the homomorphisms f and p be the same as in Proposition 5.6. Then $\mathbb{Z}^l \subset f(D) \subset \overline{f(D)}$. Section 1.20 yields $\ker p = A(\mathbb{R}^l, \overline{f(D)}) \subset A(\mathbb{R}^l, \mathbb{Z}^l) \approx \mathbb{Z}^l$. Put $G = \ker p$.

Let first $\mu_1, \mu_2 \in \Gamma^s(K)$. By Proposition 5.6, $\mu_i = p(\nu_i)$, where $\nu_i \in \Gamma^s(\mathbb{R}^l)$, $i = 1, 2$. Put $L_i = \sigma(\nu_i)$, $i = 1, 2$.

If $L_1 = L_2$, then the distributions ν_1 and ν_2 are mutually absolutely continuous, as well as the distributions μ_1 and μ_2. Let $L_1 \neq L_2$. Then the distributions ν_1 and ν_2 are mutually singular. Assume for definiteness that $L = L_1 \cap L_2$ is a proper subspace of L_2. Also note that $p^{-1}(p(L_1)) = L_1 + G$. By the assumption, $\mu_1(p(L_1)) = 1$. On the other hand, we have $\mu_2(p(L_1)) = \nu_2(p^{-1}(p(L_1))) = \nu_2(L_1 + G) = \nu_2((L_1 + G) \cap L_2) = 0$, since $(L_1 + G) \cap L_2$ is a proper subgroup of L_2. Hence μ_1 and μ_2 are mutually singular.

Consider now the general case, $\mu_i = \mu_i' * E_{x_i}$, $\mu_i' \in \Gamma^s(K)$, $i = 1, 2$. Obviously, without loss of generality, one can suppose that $x_2 = 0$. If $x_1 \notin p(\mathbb{R}^l)$, then $p(\mathbb{R}^l) \cap (x_1 + p(\mathbb{R}^l)) = \varnothing$; therefore, $\sigma(\mu_1) \cap \sigma(\mu_2) = \varnothing$, i.e., μ_1 and μ_2 are mutually singular. But if $x_1 = p(t_1)$, $t_1 \in \mathbb{R}^l$, then $\mu_1 = p(\nu_1)$, where $\nu_1 = \nu_1' * E_{t_1}$, $\nu_1' \in \Gamma^s(\mathbb{R}^l)$. The reasoning for this case is the same as in that for symmetric μ_1 and μ_2. $\square$

We now examine the case of an infinite-dimensional group X.

5.17. LEMMA. *Let X be a connected infinite-dimensional group, and let the homomorphism $f\colon Y \to \mathbb{R}_0^\infty$ be constructed as in §5.9. If the group X does not contain a subgroup topologically isomorphic to $\mathbb{T}$, then $\overline{f(Y)} = \mathbb{R}_0^\infty$.*

PROOF. By Theorem 1.15, $X \approx \mathbb{R}^n + K$, where $n \geq 0$ and K is a connected infinite-dimensional compact group. By Corollary 1.10 and Theorems 1.11, 1.13, we have $Y \approx \mathbb{R}^n + D$, where D is a countable torsion-free discrete group of finite rank. Since the restriction of f to $\mathbb{R}^n$ has the form $f(s_1, \ldots, s_n; 0) = (s_1, \ldots, s_n, 0)$, it suffices to prove the lemma for the case $X = K$, $Y = D$.

Consider the group $\mathbb{Z}^n$ naturally embedded into $\mathbb{R}^n$, and also the natural embeddings $\mathbb{R} \subset \mathbb{R}^2 \subset \cdots \subset \mathbb{R}^n \subset \mathbb{R}_0^\infty$. Put $H_n = \overline{f(D)} \cap \mathbb{R}^n$. Then H_n is a closed subgroup of $\mathbb{R}^n$. By Theorem 1.23(a), $H_n \approx \mathbb{R}^{l_n} + \mathbb{Z}^{k_n}$, where $l_n + k_n = n$, since $\mathbb{Z}^n \subset H_n$. The lemma will be proved if we verify that the equality $k_n = 0$ holds for any natural number n. Assume the contrary is valid, i.e., $k_n > 0$ for some n. Accounting for the fact that $H_n = H_{n+1} \cap \mathbb{R}^n$, Theorem 1.23(b) implies that H_n is a topological direct summand in H_{n+1}. Therefore the group $\overline{f(D)}$ contains a subgroup F, $F \approx \mathbb{Z}$, as a direct summand, i.e., $\overline{f(D)} = F + H$. Let e be a generator of the group F. Take an arbitrary element $a_0 = e + h_0 \in (e + H) \cap f(D)$, $h_0 \in H$. Let L be the subgroup of D generated by the element a_0. Clearly, $f(D) = L + (f(D) \cap H)$, i.e., the group $f(D)$ contains, as a direct summand, a group isomorphic to $\mathbb{Z}$. Hence the group D also contains, as a direct summand, a subgroup isomorphic to $\mathbb{Z}$ and the group K contains a subgroup topologically isomorphic to $\mathbb{T}$, contrary to the assumption that $k_n > 0$. $\square$

5.18. REMARK. Let X be a connected group of finite dimension l and $f: Y \to \mathbb{R}^l$ the homomorphism constructed in §5.6. If the group X does not contain a subgroup topologically isomorphic to $\mathbb{T}$, then $\overline{f(Y)} = \mathbb{R}^l$.

The proof is similar to that of Lemma 5.17.

5.19. PROPOSITION. *Let X be a connected infinite-dimensional group that does not contain a subgroup isomorphic to $\mathbb{T}$. Then any two Gaussian distributions on X are either mutually absolutely continuous or mutually singular.*

PROOF. By Theorem 1.15, $X \approx \mathbb{R}^n + K$, where $n \geq 0$ and K is a connected compact group. As in Proposition 5.16, we restrict ourselves to the case $X = K$, $\dim K = \infty$. Put $D = K^*$. Let f and p be the same homomorphisms as in Proposition 5.9. According to Lemma 5.17, $\overline{f(D)} = \mathbb{R}_0^\infty$. But then, as can easily be seen $\ker p = \{0\}$, and, hence, the homomorphism p realizes a one-to-one correspondence between the distributions on $\mathbb{R}^\infty$ and the distributions on K, concentrated on $p(\mathbb{R}^\infty)$.

Let $\mu_1, \mu_2 \in \Gamma^s(K)$. Then Proposition 5.9 implies $\mu_i = p(\nu_i)$, where $\nu_i \in \Gamma^s(\mathbb{R}^\infty)$, $i = 1, 2$. But any two Gaussian distributions in the space $\mathbb{R}^\infty$ are either mutually absolutely continuous or mutually singular [Ga]. Since the mapping p is one-to-one, the Gaussian distributions μ_1 and μ_2 are either mutually absolutely continuous or mutually singular as well.

The general case $\mu_i = \mu_i' * E_{x_i}$, $\mu_i' \in \Gamma^s(K)$, $i = 1, 2$, is treated in exactly the same way as in Proposition 5.16. $\square$

We now turn to consideration of arithmetic properties of Gaussian distributions. We recall the formulation of the Cramér theorem [LinO, Chapter II, §1].

CRAMÉR THEOREM. *If $\mu \in \Gamma(\mathbb{R}^n)$, then all divisors of μ are also Gaussian distributions.*

On the other hand, as Marcinkiewicz has noticed, such an assertion ceases to be true for the group $X = \mathbb{T}$.

5.20. THEOREM. *Any Gaussian distribution on the group $X = \mathbb{T}$ has non-Gaussian divisors.*

PROOF. Obviously, any distribution $\mu \in \Gamma(\mathbb{T})$ satisfies condition 4.16(i). Therefore decomposition 4.16(ii) is valid, and the divisors μ in this distribution are non-Gaussian. $\square$

Before giving a complete description of the groups X to which the Cramér theorem can be extended, we shall prove the following.

5.21. PROPOSITION. *Let $\mu \in \Gamma(X)$. If $\mu \in I_0$, then μ has only Gaussian divisors.*

PROOF. Let $\mu = \mu_1 * \mu_2$, where $\mu_i \in \mathcal{M}^1(X)$, $i = 1, 2$. Theorem 4.3 implies that μ_i are infinitely divisible distributions having no nondegenerate idempotent divisors. Let the characteristic function of the distribution μ have the representation $(x, 0, \varphi)$, and the characteristic functions of the distributions μ_i have the representations (x_i, Φ_i, φ_i) (Remark 2.22). Since φ in representation 2.22(i) is unique, it follows that $\varphi_1 + \varphi_2 = \varphi$; hence, we have

$$E_x = \nu_1 * \nu_2, \tag{1}$$

where ν_i, $i = 1, 2$, is an infinitely divisible distribution whose characteristic function has the representation $(x_i, \Phi_i, 0)$. From (1) we get that $\nu_i = E_{t_i}$, $t_i \in X$, $i = 1, 2$, and consequently, $\mu_i \in \Gamma(X)$, $i = 1, 2$. $\square$

The above reasoning also proves that in the class of infinitely divisible distributions the Gaussian one has only Gaussian divisors.

Now let us prove a group analogue of the Cramér theorem.

5.22. THEOREM. *For every Gaussian distribution on a group X to have only Gaussian divisors, it is necessary and sufficient that the group X contain no subgroup topologically isomorphic to $\mathbb{T}$.*

PROOF. The necessity immediately follows from Theorem 5.20. Assume that the group X contains no subgroup topologically isomorphic to $\mathbb{T}$ and $\mu \in \Gamma(X)$. Without loss of generality one may consider $\mu \in \Gamma^s(X)$. Proposition 5.5 implies that $\sigma(\mu) = G$, where G is a connected subgroup of X. Thereby $\mu \in \Gamma^s(G)$ and the group G also does not contain a subgroup topologically isomorphic to $\mathbb{T}$. Thus the group X could be assumed connected from the outset.

Let $\dim X = l < \infty$, and let f and p be the same homomorphisms as in Proposition 5.6. Then Proposition 5.6 implies $\mu = p(\nu)$, where $\nu \in \Gamma^s(\mathbb{R}^l)$. By Remark 5.18, $\overline{f(Y)} = \mathbb{R}^l$, whereby, according to 1.20(b), $\ker p = \{0\}$, i.e., p is a monomorphism. Since by the Cramér theorem $\nu \in I_0(\mathbb{R}^l)$, Proposition 4.18 yields $\mu = p(\nu) \in I_0(X)$.

If the group X is infinite dimensional, then let the homomorphisms f and p be the same as in Proposition 5.9. By Proposition 5.9, $\mu = p(\nu)$, where $\nu \in \Gamma^s(\mathbb{R}^\infty)$. According to Lemma 5.17, $\overline{f(Y)} = \mathbb{R}_0^\infty$. One can easily check that $\ker p = \{0\}$, i.e., that p is a monomorphism. From the Cramér theorem for the group $X = \mathbb{R}^l$ it immediately follows that the theorem holds for the group $X = \mathbb{R}^\infty$. Therefore, $\nu \in I_0(\mathbb{R}^\infty)$, whereby Proposition 4.18 implies $\mu = p(\nu) \in I_0(X)$. Now the assertion of the theorem follows from Proposition 5.21. $\square$

Let us add the following assertion to Theorem 5.20.

5.23. THEOREM. *Let X be a connected group. In order that any Gaussian distribution μ on X, such that $\sigma(\mu) = X$, have non-Gaussian divisors, it is necessary and sufficient that the group X be topologically isomorphic to the group*

$$\text{(i)} \quad \mathbb{R}^n + \mathbb{T}, \qquad n \geq 0.$$

PROOF. *Necessity.* Theorem 1.15 implies $X \approx \mathbb{R}^n + K$, where $n \geq 0$ and K is a connected compact group. Then $Y \approx \mathbb{R}^n + D$, where $D = K^*$. Let the group X be topologically nonisomorphic to a group f the form (i). Then $K \ntrianglelefteq \mathbb{T}$. Assume $\dim K = \infty$ (if $\dim K < \infty$, then the reasoning would be obviously simplified). Let the homomorphism $f : D \to \mathbb{R}_0^\infty$ be the same as in Proposition 5.9. Choose an independent set of real numbers $\{a_k\}$, set $a = (a_1, \ldots, a_k, \ldots) \in \mathbb{R}^\infty$, and consider a continuous homomorphism $f_1 : Y \to \mathbb{R}^{n+1}$ defined by the formula $f_1(s_1, \ldots, s_n; d) = (s_1, \ldots, s_n; \langle f(d), a \rangle)$. Put $p_1 = \tilde{f}_1$; then $p_1 : \mathbb{R}^{n+1} \to X$. Since the set $\{a_k\}$ is independent, f_1 is a monomorphism. Then 1.20(b) implies that the image $p_1(\mathbb{R}^{n+1})$ is dense in X. On the other hand, inasmuch as the set $f_1(Y)$ is dense in $\mathbb{R}^{n+1}$, then 1.20(b) implies that p_1 is a monomorphism.

Let ν be a Gaussian distribution on $\mathbb{R}^{n+1}$ having a characteristic function $\hat{\nu}(s_1, \ldots, s_n, s_{n+1}) = \exp\{-(s_1^2 + \cdots + s_n^2 + s_{n+1}^2)\}$. By Proposition 2.11, the characteristic function of the distribution $\mu = p_1(\nu) \in \Gamma(X)$ has the form $\hat{\mu}(y) = \hat{\mu}(s_1, \ldots, s_n; d) = \exp\{-(s_1^2 + \cdots + s_n^2 + \langle f(d), a \rangle^2)\}$.

The distribution μ has the following properties. Since $p_1(\mathbb{R}^{n+1})$ is dense in X, $\sigma(\mu) = X$. Since p_1 is a monomorphism and $\nu \in I_0(\mathbb{R}^{n+1})$, by Proposition 4.17 $\mu = p_1(\nu) \in I_0(X)$. As follows from Proposition 5.21, μ has only Gaussian divisors. The necessity is proved.

Sufficiency. Assume $X = \mathbb{R}^n + \mathbb{T}$, $Y = \mathbb{R}^n + \mathbb{Z}$. Let $\mu \in \Gamma^s(X)$, and let the characteristic function of the distribution μ have the form $\hat{\mu}(y) = \exp\{-\varphi(y)\}$, where the function $\varphi(y)$ is the same as in Definition 5.1. From the proof of Proposition 5.6, it follows that the function $\varphi(y) = \varphi(s_1, \ldots, s_n; k)$ admits the representation

$$\varphi(s_1, \ldots, s_n, k) = \langle As, s \rangle + 2k\langle \beta, s \rangle + bk^2,$$

where $s = (s_1, \ldots, s_n)$, $A = (a_{ij})_{i,j=1}^n$ is a positive semidefinite matrix, and $\beta = (\beta_1, \ldots, \beta_n)$, $b \geq 0$. So far as $\sigma(\mu) = X$, then it follows from the reasoning of Proposition 5.5 that $\varphi(y) = 0$ only if $y = 0$. Accounting for the fact that $\varphi(y) \geq 0$ for all $y \in Y$ we conclude that the inequality $\langle As, s \rangle + 2s_{n+1}\langle \beta, s \rangle + bs_{n+1}^2 > 0$ holds for any $(s, s_{n+1}) \in \mathbb{R}^{n+1}$, $(s, s_{n+1}) \neq 0$. Consequently, the inequality

$$\langle As, s \rangle + 2s_{n+1}\langle \beta, s \rangle + bs_{n+1}^2 \geq \varepsilon(s_1^2 + \cdots + s_n^2 + s_{n+1}^2)$$

is valid for some $\varepsilon > 0$ and arbitrary $(s, s_{n+1}) \in \mathbb{R}^{n+1}$.

Therefore the distribution μ has a divisor μ_1 with a characteristic function

$$\hat{\mu}_1(y) = \hat{\mu}_1(s_1, \ldots, s_n, k) = \exp\{-\varepsilon k^2\}.$$

Since $\sigma(\mu_1) = \mathbb{T}$ and $\mu_1 \in \Gamma(\mathbb{T})$, Theorem 5.20 implies that μ_1 and, hence, μ have non-Gaussian divisors.

The case of a nonsymmetric distribution μ can easily be reduced to the considered one. $\square$

5.24. COROLLARY. *For every Gaussian distribution on a group X to have non-Gaussian divisors, it is necessary and sufficient that the group X be topologically isomorphic to $\mathbb{T}$.*

5.25. REMARK. Given connected group X, a distribution $\mu \in \Gamma(X)$ was constructed in Remark 5.12 such that $\sigma(\mu) = X$. The reasoning presented when proving the necessity in Theorem 5.23 give rise to another construction of such a distribution.

5.26. REMARK. Let $\mu \in \Gamma^s(\mathbb{T})$. Then Theorem 5.20 implies that μ has a non-Gaussian divisor. But one may assert even more. Specifically, $\mu = \nu^{*2}$, where $\nu \in \mathcal{M}^1(\mathbb{T})$, $\nu \notin \Gamma(\mathbb{T})$, and $\nu = \bar{\nu}$. Indeed, let the characteristic function of the distribution μ have the form $\hat{\mu}(n) = \exp\{-an^2\}$, $a > 0$, $n \in \mathbb{Z}$. Consider the function

$$\rho(t) = \sum_{n=-\infty}^{\infty} \exp\left\{-\frac{a}{2}n^2 - \text{i}nt\right\} = 1 + 2\sum_{n=1}^{\infty} \exp\left\{-\frac{a}{2}n^2\right\} \cos nt$$

on the group $\mathbb{T}$. Obviously, $\rho(t) \geq \varepsilon > 0$ for all $t \in [0, 2\pi]$. Choose n_0 in such a way that $\exp\{-\frac{a}{2}n_0^2\} \leq \frac{\varepsilon}{8}$, and consider the function

$$c(n) = \begin{cases} \exp\{\frac{-a}{2}n^2\}, & |n| \neq n_0, \\ -\exp\{\frac{-a}{2}n^2\}, & |n| = n_0, \end{cases}$$

on the group $\mathbb{Z}$. Since

$$f(t) = \sum_{n=-\infty}^{\infty} c(n) \exp\{-\text{i}nt\} = \rho(t) - 4\exp\left\{-\frac{a}{2}n_0^2\right\} \cos n_0 t \geq \frac{\varepsilon}{2},$$

$\{c(n)\}_{n=-\infty}^{\infty}$ is a characteristic function for some distribution $\nu \in \mathcal{M}^1(\mathbb{T})$ having a density $f(t)$ with respect to m_T. Apparently, ν satisfies all the requirements.

5.27. In Propositions 5.6 and 5.9, "linearization" of a Gaussian distribution on a group was constructed, i.e., a representation of the distribution as a continuous homomorphic image of a Gaussian distribution on a linear space. Such a representation is far from being unique. Let us give an example of another "linearization", which can sometimes be useful.

We restrict our reasoning by assuming that $X = K$ is an infinite-dimensional connected compact group, $D = K^*$. The changes to be made in the general case will be seen from the construction itself.

Let the homomorphism $f: D \to \mathbb{R}_0^\infty$ be the same as in Proposition 5.9. Then the characteristic function of an arbitrary distribution $\mu \in \Gamma^s(X)$ has the form

$$\hat{\mu}(d) = \exp\{-\langle Af(d), f(d)\rangle\} = \exp\left\{-\sum_{i,j=1}^\infty \alpha_{ij} \frac{k_i}{k} \frac{k_j}{k}\right\},$$

where $A = (\alpha_{ij})_{i,j=1}^\infty$ is a symmetric positive semidefinite matrix (Proposition 5.9). Reducing the quadratic form obtained to the sum of squares, we get

$$\sum_{i,j=1}^\infty \alpha_{ij} \frac{k_i}{k} \frac{k_j}{k} = \sum_{i=1}^\infty \left(\sum_{j=1}^\infty \beta_{ij} \frac{k_j}{k}\right)^2. \tag{1}$$

Furthermore, changing, if needed, the numeration of d_j, we may assume that $\beta_{ii} \neq 0$ for all $i = 1, 2, 3, \ldots$.

Assuming that we have squared infinitely many linear forms, we define the homomorphisms $f_1: D \to \mathbb{R}_0^\infty$, $p_1: \mathbb{R}^\infty \to K$, taking

$$f_1(d) = \left(\sum_{j=1}^\infty \beta_{1j} \frac{k_j}{k}, \ldots, \sum_{j=n}^\infty \beta_{jn} \frac{k_j}{k}, 0, \ldots\right), \qquad p_1 = \tilde{f}_1.$$

Let us consider a distribution $\nu_0 \in \Gamma^s(\mathbb{R}^\infty)$ on the group $\mathbb{R}^\infty$ having the characteristic function

$$\hat{\nu}_0(s) = \hat{\nu}_0(s_1, \ldots, s_n, 0) = \exp\{-(s_1^2 + \cdots + s_n^2 + \cdots)\}.$$

Proposition 2.11 implies that the distributions μ and $p_1(\nu_0)$ have the same characteristic function. Thus according to 2.10(b), $\mu = p_1(\nu_0)$.

If there is a finite number n of linear forms in (1), then we define homomorphisms $f_2: D \to \mathbb{R}^n$, $p_2: \mathbb{R}^n \to K$, taking

$$f_2(d) = \left(\sum_{j=1}^\infty \beta_{1j} \frac{k_j}{k}, \ldots, \sum_{j=n}^\infty \beta_{nj} \frac{k_j}{k}\right), \qquad p_2 = \tilde{f}_2.$$

Let us consider a distribution $\nu_0 \in \Gamma^s(\mathbb{R}^n)$ on the group $\mathbb{R}^n$ having the characteristic function

$$\hat{\nu}_0(s) = \hat{\nu}_0(s_1, \ldots, s_n) = \exp\{-(s_1^2 + \cdots + s_n^2)\}.$$

Proposition 2.11 implies that the distributions μ and $p_2(\nu_0)$ have the same characteristic function. Consequently, by 2.10(b), $\mu = p_2(\nu_0)$.

We now use Propositions 5.6 and 5.9 to find conditions for a distribution $\mu \in \Gamma(X)$ to have only Gaussian divisors. Obviously, the only interesting case is when the group X contains a subgroup topologically isomorphic to $\mathbb{T}$. Otherwise by Theorem 5.22 any Gaussian distribution on X has only Gaussian divisors.

5.28. PROPOSITION. *Let X be a connected finite-dimensional group. For a symmetric Gaussian distribution μ to have only Gaussian divisors, it is necessary and sufficient that the distribution μ be a continuous monomorphic image of some Gaussian distribution ν in a finite-dimensional linear space.*

PROOF. *Necessity.* By Theorem 1.15 $X \approx \mathbb{R}^n + K$, where $n \geq 0$ and K is a connected compact group. Let us restrict ourselves to the case $X = K$. It differs from the general one only in notation.

Let $\dim K = l$, $D = K^*$, let the homomorphisms f and p be the same as in Proposition 5.6 and let $\mu \in \Gamma^s(X)$. As was proved in Proposition 5.6, the characteristic function $\hat{\mu}(d)$ has the form

$$\hat{\mu}(d) = \exp\{-\langle Af(d), f(d)\rangle\}, \qquad d \in D,$$

where $A = (\alpha_{ij})_{i,j=1}^l$ is a symmetric positive semidefinite matrix, $\mu = p(\nu)$, $\nu \in \Gamma^s(\mathbb{R}^l)$, and the characteristic function $\hat{\nu}(s)$ has the form

$$\hat{\nu}(s) = \exp\{-\langle As, s\rangle\}, \qquad s \in \mathbb{R}^l.$$

Set $\sigma(\nu) = G$. Then $G = A(\mathbb{R}^l, \ker A)$. (Here we regard A as a linear operator from $\mathbb{R}^l$ into $\mathbb{R}^l$.) Let us prove that $G \cap \ker p = \{0\}$ if μ has only Gaussian divisors, whereby the necessity would be proven.

The kernel $\ker p$ is a closed subgroup of $\mathbb{R}^l$. By Theorem 1.23(a), $\ker p = F + S$, where $F \approx \mathbb{R}^n$, $S \approx \mathbb{Z}^m$. We verify that $n = 0$. Indeed, if for some $t_0 \in \ker p$ we have $\lambda t_0 \in \ker p$ for all $\lambda \in \mathbb{R}$, then

$$(\lambda t_0, f(d)) = (p(\lambda t_0), d) = 1$$

for all $d \in D$, which is impossible since the linear hull of the set $f(D)$, is, by construction, the whole of $\mathbb{R}^l$. So, $\ker p = S \approx \mathbb{Z}^m$. Let there exist $y_0 \in G \cap \ker p$. We may assume that the element y_0 is chosen in such a way that $\lambda y_0 \notin G \cap \ker p$ for all $\lambda \in]0, 1[$. Let us note now that for some $\varepsilon > 0$ the inequality

$$\langle As, s\rangle \geq \varepsilon \langle s, y_0\rangle^2, \qquad s \in \mathbb{R}^l, \tag{1}$$

holds, since the bilinear form $\langle A\cdot, \cdot\rangle$ defines an inner product on the factor-space $\mathbb{R}^l / \ker A$, and $\langle \cdot, y_0\rangle$ is a linear functional on this space.

It follows from inequality (1) that the distribution $\nu_1 \in \Gamma(\mathbb{R}^l)$ having the characteristic function

$$\hat{\nu}_1(s) = \exp\{-\varepsilon\langle s, y_0\rangle^2\}, \qquad s \in \mathbb{R}^l, \tag{2}$$

is a divisor of ν. Consequently, the distribution $\mu_1 = p(\nu_1)$ is a divisor of μ. Equation (2) yields $H = \sigma(\nu_1) = \{t \in \mathbb{R}^l : t = \lambda t_0, \lambda \in \mathbb{R}\}$, i.e., $H \approx \mathbb{R}$. By construction $p(H) \approx \mathbb{T}$. By Theorem 5.20, the distribution μ_1 and, hence, μ have non-Gaussian divisors. The contradiction obtained indicates that $G \cap \ker p = \{0\}$.

Sufficiency. Let $\mu = q(\nu)$, where $q: \mathbb{R}^n \to X$ is a continuous monomorphism and $\nu \in \Gamma(\mathbb{R}^n)$. By the Cramér theorem $\nu \in I_0(\mathbb{R}^n)$. By Proposition 4.17, $\mu \in I_0(X)$ as well. Applying Proposition 5.21, we obtain the desired statement. $\square$

On can prove a similar statement for some class of infinite-dimensional groups.

5.29. PROPOSITION. *Let X be an infinite-dimensional connected group containing no subgroup topologically isomorphic to the group $\mathbb{T}^\infty$. For a symmetric Gaussian distribution μ on X to have only Gaussian divisors, it is necessary and sufficient that the distribution μ be a continuous monomorphic image of some Gaussian distribution ν concentrated on a linear subspace $G \subset \mathbb{R}^\infty$.*

PROOF. *Necessity.* By Theorem 1.15 $X \approx \mathbb{R}^n + K$, where $n \geq 0$ and K is a connected compact group. As in the proof of Proposition 5.28, we restrict ourselves to the case $X = K$. Put $D = K^*$. Then by Corollary 1.10 and Theorems 1.11 and 1.13, D is an infinite-rank countable discrete torsion-free group. First check that there exists an m such that the group K contains a subgroup topologically isomorphic to $\mathbb{T}^m$ and does not contain a subgroup topologically isomorphic to $\mathbb{T}^{m+1}$. According to the Stein theorem [Fuchs, §19], each countable abelian group can be represented as a direct sum

$$D = A + B, \tag{1}$$

where A is a free group and the group B has no free factor-groups. (A countable discrete abelian group is called a free group if it is topologically isomorphic either to $\mathbb{Z}_0^\infty$ or to $\mathbb{Z}^m$ for some m.) The subgroup B is uniquely determined by the group D. Since $(\mathbb{T}^\infty)^* \approx \mathbb{Z}_0^\infty$ and the group K does not contain a subgroup topologically isomorphic to $\mathbb{T}^\infty$, we have $A \not\approx \mathbb{Z}_0^\infty$. Then let $A \approx \mathbb{Z}^m$. Assume that there exists the expansion

$$D = A_1 + B_1, \tag{2}$$

where $A_1 \approx \mathbb{Z}^{m+1}$. Applying the Stein theorem to the group B_1 and taking into account that the summand B in (1) is uniquely defined, we conclude that expansion (2) is impossible. By Proposition 1.22, a subgroup of K that is topologically isomorphic to $\mathbb{T}^n$ is a topological direct summand of K. Therefore, it follows from the above said that K does not contain a subgroup topologically isomorphic to $\mathbb{T}^{m+1}$.

Using the distribution μ, let us construct either the homomorphisms f_1 and p_1 or f_2 and p_2 in the same way as in §5.27. According to that section, two cases are possible.

1. $\mu = p_1(\nu_0)$, $\nu_0 \in \Gamma^s(\mathbb{R}^\infty)$. Let us study the kernel $\ker p_1$. First check that the $\ker p_1$ does not contain a subgroup topologically isomorphic to $\mathbb{R}$. The intersection $H_n = \overline{f_1(D)} \cap \mathbb{R}^n$ is a closed subgroup in $\mathbb{R}^n$. By Theorem

1.23(a), $H_n \approx \mathbb{R}^{l_n} + \mathbb{Z}^{k_n}$, $l_n + k_n = n$, since $f_1(d_j) = (\beta_{1j}, \ldots, \beta_{jj}, 0, \ldots)$, and $\beta_{jj} \neq 0$.

It should be noted that $\ker p_1 = \{t \in \mathbb{R}^\infty : (t, f_1(d)) = 1 \text{ for all } d \in D\}$. Therefore, if the kernel $\ker p_1$ contains a subgroup topologically isomorphic to $\mathbb{R}$, then for some $a = (a_1, \ldots, a_n, \ldots) \in \mathbb{R}^\infty$ the equality $(\lambda a, f_1(d)) = 1$ holds for all $\lambda \in \mathbb{R}$, $d \in D$. In particular, when $d = d_k$, we have $(\lambda a, f_1(d_k)) = 1$, $k = 1, 2, \ldots$. For $k = 1$ we obtain $(\lambda a, f_1(d_1)) = 1$, i.e., $\lambda a_1 \beta_{11} \equiv 0 \pmod{2\pi}$. Hence $\lambda a_1 \beta_{11} = 2k_\lambda \pi$, $k_\lambda \in \mathbb{Z}$. Since k_λ depends continuously on λ, we get $\lambda a_1 \beta_{11} = 2k\pi$. The latter is possible only if $a_1 = 0$. Taking into consideration that $a_1 = 0$ and $\beta_{22} \neq 0$, we prove in a similar way that $a_2 = 0$ and so on. Hence $a = 0$.

As has been shown above, $K \approx \mathbb{T}^m + K_1$, where the group K_1 already does not contain a subgroup topologically isomorphic to $\mathbb{T}$. We prove that $\ker p_1 \approx \mathbb{Z}^l$ for some $l \leq m$. Indeed, let us choose l linearly independent vectors $\{c_1, \ldots, c_l\}$ in the kernel $\ker p_1$, and consider the subspace E_l they generate. Then $H_l = E_l \cap \ker p_1$ is a closed subgroup in E_l. By Theorem 1.23(a), $H_l \approx \mathbb{R}^p + \mathbb{Z}^q$, $p + q \leq l$. Since the kernel $\ker p_1$ does not contain subgroups topologically isomorphic to $\mathbb{R}$ and H_l contains l linearly independent vectors, it follows that $H_l \approx \mathbb{Z}^l$ and $p_1(E_l) \approx \mathbb{T}^l$. Hence $l \leq m$. Thus the kernel $\ker p_1$ contains no more than m linearly independent vectors, i.e., $\ker p_1 \approx \mathbb{Z}^l$, $l \leq m$.

Denote by l^2 the subspace in $\mathbb{R}^\infty$ consisting of $t = (t_1, \ldots t_n, \ldots) \in \mathbb{R}^\infty$, such that $\sum_{n=1}^\infty t_n^2 < \infty$, and prove that $\ker p_1 \cap l^2 = \{0\}$.

Let $b = (b_1, \ldots, b_n, \ldots) \in \ker p_1 \cap l^2$. Then the inequality

$$(b_1 s_1 + \cdots + b_n s_n + \cdots)^2 \leq \left(\sum_{n=1}^\infty b_n^2 \right) \left(\sum_{n=1}^\infty s_n^2 \right) \tag{3}$$

holds for any $s \in \mathbb{R}_0^\infty$. It follows from (3) that

$$\sum_{n=1}^\infty s_n^2 \geq \delta(b_1 s_1 + \cdots + b_n s_n + \cdots)^2,$$

where $\delta = \left(\sum_{n=1}^\infty b_n^2 \right)^{-1} > 0$. Therefore, the distribution $\nu_1 \in \Gamma^s(\mathbb{R}^\infty)$ having the characteristic function

$$\hat{\nu}_1(s) = \hat{\nu}_1(s_1, \ldots, s_n, 0, \ldots) = \exp\{-\delta(b_1 s_1 + \cdots + b_n s_n + \cdots)^2\}$$

is a divisor of the distribution ν_0. Apparently, the support $\sigma(\nu_1) = L = \{t \in \mathbb{R}^\infty : t = \lambda b, \lambda \in \mathbb{R}\}$. Note that the distribution $\mu_1 = p_1(\nu_1) \in \Gamma^s(K)$ is a divisor of μ. The support of the distribution μ_1 is the subgroup $p_1(L) \approx \mathbb{T}$, since the kernel $\ker p_1$ contains (as was proved above) no subgroups topologically isomorphic to $\mathbb{R}$. By Theorem 5.20, the distribution μ_1 and, hence, μ has non-Gaussian divisors. The contradiction obtained implies that $\ker p_1 \cap l^2 = \{0\}$.

Now we can construct the derived subspace G. It is known that the intersection of all measurable subspaces $E \subset \mathbb{R}^\infty$ such that $\nu_0(E) = 1$ coincides with l^2 [VSu]. Since, as proved above, the kernel $\ker p_1$ is countable and $\ker p_1 \cap l^2 = \{0\}$, there exists a measurable subspace $G \subset \mathbb{R}^\infty$ such that $\nu_0(G) = 1$ and $G \cap \ker p_1 = \{0\}$. The necessity for the case 1 is proven.

2. $\mu = p_2(\nu_0)$, $\nu_0 \in \Gamma^s(\mathbb{R}^n)$. Just as in the case 1 we verify that the kernel $\ker p_2$ does not contain a group topologically isomorphic to $\mathbb{R}$. Then it immediately follows from Theorem 1.23(a) that $\ker p_2 \approx \mathbb{Z}^l$, $l \le n$. The argument which resulted in $\ker p_1 \cap l^2 = \{0\}$ for the case 1 in the situation at hand implies that $\ker p_2 = \{0\}$, i.e., p_2 is a monomorphism and $G = \mathbb{R}^n$. The necessity is proven completely. The sufficiency can be proved as in Theorem 5.22. □

We shall apply Propositions 5.28 and 5.29 to the study of convolutions of Gaussian and Poisson distributions in §7.

The main problem, to the solution of which the rest of this section is devoted, is to provide a complete description of groups each distribution of which, transformed by every character into a Gaussian distribution on the group $\mathbb{T}$, is itself Gaussian. The solution is given in Theorem 5.36. This result, along with Theorem 5.22, will be used in Chapter III in the study of characterization problems on groups.

The following assertion is needed.

5.30. PROPOSITION. *A distribution μ on a group X is Gaussian if and only if it satisfies the following conditions*:

   (i) *μ is an infinitely divisible distribution.*
   (ii) *If $\mu = e(\Phi) * \alpha$, where $\Phi \in \mathscr{M}_+(X)$, and α is an infinitely divisible distribution, then the measure Φ is degenerate at zero.*

The proof follows directly from Theorem 2.21 and Remark 2.22.

5.31. PROPOSITION. *A distribution μ on a group X is Gaussian if and only if it satisfies the following conditions*:

   (i) *μ is an infinitely divisible distribution.*
   (ii) *Any character $y \in Y$ transforms the distribution μ into a Gaussian distribution on the group $\mathbb{T}$.*

PROOF. A Gaussian distribution μ obviously satisfies conditions (i) and (ii). Let μ be an infinitely divisible distribution and (ii) holds. Suppose that $\mu = e(\Phi) * \nu$, where $\Phi \in \mathscr{M}_+(X)$ and ν is an infinitely divisible distribution. Then $y(\mu) = e(y(\Phi)) * y(\nu)$. But $y(\nu)$ is an infinitely divisible distribution on $\mathbb{T}$ and by condition (ii) $y(\mu) \in \Gamma(\mathbb{T})$. Proposition 5.30 implies that the measure $y(\Phi)$ is degenerate at zero for any $y \in Y$. Then the measure Φ is also degenerate at zero. Applying Proposition 5.30 once more, we obtain that $\mu \in \Gamma(X)$. □

5.32. DEFINITION. A distribution γ on a group X is called Gaussian in the sense of Urbanik if $y(\gamma) \in \Gamma(\mathbb{T})$ for any $y \in Y$.

Denote by $\Gamma_U(X)$ the set of Gaussian in the sense of Urbanik distributions on the group X. Obviously, $\Gamma(X) \subset \Gamma_U(X)$. We proceed with a complete description of those groups X for which $\Gamma(X) = \Gamma_U(X)$.

5.33. DEFINITION. An element $x_0 \in X$ is called infinitely divisible if the equation $nx = x_0$ has a solution in X for arbitrarily large natural n.

It should be noted that the definition of an infinitely divisible element differs from that of an infinitely divisible distribution (§2.15.), in which the corresponding equation has to be solvable for *any* natural n.

Before turning to the formulation and proof of the principal theorem, we prove several lemmas.

5.34. LEMMA. *Assume that the group X does not contain nonzero infinitely divisible elements. Then X is a discrete torsion-free group.*

PROOF. By Theorem 1.14, $X \approx \mathbb{R}^n + G$, where $n \geq 0$ and the group G contains an open compact group K. If X does not contain nonzero infinitely divisible elements, then $n = 0$ and G is a torsion-free group. In particular, K is also a torsion-free group.

Applying Theorem 1.16 and noticing that all elements of the groups $\Sigma_{\mathbf{a}}$ and Δ_p are infinitely divisible we infer that $K = \{0\}$, i.e., X is a discrete torsion-free group. □

5.35. LEMMA. *There exists a distribution μ_0 on the group $X = \mathbb{R} + \mathbb{T}$ such that $\mu_0 \in \Gamma_U(X)$, $\mu_0 \notin \Gamma(X)$, and $\hat{\mu}_0(y) > 0$ for all $y \in Y$.*

PROOF. Since $Y \approx \mathbb{R} + \mathbb{Z}$, we may assume $Y = \mathbb{R} + \mathbb{Z}$ and denote by $y = (s, n)$, $s \in \mathbb{R}$, $n \in \mathbb{Z}$ the elements of the group Y. Let $a > 0$, $b > 0$, $c > 0$. Put

$$\varphi(s, n) = \begin{cases} as^2, & n = 0, \\ bs^2 + cn^2, & n \neq 0. \end{cases}$$

Denote by α_a the Gaussian distribution on the group $\mathbb{R}$ with the characteristic function $\hat{\alpha}_a(s) = \exp\{-as^2\}$ and by β_c the Gaussian distribution on the group $\mathbb{T}$ with the characteristic function $\hat{\beta}_c(n) = \exp\{-cn^2\}$. Choose a, b, and c in such a way that $2\alpha_a \geq \alpha_b$, $2\beta_c \geq m_T$. As $\mathbb{R}$ and $\mathbb{T}$ are subgroups of X, we may regard all the mentioned distributions as distributions on X and set

$$\mu_0 = \alpha_a * m_T - \alpha_b * m_T + \beta_c * \alpha_b. \tag{2}$$

Since

$$\mu_0 = \left(\alpha_a - \frac{1}{2}\alpha_b\right) * m_T + \left(\beta_c - \frac{1}{2}m_T\right) * \alpha_b,$$

$\mu_0 \in \mathcal{M}^1(X)$. It follows from (2) that $\hat{\mu}_0(s, n) = \exp\{-\varphi_1(s, n)\}$. Since the equality $\varphi_1(ky) = k^2\varphi_1(y)$, $k \in \mathbb{Z}$, $y \in Y$ holds for arbitrary a, b, and c, then $\mu_0 \in \Gamma_U(X)$. Choosing $a \neq b$, we obtain that $\mu_0 \notin \Gamma(X)$. □

Let X be a discrete torsion-free group. We shall call an element $x \in X$ dependent on elements $x_1, \ldots, x_l \in X$ if there are $n, n_1, \ldots, n_l \in \mathbb{Z}$ such that $nx = n_1 x_1 + \cdots + n_l x_l$. Denote by L_x the subgroup of elements dependent on x. The group L_x is topologically isomorphic to a subgroup of $\mathbb{Q}$.

5.36. THEOREM. *The equality*

$$\text{(i)} \quad \Gamma(X) = \Gamma_U(X),$$

holds on a group X, if and only if the group X satisfies one of the conditions:

(ii) *Any nonzero factor-group of the group Y contains a nonzero infinitely divisible element.*

(iii) *The factor-group of the group Y by the subgroup of all compact elements is topologically isomorphic to the group $\mathbb{Z}$.*

PROOF. *Necessity*. Assume that there exits a subgroup $Y_1 \subset Y$ such that the factor-group $Y/Y_1 = H$ does not contain nonzero infinitely divisible elements. According to Lemma 5.34, H is a discrete torsion-free group. There are the two possibilities.

1. $H \neq \mathbb{Z}$. In this case $r(H) > 1$, since otherwise the group H would be topologically isomorphic to a subgroup B of $\mathbb{Q}$, $B \neq \mathbb{Z}$, and then all elements of H would be infinitely divisible. Ascribe to each element $h \in H$, $h \neq 0$, the subgroup L_h in H consisting of all elements dependent on h. Since the group L_h is discrete and $r(L_h) = 1$, $L_h \approx \mathbb{Z}$ for all $h \in H$. Denote by h_k a generator of the group L_h. Since the group H is countable, we obtain no more than a countable set of pairwise intersecting (only at zero) different subgroups L_{h_k} such that $H = \bigcup_k L_{h_k}$. Any element $h \in H$, $h \neq 0$, can be uniquely expressed in the form $h = mh_k$, $m \in \mathbb{Z}$, $h_k \in L_{h_k}$. On the group H let us define the function

$$\varphi_0(h) = \begin{cases} a_k m^2 & \text{if } h = mh_k, \ h \neq 0, \\ 0 & \text{if } h = 0, \end{cases}$$

where the numbers a_k are chosen so that

$$\sum_{k=1}^{\infty} \sum_{m \neq 0} \exp\{-a_k m^2\} < 1.$$

On the group $G = H^*$ consider the continuous nonnegative function

$$\rho(g) = 1 + \sum_{h \in H, \ h \neq 0} \exp\{-\varphi_0(h)\}\overline{(g, h)}.$$

The characteristic function of the distribution $\mu \in \mathscr{M}^1(G)$ having the density $\rho(g)$ with respect to m_G has the form

$$\hat{\mu}(h) = \exp\{-\varphi_0(h)\}.$$

Since $r(H) > 1$, the numbers a_k can always be chosen in such a way that $\mu \notin \Gamma(G)$. On the other hand, since $\varphi_0(kh) = k^2 \varphi_0(h)$ for any $k \in \mathbb{Z}$, $h \in H$, then $\mu \in \Gamma_U(G)$. It remains to note that, by Theorem 1.6, $G = A(X, Y_1) \subset X$, i.e., $\mu \in \Gamma_U(X)$ and $\mu \notin \Gamma(X)$.

2. $H \approx \mathbb{Z}$. In this case it follows from Theorem 1.6 and Proposition 1.22 that $Y \approx Y_1 + \mathbb{Z}$. Now it should be noted that the subgroup Y_1 contains a noncompact element. Otherwise the group X would satisfy condition (iii). Therefore $G = C_{Y_1^*} \neq \{0\}$ (Corollary 1.10). So the group X contains a subgroup $X_1 \approx G + \mathbb{T}$, where G is a connected group. By Theorem 1.15 $G \approx \mathbb{R}^n + K$, where $n \geq 0$ and K is a connected compact group. If $n > 0$, then the group X contains a subgroup $G_1 \approx \mathbb{R} + \mathbb{T}$, and Lemma 5.35 allows one to construct a distribution $\mu \in \Gamma_U(X)$ such that $\mu \notin \Gamma(X)$.

If $n = 0$, then $X_1 \approx K + \mathbb{T}$. Assume for definiteness that $\dim K = \infty$. Let $D = K^*$, and let $f: D \to \mathbb{R}_0^\infty$ be the same homomorphism as in Proposition 5.9. Denote the elements of the group $X_1^* \approx D + \mathbb{Z}$ by (d, n), $d \in D$, $n \in \mathbb{Z}$.

Define the homomorphism $f_1: X_1^* \to \mathbb{R} + \mathbb{Z}$ by the formula $f_1(d, n) = (k_1/k, n)$ if $f(d) = (k_1/k, \ldots, k_1/k, 0, \ldots)$. Put $\tilde{p}_1 = f_1$, $p_1: \mathbb{R} + \mathbb{T} \to X_1$ and $\mu = p_1(\mu_0)$, where μ_0 is the distribution constructed in Lemma 5.35. Proposition 2.11 yields $\hat{\mu}(d, n) = \widehat{p_1(\mu_0)}(d, n) = \mu_0(f_1(d, n)) = \exp\{-\varphi_1(f_1(d, l))\}$. It follows from (1) of Lemma 5.35 that $\mu \in \Gamma_U(X_1)$ and $\mu \notin \Gamma(X_1)$. Consequently, $\mu \in \Gamma_U(X)$ and $\mu \notin \Gamma(X)$.

Sufficiency. Let $\mu \in \Gamma_U(X)$. We verify that $\mu \in \Gamma(X)$. First assume that condition (ii) holds. In particular, it follows from (ii) that a group topologically isomorphic to $\mathbb{Z}$ cannot be a factor-group of the group Y. By Theorem 1.6, the group X does not contain a group topologically isomorphic to $\mathbb{T}$. Therefore, according to Theorem 5.22, it is enough to check that the distribution $\nu = \mu * \overline{\mu} \in \Gamma(X)$.

As can easily be seen, if $\gamma \in \Gamma_U(X)$, then $\hat{\gamma}(y) \neq 0$ for all $y \in Y$; 2.10(d),(e) imply that $\hat{\nu}(y) = |\hat{\mu}(y)|^2 > 0$ for all $y \in Y$. Consequently, the function $\hat{\nu}(y)$ admits the representation

$$\hat{\nu}(y) = \exp\{-\varphi(y)\},$$

where $\varphi(y)$ is a continuous nonnegative function on Y satisfying the equation

$$\varphi(ky) = k^2 \varphi(y) \tag{1}$$

for arbitrary $y \in Y$, $k \in \mathbb{Z}$.

Let A be the set of all elements $y \in Y$ for which a sequence $\{y_n\}$ of elements in Y exists such that y_n is divisible by n and $\varphi(y - y_n) \to 0$ (an element $x_0 \in X$ is divisible by n if the equation $nx = x_0$ has a solution in X).

Let us prove that A is a closed subgroup in Y. Let $u, v \in A$, $\{u_n\}$ and $\{v_n\}$ being the relevant sequences of elements. Using inequality 2.10(c), we have

$$|1 - \hat{\nu}(u - v - u_n + v_n)|$$
$$\leq |\hat{\nu}(u - v - u_n + v_n) - \hat{\nu}(v - v_n)| + |1 - \hat{\nu}(v - v_n)|$$
$$\leq \sqrt{2}|1 - \hat{\nu}(u - u_n)|^{1/2} + |1 - \hat{\nu}(v - v_n)|.$$

Since for each sequence $\{y_n\} \subset Y$ the relation $\varphi(y_n) \to 0$ holds if and only if $\hat{\nu}(y_n) \to 1$, we infer that $\varphi(u - v - u_n + v_n) \to 0$. Clearly, $u_n - v_n$ can be divided by n. So, $u - v \in A$, providing the proof that A is an algebraic subgroup of Y.

A is a closed subgroup. Indeed, let $u^{(j)} \in A$, $u^{(j)} \to u$, and $\{u_n^{(j)}\}$ a sequence corresponding to the element $u^{(j)}$. By 2.10(c), we get

$$|1 - \hat{\nu}(u - u_n^{(j)})| \leq |\hat{\nu}(u - u_n^{(j)}) - \hat{\nu}(u^{(j)} - u_n^{(j)})| + |1 - \hat{\nu}(u^{(j)} - u_n^{(j)})|$$
$$\leq \sqrt{2}|1 - \hat{\nu}(u - u^{(j)})|^{1/2} + |1 - \hat{\nu}(u^{(j)} - u_n^{(j)})|. \tag{2}$$

As already noted above, $\varphi(y_n) \to 0$ if and only if $\hat{\nu}(y_n) \to 1$. Therefore (2) implies that A is closed.

We now prove that the factor-group Y/A does not contain infinitely divisible nonzero elements. Admit the contrary. Then there exist an element $u \notin A$, an unbounded sequence of numbers $\{p_n\} \subset \mathbb{Z}$, and a sequence $\{y_n\} \subset Y$ such that

$$u - p_n y_n \in A$$

for all positive integers n. Without loss of generality we may assume that $p_n \geq n^2$. Therefore there exists a sequence of integers $\{qn\}$ such that $nq_n/p_n \to 1$.

As follows from the definition of the set A, there exists an element $v_n \in Y$ which can be divided by p_n and such that $\varphi(u - p_n y_n - v_n)$ is arbitrarily close to zero. Hence there exists a sequence $\{z_n\} \subset Y$ such that z_n can be divided by p_n and $\varphi(u - z_n) \to 0$. Put $u_n = nq_n z_n/p_n$, $n = 1, 2, 3, \ldots$. In view of 2.10(c), we have

$$|1 - \hat{\nu}(u - u_n)| \leq |\hat{\nu}(u - u_n) - \hat{\nu}(u - z_n)| +$$
$$|1 - \hat{\nu}(u - z_n)| \leq \sqrt{2}|1 - \hat{\nu}(u_n - z_n)|^{1/2} + |1 - \hat{\nu}(u - z_n)|. \tag{3}$$

Since $\varphi(u - z_n) \to 0$, then $\hat{\nu}(u - z_n) \to 1$. On the other hand, accounting for (1), we obtain

$$\varphi(u_n - z_n) = \varphi\left(\frac{(nq_n - p_n)z_n}{p_n}\right) = \left(\frac{nq_n}{p_n} - 1\right)^2 \varphi(z_n) \to 0.$$

So, $\hat{\nu}(u_n - z_n) \to 1$. Consequently we find from (3) that $\hat{\nu}(u - u_n) \to 1$, i.e., $\varphi(u - u_n) \to 0$. Since u_n can be divided by n, we have come to a contradiction with the assumption $u \notin A$.

So far as we have assumed that the condition (ii) is satisfied, it follows from the above said that $Y = A$. Let $u, v \in Y$. Then one can find sequences

$\{u_n\}$, $\{v_n\}$ such that u_n and v_n can be divided by n and $\varphi(u - u_n) \to 0$, $\varphi(v - v_n) \to 0$. Then it follows from 2.10(c) that both $\varphi(u_n) \to \varphi(u)$ and $\varphi(v_n) \to \varphi(v)$ and $\varphi(u_n + v_n) \to \varphi(u + v)$ and $\varphi(u_n - v_n) \to \varphi(u - v)$.

Since, by Theorem 2.8 $\exp\{-\varphi(y)\}$ is a positive definite function on Y, the inequality 2.8(i) holds for any $y_1, \ldots, y_k \in Y$, $\xi_1, \ldots, \xi_k \in \mathbb{C}$. In our case this inequality takes the form

$$\sum_{i,j=1}^{k} \exp\{-\varphi(y_i - y_j)\}\xi_i\bar{\xi}_j \geq 0.$$

Setting $k = 4$, $y_1 = -y_2 = u_n/n$, $y_3 = -y_4 = v_n/n$, $\xi_1 = \xi_2 = -\xi_3 = -\xi_4 = n$, here we obtain

$$\left[2\exp\left\{ -\frac{4}{n^2}\varphi(u_n) \right\} + 2\exp\left\{ -\frac{4}{n^2}\varphi(v_n) \right\} \right.$$
$$\left. - 4\exp\left\{ -\frac{1}{n^2} \times \varphi(u_n - v_n) \right\} - 4\exp\left\{ -\frac{1}{n^2}\varphi(u_n + v_n) \right\} + 4 \right] n^2 \geq 0.$$

Passing to the limit as $n \to \infty$, we find

$$4\varphi(u + v) + 4\varphi(u - v) - 8\varphi(u) - 8\varphi(v) \geq 0$$

or

$$2[\varphi(u) + \varphi(v)] \leq \varphi(u + v) + \varphi(u - v). \tag{4}$$

Replacing u by $u + v$ and v by $u - v$ here and accounting for (1), we obtain

$$\varphi(u + v) + \varphi(u - v) \leq 2[\varphi(u) + \varphi(v)]. \tag{5}$$

It follows from (4) and (5) that the function $\varphi(y)$ satisfies equation 5.1(ii), i.e., $\nu \in \Gamma(X)$. So the sufficiency of condition (ii) is proven.

Before proving the sufficiency of condition (iii) note that the support of an arbitrary distribution $\mu \in \Gamma_U(X)$ is contained in a conjugacy class of the subgroup C_X. Indeed, let $\nu = \mu * \bar{\mu}$. According to Proposition 2.5, it is enough to verify that $\sigma(\nu) \subset C_X$. Let $y_0 \in Y_0$. Then for some sequence $n_l \to \infty$ we have $n_l y_0 \to y'$. Taking (1) into account, we get $\varphi(y_0) = \lim_{n_l \to \infty}(\varphi(y')/n_l^2) = 0$. Hence $\varphi(y) \equiv 0$ on Y_0, i.e., $\hat{\nu}(y) \equiv 1$ on Y_0. Therefore, we obtain from Proposition 2.13 and Theorem 1.9 that $\sigma(\nu) \subset A(X, Y_0) = C_X$.

Now it is easy to prove the sufficiency of condition (iii). Replace the distribution μ by its shift $\mu' = \mu * E_x$ in such a way that $\sigma(\mu') \subset C_X$. But if the group X satisfies condition (iii), then Theorems 1.6 and 1.9 imply

$C_X \approx \mathbb{T}$. Since $\mu' \in \Gamma_u(C_x)$, then, obviously, $\mu' \in \Gamma(C_X)$. Hence $\mu \in \Gamma(X)$. □

The groups X satisfying condition (ii) of Theorem 5.36 can easily be constructed by virtue of the following assertion.

5.37. PROPOSITION. *For any nonzero factor-group Y to contain a nonzero infinitely divisible element, it is sufficient that the group X satisfy one of the conditions:*

(i) $C_X = \{0\}$.

(ii) *The group C_X^* possesses a maximal independent system consisting of infinitely divisible elements.*

PROOF. Consider the subgroup Y_0 and assume that for a subgroup $Y_1 \subset Y$ the factor-group Y/Y_1 does not contain nonzero infinitely divisible elements. According to Lemma 5.34, Y/Y_1 is a discrete torsion-free group. Hence $Y_0 \subset Y_1$.

Let condition (i) hold. Then in view of Theorem 1.3 and Corollary 1.10 $Y = Y_0$. It follows immediately from the above said that any nonzero factor-group of the group Y contains a nonzero infinitely divisible element.

Let condition (ii) hold. Suppose that for a subgroup $Y_1 \subset Y$ the factor-group Y/Y_1 does not contain nonzero infinitely divisible elements. As $Y_0 \subset Y_1$, then $Y/Y_1 \approx (Y/Y_0)/(Y_1/Y_0)$. Denote by N the set of infinitely divisible elements of the factor-group Y/Y_0, and note that Theorems 1.6 and 1.9 imply that $Y/Y_0 \approx C_X^*$. Let π be the natural homomorphism $\pi\colon Y/Y_0 \to (Y/Y_0)/(Y_1/Y_0)$. If an element $h \in N$, $h \notin Y_1/Y_0$, exists, then $\pi(h)$ is a nonzero infinitely divisible element of the factor-group Y/Y_1. The latter contradicts the choice of the subgroup Y_1. Hence $N \subset Y_1/Y_0$, and using condition (ii), we obtain that the maximal independent system of elements of the group Y/Y_0 is contained in the subgroup Y_1/Y_0. Consequently, the group Y/Y_1 is periodic, i.e., it contains a nonzero infinitely divisible element. □

§6. Decomposition of a generalized Poisson distribution

Our main purpose in this section is to prove group analogs of theorems on the membership of a generalized Poisson distribution in the class I_0.

6.1. DEFINITION. By a generalized Poisson distribution on a group X we mean a shift of the distribution

$$e(\Phi) = \exp\{-\Phi(X)\} \left(E_0 + \Phi + \frac{\Phi^{*2}}{2!} + \cdots + \frac{\Phi^{*n}}{n!} + \cdots \right),$$

where $\phi \in \mathscr{M}_+(X)$.

Let a measure $\Phi \in \mathscr{M}_+(X)$ be concentrated on a set $A \in \mathscr{B}(X)$. We always assume A to be a set from the class F_σ. This, of course, does not involve loss of generality, but enables one to avoid possible difficulties arising from the fact that the arithmetical sum $A_1 + A_2$ of two Borel sets might not be a Borel set.

Let $A \in \mathscr{B}(X)$ be an F_σ-set. Denote by $M(A)$ the set of all finite linear combinations with integer coefficients of elements of the set A, and put

$$M^+(A) = \bigcup_{k=1}^{\infty} (k)A.$$

To proceed we need the following lemmas.

6.2. LEMMA. *Let $\Phi \in \mathscr{M}_+(X)$, and let the measure Φ be concentrated on a set $A \in \mathscr{B}(X)$. Then the distribution $\mu = e(\Phi)$ is concentrated on the set $B = \{0\} \cup M^+(A)$.*

The proof of this lemma is omitted since it is fairly obvious.

6.3. LEMMA. *Let $\mu \in \mathscr{M}^1(X)$ be concentrated on a set $E \in \mathscr{B}(X)$, and let $\mu(\{0\}) > 0$. If $\mu = \mu_1 * \mu_2$, where $\mu_i \in \mathscr{M}^1(X)$, $i = 1, 2$, then one can choose shifts μ_i' of the distributions μ_i such that $\mu = \mu_1' * \mu_2'$, and the distributions μ_i' are concentrated on E.*

PROOF. For an arbitrary distribution $\mu \in \mathscr{M}^1(X)$ denote $D(\mu) = \{x \in X : \mu(\{x\}) > 0\}$ and observe that if $\mu = \mu_1 * \mu_2$, then $D(\mu) = D(\mu_1) + D(\mu_2)$. By assumption $0 \in D(\mu)$. Therefore, there exist elements $x_i \in D(\mu_i)$, $i = 1, 2$, such that $x_1 + x_2 = 0$. Set $\mu_i' = \mu_i * E_{-x_i}$. Then $\mu = \mu_1' * \mu_2'$ and $0 \in D(\mu_i')$, $i = 1, 2$. According to Lemma 4.8, the distributions μ_i' are concentrated on E. $\square$

First we prove a number of theorems concerning the membership of a generalized Poisson distribution $e(\Phi)$ generated by a discrete measure Φ to the class I_0. We need the following statement.

6.4. LEMMA. *Let G be a periodic subgroup of a group X, let $H = A(Y, G)$, and let y_0 be an arbitrary element of the group Y. If $\{x_1, \ldots, x_n\}$ is an independent set of elements of the group X, then the set of points $((x_1, y), \ldots, (x_n, y))$ when y runs through the conjugacy class $y_0 + H$ is dense in the group $\mathbb{T}^n$.*

PROOF. According to the Kronecker theorem [HR1, §26] the set of points $((x_1, y), \ldots, (x_n, y))$ when y runs through the whole group Y is dense in $\mathbb{T}^n$. Denote by τ the natural homomorphism $\tau : X \to X/G$. The set $\{\tau(x_1), \ldots, \tau(x_n)\}$ is independent in the factor-group X/G. Let $\hat{h}$ be the character of the factor-group corresponding to an element $h \in H$ under the topological isomorphism $(X/G)^* \approx A(Y, G)$ (see Theorem 1.6). Then $(\tau(x), \hat{h}) = (x, h)$ for any $x \in X$ and $h \in H$. For $y = y_0 + h \in y_0 + H$ we have $((x_1, y), \ldots, (x_n, y)) = ((x_1, y_0), \ldots, (x_n, y_0)) \cdot ((x_1, h), \ldots, (x_n, h)) = ((x_1, y_0), \ldots, (x_n, y_0)) \cdot ((\tau(x_1), \hat{h}), \ldots, (\tau(x_n), \hat{h}))$.

When $y = y_0 + h$ runs through the conjugacy class $y_0 + H$, the element h runs through the subgroup H while the element $\hat{h}$ runs through the group $(X/G)^*$. By the Kronecker theorem the set $((\tau(x_1), \hat{h}), \dots, (\tau(x_n), \hat{h}))$ is dense in $\mathbb{T}^n$, thus the set of points $((x_1, y), \dots, (x_n, y))$ is dense in $\mathbb{T}^n$ as well. $\square$

6.5. THEOREM. *Let* $\Phi = \sum_{j=0}^{n} \psi_j E_{x_j}$, *where* $\psi_j > 0$, $j = 0, 1, \dots, n$, *the set* $\{x_1, \dots, x_n\} \in X$ *is independent, and* $2x_0 = 0$. *Then* $\mu = e(\Phi) \in I_0$.

PROOF. By Lemma 6.2 the distribution μ is concentrated on the set

$$B = \{k_1 x_1 + \cdots + k_n x_n\}_{k_1, \dots, k_n \geq 0} \cup \{x_0 + k_1 x_1 + \cdots + k_n x_n\}_{k_1, \dots, k_n \geq 0}$$

and each point of the set B is uniquely representable either in the form $k_1 x_1 + \cdots + k_n x_n$ or in the form $x_0 + k_1 x_1 + \cdots + k_n x_n$.

Let $\mu = \mu_1 * \mu_2$, where $\mu_j \in \mathcal{M}^1(X)$, $j = 1, 2$. Since $\mu(\{0\}) \geq \exp\{-\Phi(X)\}$, in view of Lemma 6.3 one may assume that the distributions μ_j are concentrated on B. Let $k = (k_1, \dots, k_n)$ and

$$c_k = \mu_1(\{k_1 x_1 + \cdots + k_n x_n\}), \qquad c'_k = \mu_1(\{x_0 + k_1 x_1 + \cdots + k_n x_n\}),$$
$$d_k = \mu_2(\{k_1 x_1 + \cdots + k_n x_n\}), \qquad d'_k = \mu_2(\{x_0 + k_1 x_1 + \cdots + k_n x_n\}).$$

The characteristic functions $\hat{\mu}_1(y)$ and $\hat{\mu}_2(y)$ may be written down in the form:

$$\hat{\mu}_1(y) = \sum_{\|k\|=0}^{\infty} c_k (x_1, y)^{k_1} \cdots (x_n, y)^{k_n} \tag{1}$$

$$+ (x_0, y) \sum_{\|k\|=0}^{\infty} c'_k (x_1, y)^{k_1} \cdots (x_n, y)^{k_n},$$

$$\hat{\mu}_2(y) = \sum_{\|k\|=0}^{\infty} d_k (x_1, y)^{k_1} \cdots (x_n, y)^{k_n} \tag{2}$$

$$+ (x_0, y) \sum_{\|k\|=0}^{\infty} d'_k (x_1, y)^{k_1} \cdots (x_n, y)^{k_n}.$$

The characteristic function of the distribution μ has the form

$$\hat{\mu}(y) = \exp\left\{ \sum_{j=0}^{n} \psi_j((x_j, y) - 1) \right\}. \tag{3}$$

By 2.10(d) from (1)–(3) we obtain

$$\exp\left\{\sum_{j=0}^{n}\psi_j((x_j,y)-1)\right\}$$

$$=\left(\sum_{\|k\|=0}^{\infty}c_k(x_1,y)^{k_1}\cdots(x_n,y)^{k_n}+(x_0,y)\sum_{\|k\|=0}^{\infty}c'_k(x_1,y)^{k_1}\cdots(x_n,y)^{k_n}\right)$$

$$\times\left(\sum_{\|k\|=0}^{\infty}d_k(x_1,y)^{k_1}\cdots(x_n,y)^{k_n}\right.$$

$$\left.+(x_0,y)\sum_{\|k\|=0}^{\infty}d'_k(x_1,y)^{k_1}\cdots(x_n,y)^{k_n}\right).$$

$$(4)$$

Let $\zeta_j=(x_j,y)$, $j=1,2,\ldots$, $\zeta=(\zeta_1,\ldots,\zeta_n)$,

$$f_+(\zeta)=\sum_{\|k\|=0}^{\infty}(c_k+c'_k)\zeta^k,\qquad f_-(\zeta)=\sum_{\|k\|=0}^{\infty}(c_k-c'_k)\zeta^k,$$

$$g_+(\zeta)=\sum_{\|k\|=0}^{\infty}(d_k+d'_k)\zeta^k,\qquad g_-(\zeta)=\sum_{\|k\|=0}^{\infty}(d_k-d'_k)\zeta^k.$$

Let G be the subgroup of X, generated by the element x_0, $G\approx\mathbb{Z}(2)$. Set $H=A(Y,G)$. Then $Y=H\cup(y_1+H)$, where $(x_0,y_1)=-1$. Consider equality (4) on the conjugacy classes H and y_1+H. We obtain the following equalities:

$$\exp\left\{\sum_{j=1}^{n}\psi_j(\zeta_j-1)\right\}=f_+(\zeta)g_+(\zeta),\qquad(5)$$

$$\exp\left\{-2\psi_0+\sum_{j=1}^{n}\psi_j(\zeta_j-1)\right\}=f_-(\zeta)g_-(\zeta).\qquad(6)$$

By virtue of Lemma 6.4 these equalities are fulfilled for ζ belonging to a dense set in $\mathbb{T}^n$ and thereby for all $\zeta\in\mathbb{T}^n$. Comparing the coefficients at equal powers of ζ on the right- and left-hand sides of (5) and taking into account the fact that c_k, c'_k, d_k, d'_k are nonnegative, using condition 3.7(i) we easily find that the functions $f_+(\zeta)$ and $g_+(\zeta)$ are entire functions in $\mathbb{C}^n$ and that equalities (5) and (6) hold for any $\zeta\in\mathbb{C}^n$.

It follows from (5) that the functions $f_+(\zeta)$ and $g_+(\zeta)$ do not vanish in $\mathbb{C}^n$. By Theorem 3.8 they may be represented in the form

$$f_+(\zeta)=\exp\{a_+(\zeta)\},\qquad g_+(\zeta)=\exp\{b_+(\zeta)\},$$

where $a_+(\zeta)$ and $b_+(\zeta)$ are entire functions.

Relation (5) implies (see Theorem 3.8)

$$M_{f_+}(r)M_{g_+}(r) = \exp\left\{\sum_{j=1}^{n}\psi_j(r-1)\right\}.$$

Hence for $r > 1$ we have

$$A_{a_+}(r) \le \sum_{j=1}^{n}\psi_j(r-1). \tag{7}$$

It follows from (7) that $A_{a_+}(r) = O(r)$ as $r \to \infty$. Applying Theorem 3.9 we obtain that $a_+(\zeta)$ is a linear function (maybe constant). Similar reasoning shows that $b_+(\zeta)$ is also linear. Since $f_+(1, \ldots, 1) = g_+(1, \ldots, 1) = 1$, we may assume that the functions $a_+(\zeta)$, $b_+(\zeta)$ are chosen in such a way that $a_+(1, \ldots, 1) = b_+(1, \ldots, 1) = 0$ and we have

$$f_+(\zeta) = \exp\left\{\sum_{j=1}^{n}\psi_j^{(1)}(\zeta_j - 1)\right\}, \qquad g_+(\zeta) = \exp\left\{\sum_{j=1}^{n}\psi_j^{(2)}(\zeta_j - 1)\right\}, \tag{8}$$

where

$$\psi_j^{(1)} + \psi_j^{(2)} = \psi_j, \qquad j = 1, \ldots, n. \tag{9}$$

From the fact that c_k, c_k', d_k, d_k' are nonnegative we obtain $\psi_j^{(1)}$, $\psi_j^{(2)} \ge 0$, $j = 1, \ldots, n$.

Since $|c_k - c_k'| \le c_k + c_k'$ and $|d_k - d_k'| \le d_k + d_k'$, the functions $f_-(\zeta)$ and $g_-(\zeta)$ are also entire, and equality (6) holds for all $\zeta \in \mathbb{C}^n$. Arguing as above we obtain the representation

$$f_-(\zeta) = \exp\left\{-2\psi_0^{(1)} + \sum_{j=1}^{n}\gamma_j(\zeta_j - 1)\right\},$$

$$g_-(\zeta) = \exp\left\{-2\psi_0^{(2)} + \sum_{j=1}^{n}\delta_j(\zeta_j - 1)\right\}, \tag{10}$$

where $\psi_0^{(1)}$, $\psi_0^{(2)} \ge 0$, $\psi_0^{(1)} + \psi_0^{(2)} = \psi_0$, γ_j, $\delta_j \in \mathbb{R}$, $j = 1, \ldots, n$, and

$$\gamma_j + \delta_j = \psi_j, \qquad j = 1, \ldots, n. \tag{11}$$

Obviously, $\gamma_j \le \psi_j^{(1)}$ and $\delta_j \le \psi_j^{(2)}$, $j = 1, \ldots, n$. Therefore (9) and (11) imply $\gamma_j = \psi_j^{(1)}$ and $\delta_j = \psi_j^{(2)}$, $j = 1, \ldots, n$. Using (8) and (11), one can reduce the characteristic functions $\hat{\mu}_1(y)$ and $\hat{\mu}_2(y)$ to the form

$$\hat{\mu}_1(y) = \exp\left\{\sum_{j=0}^{n}\psi_j^{(1)}((x_j, y) - 1)\right\},$$

$$\hat{\mu}_2(y) = \exp\left\{\sum_{j=0}^{n}\psi_j^{(2)}((x_j, y) - 1)\right\},$$

i.e., they are the characteristic functions of generalized Poisson distributions and so $\mu \in I_0$. □

6.6. PROPOSITION. *Let* $\mu = e(\Phi)$, *where* $\Phi = \psi E_x$ *for some* $\psi > 0$ *and* $x \in X$. *If the order of the element* x *is either infinity or two, then* $\mu \in I_0$. *If* x *is an element of finite order* $p > 2$, *then* $\mu \notin I_0$.

PROOF. In the case, when the order of the element x is either infinity or two, the proposition is a direct consequence of Theorem 6.5. If x is an element of order $p > 2$, then, using Proposition 2.5, the proof is reduced to the case $X = \mathbb{Z}(p)$. Let $\{1, \zeta, \ldots, \zeta^{p-1}\}$ be the elements of the group $\mathbb{Z}(p)$, $\zeta = \exp\{2\pi i/p\}$. Then we have

$$\mu(\{\zeta^l\}) = \exp\{-\psi\} \sum_{n \equiv l \pmod{p}} \frac{\psi^n}{n!}, \qquad l = 0, 1, \ldots, p-1.$$

Therefore μ satisfies the condition 4.16(i) and, by Theorem 4.17, we obtain $\mu \in I_0$. □

The convolution of two arbitrary Poisson distributions is a generalized Poisson distribution of the form $\mu = e(\Phi)$, where the measure Φ is concentrated on a set $A = \{x_1, x_2\}$. Let us now turn to the question of whether this convolution belongs to the class I_0.

6.7. THEOREM. *Let* $\Phi = \psi_1 E_{x_1} + \psi_2 E_{x_2}$, *where* $\psi_j > 0$, $j = 1, 2$, *and* $x_1, x_2 \in X$ *are elements of infinite order. Let* $A = \{x_1, x_2\}$. *If*

$$(i) \quad 0 \notin M^+(A),$$

then $\mu = e(\Phi) \in I_0$.

PROOF. If the elements x_1 and x_2 are independent, then by Theorem 6.5, $\mu \in I_0$. Let the elements x_1 and x_2 be dependent, and consider the group $M(A)$ with the discrete topology. It follows from Theorem 1.24 that either $M(A) \approx \mathbb{Z}$, or $M(A) \approx \mathbb{Z} + \mathbb{Z}(n)$. Thus

$$\begin{aligned} x_1 &= l_1 h + t_1 e, \\ x_2 &= l_2 h + t_2 e, \end{aligned} \tag{1}$$

for some $l_1, l_2 \in \mathbb{Z}$, where $h \in X$ is an element of infinite order, $t_1, t_2 \in \{0, 1 \ldots, n-1\}$, and $e \in X$ is an element of order n. Since (i) is fulfilled, both l_1 and l_2 have the same sign. Therefore, without loss of generality, we assume that in (1) $0 < l_1 \le l_2$ and l_1, l_2 are relatively prime. Every element of the group $M(A)$ is uniquely represented in the form $lh + te$, where $l \in \mathbb{Z}$, $t \in \{0, 1, \ldots, n-1\}$.

Consider the group G generated by the element e, i.e., $G = \{0, e, \ldots, (n-1)e\}$, and denote $e_t = te$, $t \in \{0, 1, \ldots, n-1\}$. Now choose a character $y_1 \in Y$, such that $(e_1, y_1) = \exp\{2\pi i/n\}$. Let $H = A(Y, G)$. As the groups

G^* and Y/H are isomorphic (see Theorem 1.6) the element $y_1 + H$ is a generator of the group Y/H and

$$Y = \bigcup_{s=0}^{n-1} (y_s + H),$$

where $y_s = sy_1$, $s \in \{0, 1, \ldots, n-1\}$. Let us identify the elements $y_s \in Y$ and the corresponding characters of the group G. Since $(e_t, y_s) = \exp\{2\pi i t s / n\}$, $t, s \in \{0, 1, \ldots, n-1\}$, the matrix $\Gamma = (e_t, y_s)_{t,s=0}^{n-1}$ has the following properties:

$$\Gamma^{-1} = \frac{1}{n}\Gamma^*, \tag{2}$$

$$(e_t, y_s) = (e_s, y_t), \tag{3}$$

$$\sum_{s=0}^{n-1}(e_t, y_s) = \begin{cases} n & \text{if } t = 0, \\ 0 & \text{if } t \neq 0. \end{cases} \tag{4}$$

By Lemma 6.2 the distribution μ is concentrated on the set $B = \{0\} \cup M^+(A) \subset \bigcup_{t=0}^{n-1}\{lh + e_t\}_{l=0}^{\infty}$. Let $\mu = \mu_1 * \mu_2$, $\mu_j \in \mathscr{M}^1(X)$, $j = 1, 2$. Since $\mu(\{0\}) \geq \exp\{-\Phi(X)\}$, then in view of Lemma 6.3 one may assume that the distributions μ_j are concentrated on B.

Set $c_l^{(t)} = \mu_1(\{lh + e_t\})$, $d_l^{(t)} = \mu_2(\{lh + e_t\})$, $t \in \{0, 1, \ldots, n-1\}$, $l = 0, 1, 2 \ldots$.

The characteristic functions $\hat{\mu}_1(y)$ and $\hat{\mu}_2(y)$ may be represented as

$$\hat{\mu}_1(y) = \sum_{t=0}^{n-1}\sum_{l=0}^{\infty} c_l^{(t)}(h, y)^l(e_t, y), \qquad \hat{\mu}_2(t) = \sum_{t=0}^{n-1}\sum_{l=0}^{\infty} d_l^{(t)}(h, y)^l(e_t, y). \tag{5}$$

The characteristic function of the distribution μ has the form

$$\hat{\mu}(y) = \exp\{\psi_1[(e_{t_1}, y)(h, y)^{l_1} - 1] + \psi_2[(e_{t_2}, y)(h, y)^{l_2} - 1]\}. \tag{6}$$

By virtue of 2.10(d) from (5) and (6) we obtain

$$\exp\{\psi_1[(t_{t_1}, y)(h, y)^{l_1} - 1] + \psi_2[(e_{t_2}, y)(h, y)^{l_1} - 1]\}$$

$$= \left(\sum_{t=0}^{n-1}(e_t, y)\sum_{l=0}^{\infty} c_l^{(t)}(h, y)^l\right)\left(\sum_{t=0}^{n-1}(e_t, y)\sum_{l=0}^{\infty} d_l^{(t)}(h, y)^l\right). \tag{7}$$

For $t \in \{0, 1, \ldots, n-1\}$, set $\zeta = (h, y)$, $f_t(\zeta) = \sum_{l=0}^{\infty} c_l^{(t)}\zeta^l$, and $g_t(\zeta) = \sum_{l=0}^{\infty} d_l^{(t)}\zeta^l$.

We consider equality (7) on every conjugacy class $y_s + H$, $s \in \{0, 1, \ldots, n-1\}$, and obtain the set of n equalities

$$\exp\{\psi_1[(e_{t_1}, y_s)\zeta^{l_1} - 1] + \psi_2[(e_{t_2}, y_s)\zeta^{l_2} - 1]\}$$

$$= \left(\sum_{t=0}^{n-1}(e_t, y_s)f_t(\zeta)\right)\left(\sum_{t=0}^{n-1}(e_t, y_s)g_t(\zeta)\right),$$

$$s \in \{0, 1, \ldots, n-1\}. \tag{8}$$

By Lemma 6.4 these equalities are fulfilled for all ζ belonging to a dense set in $\mathbb{T}$ and therefore for all $\zeta \in \mathbb{T}$.

Setting $s = 0$ in (8) we obtain

$$\exp\{\psi_1[\zeta^{l_1} - 1] + \psi_2[\zeta^{l_2} - 1]\} = \left(\sum_{t=0}^{n-1} f_t(\zeta)\right) \left(\sum_{t=0}^{n-1} g_t(\zeta)\right). \tag{9}$$

Set

$$f(\zeta) = \sum_{t=0}^{n-1} f_t(\zeta), \qquad g(\zeta) = \sum_{t=0}^{n-1} g_t(\zeta).$$

The Taylor coefficients of the functions $f(\zeta)$ and $g(\zeta)$ are nonnegative. By comparing coefficients at equal powers of ζ on the right- and left-hand sides of (9) (see 3.1) we see that $f(\zeta)$ and $g(\zeta)$ are entire functions. It also follows from equality (9) that

$$M_f(r)M_g(r) = \exp\{\psi_1[r^{l_1} - 1] + \psi_2[r^{l_2} - 1]\}. \tag{10}$$

Hence the orders of the functions $f(\zeta)$ and $g(\zeta)$ do not exceed l_2. Since the coefficients $c_l^{(t)}$ and $d_l^{(t)}$ are nonnegative, the functions $f_t(\zeta)$ and $g_t(\zeta)$ are entire as well and, besides,

$$M_{f_t}(r) \le M_f(r), \qquad M_{g_t}(r) \le M_g(r). \tag{11}$$

Thus the order of $f_t(\zeta)$ and $g_t(\zeta)$ does not exceed l_2 either and the sums

$$\sum_{t=0}^{n-1} (e_t, y_s)f_t(\zeta), \quad \sum_{t=0}^{n-1} (e_t, y_s)g_t(\zeta), \qquad s \in \{0, 1, \dots, n-1\},$$

represent entire functions of order not exceeding l_2. By virtue of (8) these functions do not vanish in $\mathbb{C}$ and, by Theorem 3.2,

$$\sum_{t=0}^{n-1} (e_t, y_s)f_t(\zeta) = e^{P_s(\zeta)}, \quad \sum_{t=0}^{n-1} (e_t, y_s)g_t(\zeta) = e^{Q_s(\zeta)}, \tag{12}$$

where

$$P_s(\zeta) = a_0^{(s)} + a_1^{(s)}\zeta + \cdots + a_{l_2}^{(s)}\zeta^{l_2},$$

$$Q_s(\zeta) = b_0^{(s)} + b_1^{(s)}\zeta + \cdots + b_{l_2}^{(s)}\zeta^{l_2},$$

$s \in \{0, 1, \dots, n-1\}$. We have to prove that

$$P_s(\zeta) = \psi_1^{(1)}[(e_{t_1}, y_s)\zeta^{l_1} - 1] + \psi_2^{(1)}[(e_{t_2}, y_s)\zeta^{l_2} - 1], \qquad s \in \{0, 1, \dots, n-1\}, \tag{13}$$

where $\psi_1^{(1)}, \psi_2^{(1)} \ge 0$, and to obtain a similar representation for $Q_s(\zeta)$. Then μ_1 and μ_2 would be generalized Poisson distributions, and therefore $\mu \in I_0$.

We restrict ourselves to obtaining the desired representation for $P_s(\zeta)$. The corresponding representation for $Q_s(\zeta)$ would be obtained in the same way.

Equations (12), together with (2) and (3), imply

$$f_t(\zeta) = \frac{1}{n} \sum_{s=0}^{n-1} \overline{(e_t, y_s)} e^{P_s(\zeta)}, \qquad t \in \{0, 1, \ldots, n-1\}. \qquad (14)$$

Equating the free terms in (14), we obtain

$$c_0^{(t)} = \frac{1}{n} \sum_{s=0}^{n-1} \overline{(e_t, y_s)} e^{a_0^{(s)}}, \qquad t \in \{0, 1, \ldots, n-1\}. \qquad (15)$$

Since $e_t \notin B$ for $t \neq 0$, we have $c_0^{(t)} = 0$ for $t \neq 0$ and from (15) it follows that $\exp\{a_0^{(s)}\} = c_0^{(0)} > 0$, $s \in \{0, 1, \ldots, n-1\}$. Therefore we may assume

$$a_0^{(s)} = a_0^{(0)}, \qquad s \in \{0, 1, \ldots, n-1\}. \qquad (16)$$

Suppose that $l_2 > 1$. To obtain the next coefficients $a_k^{(s)}$ we compare the coefficients at equal powers of ζ in system (14). Let $l_1 > 1$. For $l = 1$ we obtain

$$c_1^{(t)} = \frac{c_0^{(0)}}{n} \sum_{s=0}^{n-1} \overline{(e_t, y_s)} a_1^{(s)}, \qquad t \in \{0, 1, \ldots, n-1\}. \qquad (17)$$

Since $h + e_t \notin B$, we have $c_1^{(t)} = 0$, for $t \in \{0, 1, \ldots, n-1\}$. The matrix Γ is nondegenerate, and hence by (17) we obtain $a_1^{(s)} = 0$ for $s \in \{0, 1, \ldots, n-1\}$. In the same way step by step we obtain that $a_k^{(s)} = 0$, $s \in \{0, 1, \ldots, n-1\}$, for $k < l_1$.

Let now $l_1 = 1$. To determine the coefficients $a_{l_1}^{(s)}$ for $s \in \{0, 1, \ldots, n-1\}$ consider the system

$$c_{l_1}^{(t)} = \frac{c_0^{(0)}}{n} \sum_{s=0}^{n-1} \overline{(e_t, y_s)} a_{l_1}^{(s)}, \qquad t \in \{0, 1, \ldots, n-1\}. \qquad (18)$$

Since $l_1 h + e_t \notin B$ for $t \neq t_1$, we have $c_{l_1}^{(t)} = 0$ for $t \neq t_1$ and the solution of (18) has the form

$$a_{l_1}^{(s)} = \frac{c_{l_1}^{(t_1)}}{c_0^{(0)}} (e_{t_1}, y_s), \qquad s \in \{0, 1, \ldots, n-1\}. \qquad (19)$$

Denoting $c_{l_1}^{(t_1)}/c_0^{(0)} = \psi_1^{(1)}$ we clearly have $\psi_1^{(1)} \geq 0$.

The proof of the fact that the coefficients $a_k^{(s)}$ vanish for $l_1 < k < 2l_1$ is carried out in the same way as for $0 < k < l_1$.

Suppose that we have already proved that $a_k^{(s)} = 0$ for $0 < k < k'l_1$, $k \neq l_1$, $s \in \{0, 1, \ldots, n-1\}$. Let us prove that $a_{k'l_1}^{(s)} = 0$, $s \in \{0, 1, \ldots, n-1\}$. To obtain $a_{k'l_1}^{(s)}$ we have the system of equations

$$c_{k'l_1}^{(s)} = \frac{c_0^{(0)}}{n} \sum_{s=0}^{n-1} \overline{(e_t, y_s)} \left[a_{k'l_1}^{(s)} + \frac{1}{k'!} (a_{l_1}^{(s)})^{k'} \right], \qquad t \in \{0, 1, \ldots, n-1\}. \quad (20)$$

The inequality $c_{k'l}^{(t)} \neq 0$ holds only for one value of t, namely, for $t \equiv k't_1 \pmod{n}$. Indeed, otherwise there exist integers $m', n' \geq 0$ such that $m'x_1 + n'x_2 = k'l_1 h + e_t$. Together with (1) this implies

$$m'l_1 + n'l_2 = k'l_1, \tag{21}$$

$$m't_1 + n't_2 \equiv t \pmod{n}. \tag{22}$$

Since $k' < l_2/l_1$, equality (21) leads to $n' = 0$, $m' = k'$, and by equality (22) we have $k't_1 \equiv t \pmod{n}$. This congruence clearly has a unique solution for t in the set $\{0, 1, \ldots, n-1\}$. Let us consider those equations from system (20) for which $t \not\equiv k't_1 \pmod{n}$. Then $c_{k'l_1}^{(t)} = 0$ and the right-hand sides of the equations can be simplified. Actually, accounting for (19) we obtain

$$\sum_{s=0}^{n-1} \overline{(e_t, y_s)}(a_{l_1}^{(s)})^{k'} = \sum_{s=0}^{n-1} \overline{(e_t, y_s)}[\psi_1^{(1)}(e_{t_1}, y_s)]^{k'}$$

$$= (\psi_1^{(1)})^{k'} \sum_{s=0}^{n-1} \overline{(e_t, y_s)}(e_{t_1 k' \pmod n}, y_s)$$

$$= (\psi_1^{(1)})^{k'} \sum_{s=0}^{n-1} \overline{(e_{t-t_1 k' \pmod n}, y_s)} = 0$$

by virtue of (4) and because $t - k't_1 \not\equiv 0 \pmod{n}$. Therefore system (20) yields the following:

$$0 = \sum_{s=0}^{n-1} \overline{(e_t, y_s)} a_{k'l_1}^{(s)}, \qquad t \in \{0, 1, \ldots, n-1\}, \ t \not\equiv k't_1 \pmod{n}. \tag{23}$$

From (8) and (12) for $s = 0$ we have

$$\exp\{\psi_1[\zeta^{l_1} - 1] + \psi_2[\zeta^{l_2} - 1]\} = \exp\{P_0(\zeta) \cdot \exp\{Q_0(\zeta)\}, \tag{24}$$

where the Taylor coefficients of the entire functions $\exp\{P_0(\zeta)\}$ and $\exp\{Q_0(\zeta)\}$ are nonnegative.

Since the theorem is true for the group $\mathbb{R}$ ([Lévy2], see also [Fr]), it follows from (24) that in particular

$$a_{l_2}^{(0)} \geq 0, \qquad b_{l_2}^{(0)} \geq 0, \tag{25}$$

and also $a_k^{(0)} = b_k^{(0)} = 0$ for $k \neq 0$, $k \neq l_1$, $k \neq l_2$. Hence $a_{k'l_1}^{(0)} = 0$ as well. Substituting zero instead of $a_{k'l_1}^{(0)}$ in (23) one obtains a system for the determination of $a_{k'l_1}^{(s)}$, $s \in \{1, 2, \ldots, n-1\}$. This system consists of $n-1$ equations with $n-1$ unknowns and has nonzero determinant. (To verify this it suffices to replace the row with the number $k'l_1 \pmod{n}$ in the matrix by the sum of all rows and take into account that $\det \Gamma \neq 0$). So $a_{k'l_1}^{(s)} = 0$, $s \in \{1, 2, \ldots, n-1\}$.

To prove that $a_k^{(s)} = 0$ for $k'l_1 < k < (k'+1)l_1 < l_2$ we apply the same argument as for $0 < k < l_1$.

To obtain $a_{l_2}^{(s)}$ we substitute (12) in (8) and then use the expression for $a_k^{(s)}$ and $b_k^{(s)}$ with $k < l_2$. We have

$$\exp\{\psi_1[(e_{t_1}, y_s)\zeta^{l_1} - 1] + \psi_2[(e_{t_1}, y_s)\zeta^{l_2} - 1]\}$$
$$= \exp\{a_0^{(s)} + \psi_1^{(1)}(e_{t_1}, y_s)\zeta^{l_1} + a_{l_2}^{(s)}\zeta^{l_2}\}$$
$$\times \exp\{b_0^{(s)} + \psi_1^{(2)}(e_{t_1}, y_s)\zeta^{l_1} + b_{l_2}^{(s)}\zeta^{l_2}\},$$

and hence

$$\psi_2(e_{t_2}, y_s) = a_{l_2}^{(s)} + b_{l_2}^{(s)}, \qquad s \in \{0, 1, 2, \ldots, n-1\}. \tag{26}$$

By (12) we have

$$M_{e^{P_s}}(r) \le M_{e^{P_0}}(r), \qquad M_{e^{Q_s}}(r) \le M_{e^{Q_0}}(r) \le M_{e^{Q_0}}(r),$$
$$s \in \{0, 1, 2, \ldots, n-1\}.$$

Therefore, taking (25) into account, we obtain

$$|a_{l_2}^{(s)}| \le a_{l_2}^{(0)}, \qquad |b_{l_2}^{(s)}| \le b_{l_2}^{(0)}, \qquad s \in \{0, 1, 2, 3, \ldots, n-1\}. \tag{27}$$

Let $a_{l_2}^{(0)} = \psi_2^{(1)}$. It follows from (26) and (27) that

$$a_{l_2}^{(s)} = \psi_2^{(1)}(e_{t_2}, y_s), \qquad s \in \{0, 1, 2, \ldots, n-1\}. \tag{28}$$

So

$$P_s(\zeta) = a_0^{(s)} + \psi_1^{(1)}(e_{t_1}, y_s)\zeta^{l_1} + \psi_2^{(1)}(e_{t_2}, y_s)\zeta^{l_2}, \qquad s \in \{0, 1, 2, \ldots, n-1\}. \tag{29}$$

It follows from (12) that $e^{P_0(1)} = 1$. Therefore we may assume that the polynomial $P_0(\zeta)$ is chosen such that $P_0(1) = 0$. Setting $s = 0$ and $\zeta = 1$ in (29) and taking (16) into account, we obtain $a_0^{(s)} = -\psi_1^{(1)} - \psi_2^{(1)}$.

Now the desired statement (13) follows from (29).

The case $l_2 = 1$ is simpler and may be treated similarly. $\square$

6.8. REMARK. An analysis of the proofs of Theorems 6.5 and 6.7 demonstrates that by using Lemma 6.4 we actually reduced the case of an arbitrary group X to the case of $X = \mathbb{Z}^n + \mathbb{Z}(2)$ in Theorem 6.5 and $X = \mathbb{Z} + \mathbb{Z}(n)$ in Theorem 6.7. If using Proposition 4.18 one can accomplish this reduction without resorting to Lemma 6.4. This approach is realized below to prove that if a measure Φ is concentrated on a set $A = \{x_1, x_2\}$, where x_1, x_2 are elements of infinite order, then the generalized Poisson distribution $\mu = e(\Phi)$ belongs to the class I_0 even if condition 6.7(i) ceases to be fulfilled.

We shall need the following lemma.

6.9. LEMMA. *Let* $X_1 = \mathbb{R}^m + X$, $Y_1 X_1^* = \mathbb{R}^n + Y$. *Denote the elements of* Y_1 *by* (s, y), *where* $s \in \mathbb{R}^m$, $y \in Y$. *Let* $\mu \in \mathscr{M}^1(X)$. *If the function* $\hat{\mu}(s, 0)$ *is entire with respect to* s, *then the function* $\hat{\mu}(s, y)$ *is entire with respect to* s *for any fixed* $y \in Y$, *the representation*

$$\text{(i)} \quad \hat{\mu}(s, y) = \int_{X_1} \exp\{i\langle t, s\rangle\}(x, y)\, d\mu(t, x)$$

is valid for all $s \in \mathbb{C}^m$, $y \in Y$, *and the inequality*

$$\text{(ii)} \quad \max_{s \in B_r} |\hat{\mu}(s, y)| \leq \max_{s \in B_r} |\hat{\mu}(s, 0)|, \qquad y \in Y,$$

holds.

PROOF. By the definition of a characteristic function representation (i) is valid for all $s \in \mathbb{R}^m$ and $y \in Y$. Then

$$\hat{\mu}(s, 0) = \int_{R^m} \exp\{i\langle t, s\rangle\}\, d\mu_1(t), \qquad s \in \mathbb{R}^m, \tag{1}$$

where $\mu_1 \in \mathscr{M}^1(\mathbb{R}^m)$, $\mu_1(E) = \mu(E + X)$, and $E \in \mathscr{B}(\mathbb{R}^m)$. Since $\hat{\mu}(s, 0)$ is an entire function, by Theorem 3.14 the integral on the right-hand side of (1) converges absolutely and uniformly on every compact set in $\mathbb{C}^m$. Therefore the integral

$$I(s, y) = \int_{X_1} \exp\{i\langle t, s\rangle\}(x, y)\, d\mu(t, x), \qquad s \in \mathbb{C}^m, \; y \in Y,$$

converges absolutely and uniformly on every compact set in $\mathbb{C}^m$ and admits the estimate

$$|I(s, y)| \leq \int_{\mathbb{R}^m} \exp\left\{-\sum_{j=1}^m t_j \operatorname{Im} s_j\right\} d\mu_1(t). \tag{2}$$

By applying, for example, the Hartogs theorem one can see that for every fixed $y \in Y$ the function $I(s, y)$ is entire with respect to s and coincides with $\hat{\mu}(s, y)$ for $s \in \mathbb{R}^m$. Inequality (ii) is a direct consequence of (2). $\square$

6.10. THEOREM. *Let* $\Phi = \psi_1 E_{x_1} + \psi_2 E_{x_2}$, *where* $\psi_j > 0$, $j = 1, 2$, *and* x_1, x_2 *are elements of infinite order in the group* X. *Then* $\mu_0 = e(\Phi) \in I_0$.

PROOF. Let $A = \{x_1, x_2\}$. In view of Theorem 6.5 it suffices to consider only the case of dependent x_1 and x_2. It was mentioned in the beginning of the proof of Theorem 6.7 that in this case the group $M(A)$ is isomorphic either to $\mathbb{Z}$ or to $\mathbb{Z} + \mathbb{Z}(n)$. Let, for definiteness, $M(A) \approx \mathbb{Z} + \mathbb{Z}(n)$. Then $x_1 = l_1 h + t_1 e$ and $x_2 = l_2 h + t_2 e$, where $l_1, l_2 \in \mathbb{Z}$, $h \in X$, is an element of infinite order, $t_1, t_2 \in \{0, 1, 2, \ldots, n-1\}$, and $e \in X$ is an element of order n. Set $\zeta = \exp\{2\pi i/n\}$. Then $\mathbb{Z}(n) = \{\zeta^k : k = 0, 1, 2, \ldots, n - 1\}$. Define a continuous monomorphism $\tau \colon \mathbb{Z} + \mathbb{Z}(n) \to X$ by the relation $\tau(l, \zeta^t) = lh + te$, $l \in \mathbb{Z}$, $t \in \{0, 1, 2, \ldots, n-1\}$.

Let $P \in \mathcal{M}_+(\mathbb{Z} + \mathbb{Z}(n))$ and $P = \psi_1 E_{(l_1, \zeta^{l_1})} + \psi_2 E_{(l_2, \zeta^{l_2})}$. Obviously $\mu_0 = \tau(e(P))$, and according to Proposition 4.18 it suffices to prove that $e(P) \in I_0$. Thus we have reduced the theorem to the case $X = \mathbb{Z} + \mathbb{Z}(n)$. It is useful to consider $\mathbb{Z} + \mathbb{Z}(n)$ as a subgroup of the group $\mathbb{R} + \mathbb{Z}(n)$. So to prove the theorem in the general form it is enough to consider the case $X = \mathbb{R} + \mathbb{Z}(n)$. But first we shall introduce convenient notation. By k we shall mean the element $\zeta^k \in \mathbb{Z}(n)$; (t, k), where $t \in \mathbb{R}$, $k \in \mathbb{Z}(n)$, will denote an element of $\mathbb{R} + \mathbb{Z}(n)$, and (s, l), where $s \in \mathbb{R}$, $l \in \mathbb{Z}(n)$, will denote an element of $(\mathbb{R} + \mathbb{Z}(n))^* \approx \mathbb{R} + \mathbb{Z}(n)$. Let the measure Φ be concentrated at points (β_1, k_1) and (β_2, k_2), where $\beta_1 \cdot \beta_2 \neq 0$, and let the distribution $\mu_0 = e(\Phi)$ be decomposed as $\mu_0 = \mu_1 * \mu_2$, where $\mu_j \in \mathcal{M}^1(\mathbb{R} + \mathbb{Z}(n))$, $j = 1, 2$. Denote $\alpha_j(s, l) = \hat{\mu}_j(s, l)$, $j = 0, 1, 2$. Then

$$\ln \alpha_0(s, l) = \sum_{j=1}^{2} \psi_j \left[\exp\left\{ i \left(\beta_j s + \frac{2\pi k_j l}{n} \right) \right\} - 1 \right]. \tag{1}$$

It follows from 2.10(d) that

$$\alpha_0(s, l) = \alpha_1(s, l)\alpha_2(s, l). \tag{2}$$

To prove the theorem it suffices to check that

$$\ln \alpha_1(s, l) = \sum_{j=1}^{2} \psi_j^{(1)} \left[\exp\left\{ i \left(\beta_j s + \frac{2\pi k_j l}{n} \right) \right\} - 1 \right] + i\beta s + \ln(k, l), \tag{3}$$

where $0 \leq \psi_j^{(1)} \leq \psi_j$, $\beta \in \mathbb{R}$, and (k, l) is a character of the group $\mathbb{Z}(n)$. There is a similar representation for $\ln \alpha_2(s, l)$.

Since the proof is rather cumbersome, we split it into several steps.

1. Setting $l = 0$ in (2) and using the Lévy theorem 3.12, we obtain that $\alpha_j(s, 0)$, $j = 1, 2$, are entire functions. By Lemma 6.9 for all $l \in \mathbb{Z}(n)$ the functions $\alpha_j(s, l)$, $j = 1, 2$, are entire with respect to s and, besides, equalities (2) and

$$\alpha_j(s, l) = \int_{\mathbb{R}+\mathbb{Z}(n)} \exp\left\{ i \left(ts + \frac{2\pi k l}{n} \right) \right\} d\mu_j(t, k), \qquad j = 0, 1, 2, \tag{4}$$

hold for all $s \in \mathbb{C}$ and $l \in \mathbb{Z}(n)$.

It follows from (4) that for any fixed $y \in \mathbb{R}$ the function $\alpha_j(-iy + x, l)/\alpha_j(-iy, 0)$ is a characteristic function of variable $(x, l) \in \mathbb{R} + \mathbb{Z}(n)$. With accounting of (2) it implies

$$\left| \frac{\alpha_0(-iy + x, l)}{\alpha_0(-iy, 0)} \right| \leq \left| \frac{\alpha_j(-iy + x, l)}{\alpha_j(-iy, 0)} \right| \leq 1, \qquad j = 1, 2. \tag{5}$$

It follows from equality (2) that for any fixed $l \in \mathbb{Z}(n)$ the entire function $\alpha_1(s, l)$ does not have roots in the complex s-plane. By Theorem 3.2 it admits the representation

$$\alpha_1(s, l) = \exp\{f(s, l)\},$$

where the branches $f(s, l)$ of the entire function $\ln \alpha_1(s, l)$ are chosen in such a way that $f(s, l) = \overline{f(-s, -l)}$. Let us verify that the function $f(s, l)$ coincides with the right-hand side of equation (3).

With accounting of (1) let us rewrite system (5) in the form

$$0 \le \mathrm{Re}[f(-iy, 0) - f(-iy + x, l)] \le 2 \sum_{j=1}^{2} \psi_j e^{\beta_j y} \sin^2 \left(\frac{\beta_j x + 2\pi k_j l/n}{2} \right),$$

$$\tag{6}$$

where $x, y \in \mathbb{R}$, $l \in \mathbb{Z}(n)$. Since

$$f(-iy + x, l) = \overline{f(-iy - x, -l)}, \qquad x, y \in \mathbb{R}, \ l \in \mathbb{Z}(n),$$

inequality (6) can be rewritten in the form

$$0 \le f(-iy, 0) - \frac{1}{2}[f(-iy + x, l) + f(-iy - x, -l)]$$

$$\le 2 \sum_{j=1}^{2} \psi_j e^{\beta_j y} \sin^2 \left(\frac{\beta_j x + 2\pi k_j l/n}{2} \right), \qquad x, y \in \mathbb{R}, \ l \in \mathbb{Z}(n). \tag{7}$$

For definiteness, let us assume that $\beta_2 \ge |\beta_1|$. We carry out the proof for $\beta_2 > |\beta_1|$, since the case $\beta_2 = |\beta_1|$ is simpler and can be handled similarly.

2. Let us check that the function $f(-iy + x, l)$, $l \in \mathbb{Z}(n)$, for any fixed $x \in \mathbb{R}$ satisfies the following estimate in the whole complex y-plane:

$$f(-iy + x, l) = O(|y| \exp(\beta_2 |\mathrm{Re}\, y|)), \qquad |y| \to \infty. \tag{8}$$

Indeed, the theorem is valid for the group $\mathbb{R}$. Let $x_1, x_2 \in \mathbb{R}$. If $x_1 x_2 > 0$ and the elements x_1, x_2 are dependent, this is a result of Lévy [Lévy2]; for independent x_1, x_2 this is a result of Raĭkov [Ra]. For $x_1 x_2 < 0$ the desired statement is a consequence of the Linnik theorem [LinO, Chapter V, §3]. Hence (1) and (2) yield (for $l = 0$)

$$f(s, 0) = \sum_{j=1}^{2} \psi_j^{(1)} (e^{i\beta_j s} - 1) + i\beta s, \qquad \psi_j^{(1)} \ge 0, \ j = 1, 2, \ \beta \in \mathbb{R}. \tag{9}$$

Without loss of generality one may assume $\beta = 0$ in (9). Now from (7) and (9) we obtain that the relation

$$|\mathrm{Re}\, f(-iy + x, l)| = O(\exp(\beta_2 |\mathrm{Re}\, y|)), \qquad |y| \to \infty, \tag{10}$$

holds for any fixed $x \in \mathbb{R}$ and $l \in \mathbb{Z}(n)$. The Schwarz formula 3.6 applied to the function $f(-iy - i\xi + x, l)$ in the disk $|\xi| < 1$ shows that

$$f(-iy - i\xi + x, l) = \frac{1}{2\pi} \int_0^{2\pi} \mathrm{Re}\, f(-iy - ie^{i\varphi} + x, l) \frac{e^{i\varphi} + \xi}{e^{i\varphi} - \xi} d\varphi + iC.$$

Differentiating with respect to ξ and subsequently setting $\xi = 0$ we have

$$-f'(-iy + x, l) = \frac{1}{\pi} \int_0^{2\pi} \mathrm{Re}\, f(-iy - ie^{i\varphi} + x, l) e^{-i\varphi} d\varphi,$$

and taking account of (10) we obtain

$$|f'(-iy + x, l)| = O(\exp(\beta_2 |\operatorname{Re} y|)), \qquad |y| \to \infty. \tag{11}$$

This estimate directly implies (8).

3. Let us now introduce the function

$$\theta_l(x, y) = f(-iy, 0) - \frac{1}{2}[f(-iy + x, l) + f(-iy - x, -l)], \qquad l \in \mathbb{Z}(n),$$

which for any fixed $x \in \mathbb{R}$ is an entire function of y. It follows from (8) and (9) that

$$\theta_l(x, y) = O(|y| \exp(\beta_2 |\operatorname{Re} y|)), \qquad |y| \to \infty. \tag{12}$$

Setting $x = x_1 = -2\pi k_2 l / n\beta_2$ in (7) we obtain from (12) that

$$\theta_l(x_1, y)e^{-\beta_1 y} = O(1), \qquad y \in \mathbb{R}, \ |y| \to \infty. \tag{13}$$

We now apply the Phragmén-Lindelöf theorem 3.5 to the function $\theta_l(x_1, y)e^{-\beta_1 y}$ in each quadrant of the complex y-plane. Estimate (12) and (13) imply that the function $\theta_l(x_1, y)e^{-\beta_1 y}$ is a polynomial of degree no greater than 1, but then it is constant, since it is bounded on the real axis. Thus $\theta_l(x_1, y) = C_l e^{\beta_1 y}$, where the constant C_l is easily seen to be real.

By (9) we have

$$f(-iy, 0) = \sum_{j=1}^{2} \psi_j^{(1)}(e^{\beta_j y} - 1). \tag{14}$$

Therefore, for $y \in \mathbb{C}$

$$\frac{1}{2}[f(-iy + x_1, l) + f(-iy - x_1, -l)] = \sum_{j=1}^{2} \psi_j^{(1)}(e^{\beta_j y} - 1) - C_l e^{\beta_1 y}. \tag{15}$$

Set

$$h(-iy + x_1, l) = \sum_{j=1}^{2} \psi_j^{(1)}(e^{\beta_j y} - 1) - C_l e^{\beta_1 y}, \qquad l \in \mathbb{Z}(n). \tag{16}$$

The entire function g defined by

$$ig(-iy + x_1, l) = f(-iy + x_1, l) - h(-iy + x_1, l) \tag{17}$$

is real for real y, since

$$h(-iy + x_1, l) = \operatorname{Re} f(-iy + x_1, l)$$

for $y \in \mathbb{R}$. Let us now replace x by $x + x_1$ in inequality (7) and use (15), (16), and (17) to obtain

$$\begin{aligned}
0 &\le f(-iy, 0) - \operatorname{Re} h(-iy + x + x_1, l) + \operatorname{Im} g(-iy + x + x_1, l) \\
&= \theta_l(x + x_1, y) \\
&\le 2\psi_1 e^{\beta_1 y} \sin^2\left(\frac{\beta_1(x + x_1) + 2\pi k_1 l/n}{2}\right) + 2\psi_2 e^{\beta_2 y} \sin^2 \frac{\beta_2 x}{2},
\end{aligned} \tag{18}$$

$x, y \in \mathbb{R}$, $l \in \mathbb{Z}(n)$. Since the function g is real for real y, we have for $x, y \in \mathbb{R}$ and $l \in \mathbb{Z}(n)$

$$\operatorname{Im} g(-iy + x + x_1, l) = \frac{1}{2i}[g(-iy + x + x_1, l) - g(-iy - x + x_1, l)].$$

Therefore,

$$\theta_l(x + x_1, y) = f(-iy, 0) - \operatorname{Re}[h(-iy + x + x_1, l)]$$

$$+ \frac{1}{2i}[g(-iy + x + x_1, l) - g(-iy - x + x_1, l)], \qquad (19)$$

$$x, y \in \mathbb{R}, \qquad l \in \mathbb{Z}(n).$$

Setting $x = 2\pi/\beta_2$ in (18), for $y \in \mathbb{R}$ we obtain

$$0 \leq \theta_l\left(\frac{2\pi}{\beta_2} + x_1, y\right) \leq 2\psi_1 e^{\beta_1 y} \sin^2\left(\frac{\beta_1(2\pi/\beta_2 + x_1) + 2\pi k_1 l/n}{2}\right). \qquad (20)$$

Applying the Phragmén-Lindelöf theorem 3.5 to $\theta_l(2\pi/\beta_2 + x_1, y)e^{-\beta_1 y}$ in each quadrant of the complex y-plane, we get

$$\theta_l\left(\frac{2\pi}{\beta_2} + x_1, y\right) = B_l e^{\beta_1 y}, \qquad B_l \geq 0.$$

Now from (19), (14), and (16) we have

$$\frac{1}{2i}\left[g\left(-iy + \frac{2\pi}{\beta_2} + x_1, l\right) - g\left(-iy - \frac{2\pi}{\beta_2} + x_1, l\right)\right] = R_l e^{\beta_1 y} \qquad (21)$$

for some constants R_l.

   4. We seek a solution of each of the difference equations (21). Denote $h_1(y) = g(-iy + x_1, l)$. It follows from (21) and (8) that the function

$$\Lambda_l(y) = h_l(y) - (1/\beta_1)h_l'(y) \qquad (22)$$

is a $4\pi i/\beta_2$-periodic entire function of exponential type σ, $\sigma \leq \beta_2$. Hence, by Theorem 3.3,

$$\Lambda_l(y) = \sum_{k=-2}^{2} b_k^{(l)} e^{k\beta_2 y/2}. \qquad (23)$$

In what follows we shall carry out the proof under the assumption $\beta_2 \neq \pm 2\beta_1$. The case $\beta_2 = \pm 2\beta_1$ is similar. Substituting (23) into (22) and solving the differential equation obtained we get

$$h_l(y) = g(-iy + x_1, l) = \sum_{k=-2}^{2} a_k^{(l)} e^{k\beta_2 y/2} + A_l e^{\beta_1 y}. \qquad (24)$$

The coefficients $a_k^{(l)}$ and A_l are real, since the function $h_l(y)$ is real for

$y \in \mathbb{R}$. Substituting (24) into (17) we get

$$f(-iy + x_1, l) = \sum_{j=1}^{2} \psi_j^{(1)}(e^{\beta_j y} - 1) - C_l e^{\beta_1 y}$$

$$+ i \left(\sum_{k=-2}^{2} a_k^{(l)} e^{k\beta_2 y/2} + A_l e^{\beta_1 y} \right). \quad (25)$$

5. Using (14), (16), and (24) we rewrite inequality (17) in the form

$$0 \leq 2 \sum_{j=1}^{2} \psi_j^{(1)} e^{\beta_j y} \sin^2 \frac{\beta_j x}{2} + (C_l \cos \beta_1 x + A_l \sin \beta_1 x) e^{\beta_1 y}$$

$$+ \sum_{k=-2}^{2} a_k^{(l)} e^{k\beta_2 y/2} \sin \frac{k\beta_2 x}{2} \leq 2\psi_1 e^{\beta_1 y} \sin^2 \left(\frac{\beta_1(x + x_1) + 2\pi k_1 l/n}{2} \right)$$

$$+ 2\psi_2 e^{\beta_2 y} \sin^2 \frac{\beta_2 x}{2}, \qquad x, y \in \mathbb{R}, \ l \in \mathbb{Z}(n). \quad (26)$$

We shall use this inequality to determine the constants $a_k^{(l)}$, C_l, and A_l.

Dividing inequality (26) by $e^{\beta_2 y}$ and passing to the limit as $y \to \infty$ we obtain

$$0 \leq 2\psi_2^{(1)} \sin^2 \frac{\beta_2 x}{2} + a_2^{(l)} \sin \beta_2 x \leq 2\psi_2 \sin^2 \frac{\beta_2 x}{2}.$$

This inequality implies that

$$a_2^{(l)} = 0, \qquad l \in \mathbb{Z}(n). \quad (27)$$

Dividing inequality (26) by $e^{-\beta_2 y}$ and passing to the limit as $y \to -\infty$ we obtain $0 \leq -a_{-2}^{(l)} \sin \beta_2 x \leq 0$ and consequently

$$a_{-2}^{(l)} = 0, \qquad l \in \mathbb{Z}(n). \quad (28)$$

Substituting (27) and (28) into (26) and dividing the result by $e^{\beta_1 y}$ we obtain

$$a_{-1}^{(l)} e^{-(\beta_2/+\beta_1)y} \sin \frac{\beta_2 x}{2} + a_1^{(l)} e^{(\beta_2/2-\beta_1)y} \sin \frac{\beta_2 x}{2} = O(1)$$

as $y \to -\infty$. Since $\beta_2 \neq \pm 2\beta_1$ and $\beta_2 \neq 0$, this yields

$$a_{-1}^{(l)} e^{-(\beta_2/2+\beta_1)y} \sin \frac{\beta_2 x}{2} + a_1^{(l)} e^{(\beta_2/2-\beta_1)y} \sin \frac{\beta_2 x}{2} = o(1)$$

as $y \to -\infty$. Therefore dividing inequality (26) by $e^{\beta_1 y}$ and passing to the limit as $y \to -\infty$ we obtain

$$0 \leq 2\psi_1^{(1)} \sin^2 \frac{\beta_1 x}{2} + C_l \cos \beta_1 x + A_l \sin \beta_1 x$$

$$\leq 2\psi_1 \sin^2 \left(\frac{\beta_1(x + x_1) + 2\pi k_1 l/n}{2} \right).$$

This inequality implies that

$$2\psi_1^{(1)} \sin^2 \frac{\beta_1 x}{2} + C_l \cos \beta_1 x + A_l \sin \beta_1 x$$
$$= 2E_l \sin^2 \left(\frac{\beta_1(x + x_1) + 2\pi k_1 l/n}{2} \right). \tag{29}$$

A comparison of the zero degree coefficients of the trigonometrical polynomials in (29) shows that $E_l = \psi_1^{(1)}$. We then rewrite inequality (26) in the form

$$0 \le 2\psi_1^{(1)} e^{\beta_1 y} \sin^2 \left(\frac{\beta_1(x + x_1) + 2\pi k_1 l/n}{2} \right) + 2\psi_2^{(1)} e^{\beta_2 y} \sin^2 \frac{\beta_2 x}{2}$$
$$+ \sum_{k=-1}^{1} a_k^{(l)} e^{k\beta_2 y/2} \sin \frac{k\beta_2 x}{2}$$
$$\le 2\psi_1 e^{\beta_1 y} \sin^2 \left(\frac{\beta_1(x + x_1) + 2\pi k_1 l/n}{2} \right)$$
$$+ 2\psi_2 e^{\beta_2 y} \sin^2 \frac{\beta_2 x}{2}, \qquad x, y \in \mathbb{R}, \ l \in \mathbb{Z}(n), \tag{30}$$

and set $x = x_0 = -2\pi k_1 l/n\beta_1 - x_1 = (2\pi l/n)(k_2/\beta_2 - k_1/\beta_1)$. We obtain

$$0 \le 2\psi_2^{(1)} e^{\beta_2 y} \sin^2 \frac{\beta_2 x_0}{2} + \sum_{k=-1}^{1} a_k^{(l)} e^{k\beta_2 y/2} \sin \frac{k\beta_2 x_0}{2}$$
$$\le 2\psi_2 e^{\beta_2 y} \sin^2 \frac{\beta_2 x_0}{2}, \qquad y \in \mathbb{R}, \ l \in \mathbb{Z}(n). \tag{31}$$

Assume $\sin(\beta_2 x_0/2) \ne 0$. Dividing inequality (31) by $e^{-\beta_2 y/2}$ and passing to the limit as $y \to -\infty$ we get

$$a_{-1}^{(l)} = 0, \qquad l \in \mathbb{Z}(n). \tag{32}$$

Substituting (32) into (31), dividing the result by $e^{\beta_2 y/2}$ and passing to the limit as $y \to -\infty$ we see that $a_1^{(l)} = 0$.

If $\sin(\beta_2 x_0/2) = 0$, then by using inequality (30) we see that in a neighborhood of the point x_0 (for small x)

$$0 \le 2\psi_1^{(1)} \frac{\beta_1^2 x^2}{4} e^{\beta_1 y} + 2\psi_2^{(1)} \frac{\beta_2^2 x^2}{4} e^{\beta_2 y} + \frac{\beta_2 x}{2} \sum_{k=-1}^{1} k a_k^{(l)} e^{k\beta_2 y/2} + o(x^2)$$
$$\le 2\psi_1 \frac{\beta_1^2 x^2}{4} e^{\beta_1 y} + 2\psi_2 \frac{\beta_2^2 x^2}{4} e^{\beta_2 y} + o(x^2).$$

Hence $a_{-1}^{(l)} = a_1^{(l)} = 0$. So we have proved that $a_k^{(l)} = 0$ for $k \ne 0$ and $l \in \mathbb{Z}(n)$.

The constants C_l and A_l are determined from relation (29), namely,

$$C_l = \psi_1^{(1)} \left(1 - \cos \left(\beta_1 x_1 + \frac{2\pi k_1 l}{n} \right) \right),$$

$$A_l = \psi_2^{(1)} \sin \left(\beta_1 x_1 + \frac{2\pi k_1 l}{n} \right).$$

By substituting these values into (25) we obtain

$$f(s, l) = \sum_{j=1}^{2} \psi_j^{(1)} \left[\exp \left\{ i \left(\beta_j s + \frac{2\pi k_j l}{n} \right) \right\} - 1 \right] + i a_0^{(l)}.$$

6. To complete the proof of the theorem it remains to prove that $\exp\{ia_0^{(l)}\}$ is a character of the group $\mathbb{Z}(n)$. Indeed, the function $\exp\{f(-iy + x, l) - f(-iy, 0)\}$ is a characteristic function of the variables $(x, l) \in \mathbb{R} + \mathbb{Z}(n)$ for any fixed $y \in \mathbb{R}$. Let $x = x_0 = -2\pi k_1 / n\beta_1$, and consider the following function defined on the group $\mathbb{Z}$:

$$\beta_y(l) = \exp\{f(-iy + lx_0, l \ (\mathrm{mod} \ n)) - f(-iy, 0)\}$$

$$= \exp \left\{ \psi_2^{(1)} e^{\beta_2 y} \left[\exp \left\{ i \left(\beta_2 l x_0 + \frac{2\pi k_2 l}{n} \right) \right\} - 1 \right] \right.$$

$$\left. + i a_0^{(l(\mathrm{mod} \ n))} \right\}, \qquad l \in \mathbb{Z}. \tag{33}$$

By 2.10(h) $\beta_y(l)$ is a characteristic function for any $y \in \mathbb{R}$. Hence the function

$$\exp\{ia_0^{(l(\mathrm{mod} \ n))}\} = \lim_{y \to -\infty} \beta_y(l)$$

is characteristic as well. Therefore the function $\exp\{ia_0^{(l)}\}$, $l \in \mathbb{Z}$, is also characteristic. Inequality (5) now implies

$$|\beta_y(l)| \geq \left| \exp \left\{ \psi_2 e^{\beta_2 y} \left[\exp \left\{ i \left(\beta_2 x_0 + \frac{2\pi k_2 l}{n} \right) \right\} - 1 \right] \right\} \right|.$$

Passing here to the limit as $y \to -\infty$ we obtain $|\exp\{ia_0^{(l)}\}| \geq 1$. Thus $|\exp\{ia_0^{(l)}\}| \equiv 1$, that is, $\exp\{ia_0^{(l)}\}$ is a character of the group $\mathbb{Z}(n)$. $\square$

It should be noted that Theorem 6.7 is a corollary of Theorem 6.10.

6.11. PROPOSITION. *Let* $\Phi = \psi_1 E_{x_1} + \psi_2 E_{x_2}$, *where* $\psi_j > 0$, $x_1 \neq x_2$ *and* $x_j \neq 0$, $2x_j = 0$, *for* $= 1, 2$. *Then* $\mu = e(\Phi) \notin I_0$.

PROOF. By Proposition 2.5 it suffices to consider the case $X = (\mathbb{Z}(2))^2$. Let $x_3 = x_1 + x_2$. Then $X = \{0, x_1, x_2, x_3\}$. Without loss of generality we may assume that $\psi_1 = \psi_2 = \psi$. By direct verification we have

$$\mu(\{0\}) = \frac{1}{4}(1 + e^{-2\psi})^2, \qquad \mu(\{x_1\}) = \frac{1}{4}(1 - e^{-4\psi}),$$

$$\mu(\{x_2\}) = \frac{1}{4}(1 - e^{-4\psi}), \qquad \mu(\{x_3\}) = \frac{1}{4}(1 - e^{-2\psi})^2.$$

Therefore the distribution μ satisfies 4.16(i). By Theorem 4.17 we have $\mu \notin I_0$. □

6.12. Let $\Phi = \psi_1 E_{x_1} + \psi_2 E_{x_2}$, where $\psi_j > 0$, $x_j \neq 0$, $j = 1, 2$, and $x_1 \neq x_2$. Denote $\mu = e(\Phi)$. The following statement is a consequence of Theorems 6.5 and 6.10 and Propositions 6.6 and 6.11.

The distribution $\mu \in I_0$ if and only if either both elements x_1, x_2 have infinite order or one of them has infinite order and the other order two.

We now turn to the theorems on the membership to the class I_0 of a generalized Poisson distribution $\mu = e(\Phi)$, generated by a not necessarily discrete measure Φ. We first consider the case when the convolution powers of the measure Φ are mutually singular. As an example of such a measure one may consider a measure concentrated on an independent set. A wide class of measures with mutually singular convolution powers will be demonstrated in Theorem 6.15.

6.13. THEOREM. *Let $\Phi \in \mathcal{M}_+(X)$, and let the measure Φ^{*n} and Φ^{*m} be mutually singular for any distinct natural numbers n, m. Then $\mu = e(\Phi) \in I_0$.*

PROOF. Let $\mu = \mu_1 * \mu_2$ where $\mu_j \in \mathcal{M}^1(X)$, $j = 1, 2$. Since $\mu(\{0\}) > 0$, replacing the distributions μ_j by their shifts if necessary, one may assume $\mu_j(\{0\}) > 0$, $j = 1, 2$. This may be shown in the same way as when proving Lemma 6.3. By Lemma 6.8, we have

$$\mu_j(E) \leq K\mu(E), \qquad j = 1, 2, \tag{1}$$

for some $K > 0$ and for any $E \in \mathcal{B}(X)$.

For every $n = 1, 2, 3, \ldots$, let A_n be the set on which the measure Φ^{*n} is concentrated. By the condition of the theorem the sets A_n may be chosen pairwise disjoint. We also have $\Phi^{*n}(\{0\}) = 0$ for any $n = 1, 2, 3, \ldots$, because if $\Phi^{*n}(\{0\}) > 0$ for some n, then by Lemma 4.8 the measure Φ^{*n} would be absolutely continuous with respect to $\Phi^{*(n+1)}$. Therefore, one may assume $0 \notin A_n$ for $n = 1, 2, 3, \ldots$. Denote $A_0 = \{0\}$. The sets A_n, $n = 0, 1, 2, \ldots$, are pairwise disjoint as well.

Denote by ν_{nj}, $j = 1, 2$, $n = 0, 1, 2, \ldots$, the restriction to A_n of the distribution μ_j. According to (1) we have

$$\nu_{nj}(X) = \nu_{nj}(A_n) = \mu_j(A_n) \leq K\mu(A_n) = K \exp\{-\Phi(X)\} \frac{\Phi^{*n}}{n!}(A_n)$$
$$\leq K \exp\{-\Phi(X)\} \frac{\Phi^{*n}}{n!}(X) = K \exp\{-\Phi(X)\} \frac{(\Phi(X))^n}{n!}. \tag{2}$$

Note that

$$\mu = \mu_1 * \mu_2 = \left(\sum_{p=0}^{\infty} \nu_{p1}\right) * \left(\sum_{q=0}^{\infty} \nu_{q2}\right) = \sum_{n=0}^{\infty} \left(\sum_{p+q=n} \nu_{p1} * \nu_{q2}\right). \tag{3}$$

By (1) the measure ν_{mj} is absolutely continuous with respect to the restriction to A_m of the distribution μ, i.e., with respect to the measure Φ^{*m}. By Lemma 4.9, the measure $\nu_{p1} * \nu_{q2}$ is absolutely continuous with respect to $\Phi^{*(p+q)}$, and therefore $\nu_{p1} * \nu_{q2}$ is concentrated on the set A_{p+q}. Then by (3) we have

$$\sum_{p+q=n}^{\infty} \nu_{p1} * \nu_{q2} = \exp\{-\Phi(X)\}\frac{\Phi^{*n}}{n!}. \tag{4}$$

It follows from inequality (2) that the function

$$\varphi_j(z, y) = \sum_{n=0}^{\infty} z^n \hat{\nu}_{nj}(y), \qquad y \in Y, \quad j = 1, 2, \tag{5}$$

is an entire function of z for any fixed $y \in Y$ and satisfies the inequality

$$|\varphi_j(z, y)| \le e^{\sigma|z|}, \qquad j = 1, 2,$$

for some $\sigma > 0$ and all $|z|$ large enough. By (4) and 2.10(b), (d) we have

$$\varphi_1(z, y)\varphi_2(z, y) = \sum_{n=0}^{\infty} z^n \left(\sum_{p+q=n}^{\infty} \hat{\nu}_{p1}(y) * \hat{\nu}_{q2}(y) \right) \qquad .$$

$$= \exp\{\Phi(X)\} \sum_{n=0}^{\infty} z^n \frac{\widehat{\Phi}^n(y)}{n!} = \exp\{z(\widehat{\Phi}(y) - \widehat{\Phi}(0))\}. \tag{6}$$

Therefore, $\varphi_j(z, y) \ne 0$ for any $z \in \mathbb{C}$. By Theorem 3.2, equality (6) implies

$$\varphi_j(z, y) = \exp\{a_{0j}(y) + a_{1j}(y)z\}, \qquad j = 1, 2. \tag{7}$$

Setting $z = 0$ in (5) we obtain

$$\varphi_j(0, y) = \hat{\nu}_{0j}(y) = c_j > 0, \qquad j = 1, 2. \tag{8}$$

By comparing coefficients at same powers of z in the expansion of the right-hand side of (7) in a power series and in (5), taking account of (6), we obtain

$$c_j \frac{a_{1j}^n(y)}{n!} = \hat{\nu}_{nj}(y), \qquad j = 1, 2, \quad n = 1, 2, 3, \dots. \tag{9}$$

Set $\nu_{1j} = \Phi_j \in \mathscr{M}_+(X)$, $j = 1, 2$. Then (8), (9), and 2.10(b), (d) imply

$$\mu_j = \sum_{n=0}^{\infty} \nu_{nj} = c_j \sum_{n=0}^{\infty} \frac{\Phi^{*n}}{n!},$$

whence $c_j = \exp\{-\Phi_j(X)\}$ and $\mu_j = e(\Phi_j)$, $j = 1, 2$. $\square$

6.14. COROLLARY. *Let $\Phi \in \mathscr{M}_+(X)$, and let the measure Φ be concentrated on an independent set A. Then $\mu = e(\Phi) \in I_0$.*

PROOF. The measure Φ^{*n} is concentrated on the set $(n)A$. Since the sets $(n)A$ and $(m)A$ are disjoint when $n \ne m$, the measures Φ^{*n} and Φ^{*m} are mutually singular for any natural n, m, $n \ne m$. $\square$

Following Lin and Saeki [LiSa] we present a class of distributions on a nondiscrete group X with pairwise singular convolution powers. Suppose that there exists a neighborhood of zero in the group X that does not contain elements of infinite order. Denote by $q(X)$ the largest positive integer such that every neighborhood of zero in X contains an element of order q.

6.15. THEOREM [LiSa]. *Let* $\{U_n\}$ *be a fundamental system of compact symmetric neighborhoods of zero in* X *such that* (2) $U_{n+1} \subset U_n$, *and let* $\{(a_n, b_n, c_n)\}$ *be a sequence of triples of nonnegative numbers such that* $a_n + b_n + c_n = 1$. *Denote* $U = \prod_{n=1}^{\infty} U_n$, *and set*

$$\Phi_{\tilde{x}} = *_{n=1}^{\infty} (a_n E_0 + b_n E_{x_n} + c_n E_{-x_n})$$

for some $\tilde{x} = (x_1, x_2, x_3, \ldots) \in U$. *Suppose that for any limit point* (a, b, c) *of the sequence* $\{(a_n, b_n, c_n)\}$ *the inequality* $\max\{a, b, c\} < 1$ *is valid. Then*

(i) *if every neighborhood of zero in* X *contains either an element of infinite order or an element of the order* ≥ 4, *then for all* $\tilde{x} \in U$ *except maybe a set of the first category the distributions* $\Phi_{\tilde{x}}^{*n}$ *and* $\Phi_{\tilde{x}}^{*m} * E_x$ *are pairwise singular for any natural* n *and* m, $n \neq m$, *and for any* $x \in X$;

(ii) *the same conclusion is valid if* $q(X) = 2$ *and* $a \neq 0$, $a \neq b + c$, *or* $q(X) = 3$ *and* $2a \neq b + c$.

6.16. REMARK. Rudin has proved [Rud] that if every neighborhood of zero contains an element of infinite order, then there exists an independent perfect subset $A \subset X$. Consequently, any measure $\Phi \in \mathscr{M}_+(X)$ concentrated on A has pairwise singular convolution powers.

We need a group analog of the following result.

OSTROVSKII'S THEOREM. *Let the measure* $\Phi \in \mathscr{M}_+(\mathbb{R})$ *be such that* $\sigma(\Phi) \subset [a, b]$, *where* $0 < b < 2a$. *Then all divisors of the distribution* $\mu = e(\Phi)$ *have the form* $\mu_1 = e(\Phi_1) * E_{t_1}$, *where* $\Phi_1 \in \mathscr{M}_+(\mathbb{R})$, $\Phi_1 \leq \Phi$, *and* $t_1 \in \mathbb{R}$.

First we prove a lemma.

6.17. LEMMA. *Let* $X_1 = \mathbb{R} + X$, *let* Φ *be a charge on* X_1, *and let for any* $y \in Y$ $\widehat{\Phi}(s, y)$ *be an entire function of exponential type with respect to* s. *Let the type of this function not exceed* b. *Then* $\sigma(\Phi) \subset [-b, b] + X$.

PROOF. By the definition of a characteristic function

$$\widehat{\Phi}(s, y) = \int_{R+X} e^{its}(x, y)\, d\Phi(t, x).$$

By the hypothesis of the lemma the functions

$$\Phi_y^{(1)}(s) = \frac{1}{2}(\widehat{\Phi}(s, y) + \widehat{\Phi}(s, -y)) = \int_{\mathbb{R}+X} e^{its}\, \mathrm{Re}(x, y)\, d\Phi(t, x)$$

and

$$\Phi_y^{(2)}(s) = \frac{1}{2i}(\widehat{\Phi}(s, y) - \widehat{\Phi}(s, -y)) = \int_{\mathbb{R}+X} e^{its}\, \mathrm{Im}(x, y)\, d\Phi(t, x)$$

are, with respect to s, entire functions of exponential type σ, $\sigma \le b$, for any $y \in Y$.

Consider the charges on $\mathbb{R}$:

$$P_y^{(1)}(E) = \int_{E+X} \operatorname{Re}(x, y)\, d\Phi(t, x), \qquad P_y^{(2)}(E) = \int_{E+X} \operatorname{Im}(x, y)\, d\Phi(t, x),$$

$$E \in \mathscr{B}(\mathbb{R}).$$

Then $\widehat{P}_y^{(j)}(s) = \Phi_y^{(j)}(s)$, $j = 1, 2$. By a theorem of Pólya (see [Ram, ad. II.2]), $\sigma(P_y^{(j)}) \subset [-b, b]$, $j = 1, 2$. Now let $E \in \mathscr{B}(\mathbb{R})$ and $E \cap [-b, b] = \varnothing$. Then

$$0 = P_y^{(1)}(E) + i P_y^{(2)}(E) = \int_{E+X} (x, y)\, d\Phi(t, x)$$

$$= \int_X (x, y)\, d\Lambda_E(x) = \widehat{\Lambda}_E(y),$$

where Λ_E is a charge on X defined by the relation $\Lambda_E(A) = \Phi(E + A)$. By 2.10(b) we obtain $\Lambda_E = 0$, i.e., $\Phi(E + A) = 0$ for any $A \in \mathscr{B}(X)$. Hence, $\sigma(\Phi) \subset [-b, b] + X$. $\square$

The following result may be considered as a group analog of the Ostrovskiĭ theorem.

6.18. THEOREM. *Let $X_1 = \mathbb{R} + X$ and $\Phi \in \mathscr{M}_+(X_1)$ be given such that $\sigma(\Phi) \subset A + X$, where A is a closed subset of $\mathbb{R}$, $A \subset [a, b]$, $a > 0$. Let $\mu = e(\Phi)$, and let μ_1 be a divisor of μ. Then the characteristic function of the distribution μ_1 has the form*

(i) $\hat{\mu}_1(s, y) = e^{i t_0 s}(x_0, y) \exp \left\{ \int_{\mathbb{R}+X} [E^{its}(x, y) - 1]\, d\Lambda(t, x) \right\},$

where Λ is a charge such that $\sigma(\Lambda) \subset M^+(A) \cap [a, b]) + X$ and the restriction of Λ to the set $[a, 2a) + X$ is a measure.

PROOF. By the definition of a characteristic function we have

$$\hat{\mu}(s, y) = \exp \left\{ \int_a^b \int_X [e^{its}(x, y) - 1]\, d\Phi(t, x) \right\}.$$

As can easily be checked, the set $M^+(A)$ is closed and by Lemma 6.2 $\sigma(\mu) \subset \{0\} \cup (M^+(A) + X)$.

Let $\mu = \mu_1 * \mu_2$ for some $\mu_j \in \mathscr{M}^1(X)$, $j = 1, 2$. By 2.10(d)

$$\hat{\mu}(s, y) = \hat{\mu}_1(s, y)\hat{\mu}_2(s, y). \tag{1}$$

Setting $y = 0$ in (1) and applying the Lévy theorem 3.12 we obtain that $\hat{\mu}_1(s, 0)$ is an entire function with respect to s. According to Lemma 6.9, the function $\hat{\mu}_1(s)$ is then entire with respect to s for any fixed $y \in Y$ and equality (1) is valid for all $s \in \mathbb{C}$ and $y \in Y$. Therefore $\hat{\mu}_1(s, y) \ne 0$ for any $s \in \mathbb{C}$, $y \in Y$. Consequently, by Theorem 3.2 $\hat{\mu}_1(s, y)$ admits the representation

$$\hat{\mu}_1(s, y) = \exp\{f(s, y)\},$$

where the branch $f(s, y)$ of the entire function $\ln \hat{\mu}_1(s, y)$ is chosen in such a way that $f(s, y) = \overline{f(-s, -y)}$.

Since $\hat{\mu}(s, 0)$ is the characteristic function of an infinitely divisible distribution on the group $X = \mathbb{R}$, whose Lévy measure is concentrated on the segment $[a, b]$, then, by the Linnik theorem 3.13, $f(s, 0)$ is an entire function of exponential type σ, $\sigma \leq b$. By 3.1 the estimate

$$|f(s, 0)| \leq Ce^{(b+\varepsilon)|s|}, \qquad C > 0, \ \varepsilon > 0,$$

holds everywhere in the complex plane. Then by 6.9(ii)

$$\max_{|s| \leq r} \operatorname{Re} f(s, y) \leq \max_{|s| \leq r} \operatorname{Re} f(s, 0) \leq Ce^{(b+\varepsilon)r},$$

and from the Carathéodory inequality 3.4 it follows now that $f(s, y)$ is an entire function with respect to s, of exponential type σ, $\sigma \leq b$, for any $y \in Y$.

Since $\mu(\{0\}) > 0$, replacing the distributions μ_1 and μ_2 by their shifts in the same way as in the proof of Lemma 6.3 if necessary, we can assume $\mu_j(\{0\}) > 0$ and $\sigma(\mu_j) \subset \sigma(\mu)$, $j = 1, 2$. Thus $\mu_1 = cE_0 + \alpha$, where $c > 0$ and $\alpha \in \mathscr{M}_+(X)$, and $\sigma(\alpha) \subset M^+(A) + X \subset [a, +\infty) + X$. We have

$$\hat{\mu}_1(s, y) = c + \int_a^b \int_X e^{its}(x, y)\, d\alpha(t, x).$$

Since, for any $y \in Y$, $\hat{\mu}_1(s, y)$ is an entire function with respect to s, this representation is valid for all $s \in \mathbb{C}$. Take $\eta > 0$ such that

$$\int_a^\infty \int_X e^{-\eta t}\, d\alpha(t, x) < c,$$

and consider the measure $\alpha_\eta \in \mathscr{M}_+(X_1)$ defined by the relation

$$\alpha_\eta(E) = \int_E e^{-\eta t}\, d\alpha(t, x), \qquad E \in \mathscr{B}(X_1).$$

For any $s \in \mathbb{R}$ the inequality

$$\left| \int_a^\infty \int_X e^{its}(x, y)\, d\alpha_\eta(t, x) \right| \leq \int_a^\infty \int_X d\alpha_\eta(t, x) < c \qquad (2)$$

holds. We have

$$f(s + i\eta, y) = \ln \left\{ c + \int_a^\infty \int_X e^{its}(x, y)\, d\alpha_\eta(t, x) \right\}$$

$$= \ln c + \sum_{k=1}^\infty (-1)^{k+1} \left[\int_a^\infty \int_X e^{its}(x, y)\, d\alpha_\eta(t, x) \right]^k \frac{1}{kc^k}. \quad (3)$$

From (2) it follows that the series

$$\Lambda_\eta = \sum_{k=1}^{\infty} (-1)^{k+1} \frac{\alpha_\eta^{*k}}{kc^k} \tag{4}$$

is norm-convergent. Since $\sigma(\alpha_\eta) = \sigma(\alpha)$, we get from (4)

$$\sigma(\Lambda_\eta) \subset \overline{M^+(\sigma(\alpha))} \subset M^+(A) + X. \tag{5}$$

From (3) and (4) it follows that

$$f(s + i\eta, y) = \ln c + \int_0^\infty \int_X e^{its}(x, y) \, d\Lambda_\eta(t, x). \tag{6}$$

Since, for any $y \in Y$, $f(z, y)$ is, with respect to z, an entire function of exponential type σ not exceeding b, then by Lemma 6.16

$$\sigma(\Lambda_\eta) \subset [-b, b] + X \tag{7}$$

and the representation (6) can be rewritten in the form

$$f(s + i\eta, y) = \int_a^b \int_X e^{its}(x, y) \, d\Lambda_\eta(t, x). \tag{8}$$

The representation (8) is valid for complex s as well. Consider the charge

$$\Lambda(E) = \int_E e^{t\eta} \, d\Lambda_\eta(t, x), \qquad E \in \mathscr{B}(X_1).$$

Setting $s = \zeta - i\eta$ ($\zeta \in \mathbb{R}$) in (8), we have

$$f(\zeta, y) = \ln c + \int_a^b \int_X e^{it\zeta}(x, y) \, d\Lambda(t, x).$$

Therefore,

$$\hat{\mu}_1(s, y) = \exp\{f(s, y)\} = c \exp\left\{ \int_{X_1} e^{its}(x, y) \, d\Lambda(t, x) \right\}. \tag{9}$$

Setting $s = 0$, $y = 0$ in (9) we determine c and obtain the desired representation.

According to (5) and (7) we have $\sigma(\Lambda) = \sigma(\Lambda_\eta) \subset (M^+(A) + X) \cap ([-b, b] + X) = (M^+(A) \cap [-b, b]) + X = (M^+(A) \cap [a, b]) + X$. Denote

$$\Omega = \sum_{k=2}^{\infty} \frac{\Lambda^{*k}}{k!}.$$

We have $\mu_1 = e(\Lambda) = \exp\{-\Lambda(X)\}(E_0 + \Lambda + \Omega)$, where

$$\sigma(\Omega) \subset \overline{\bigcup_{k=2}^{\infty}(k)\sigma(\Lambda)} \subset [2a, +\infty) + X.$$

Therefore, the restriction of the charge Λ to the set $[a, 2a) + X$ is a measure. $\square$

6.19. COROLLARY. *Let* $X_1 = \mathbb{R} + X$, $\Phi \in \mathcal{M}_+(X_1)$, *and* $\sigma(\Phi) \subset [a, b] + X$, *where* $0 < b < 2a$. *Then* $\mu = e(\Phi) \in I_0$.

PROOF. Let μ_1 be an arbitrary divisor of μ. Set $A = [a, b]$, and consider the representation 6.18(i) of the characteristic function $\hat{\mu}_1$. Since $M^+(A) \cap [a, b] \subset [a, 2a)$, then $\sigma(\Lambda) \subset [a, 2a) + X$ and, hence, the charge Λ is a measure. □

To finish the study of conditions on a measure Φ, ensuring the membership of the generalized Poisson distribution $\mu = e(\Phi)$ in the class I_0, we make the following remark.

6.20. REMARK. Let $\Phi \in \mathcal{M}_+(X)$, $\mu = e(\Phi)$, and $\mu = \mu_1 * \mu_2$, where $\mu_j \in \mathcal{M}^1(X)$, $j = 1, 2$. If $\mu \in I_0$, then $\mu_j = E_{x_j} * e(\Phi_j)$, where $x_j \in X$ and $\Phi_j \in \mathcal{M}_+(X)$, $= 1, 2$.

Indeed, by Theorem 4.3, μ_j are infinitely divisible distributions without nondegenerate idempotent divisors. Let (x_j, Φ_j, φ_j) be the representation of the characteristic function of the distribution μ_j, $j = 1, 2$, in accordance with Remark 2.22. Since the function φ in 2.22(i) is unique, we have $\varphi_1 = \varphi_2 = 0$. Since $\mu(\{0\}) > 0$, then, as noted in the proof of Lemma 6.3, we can replace the distributions μ_j by their shifts if necessary and assume $\mu_j(\{0\}) > 0$. But if the characteristic function of the distribution μ_j has the representation $(x_j, \Phi_j, 0)$ and $\mu_j(\{0\}) > 0$, then (see Corollary A.3.6 below) $\Phi_j \in \mathcal{M}_+(X)$.

Now let us turn to the following problem. Let X be a nondiscrete group, $\Phi \in \mathcal{M}_+(X)$, and let the distribution $\mu = e(\Phi)$ belong to the class I_0. What is the possible structure of the measure Φ, i.e., the expansion 5.13(i)

$$\Phi = \Phi_{\mathrm{ac}} + \Phi_{\mathrm{s}} + \Phi_{\mathrm{d}},$$

subject to properties of the group X?

6.21. LEMMA. *Let* X *be a nondiscrete compact group,* $\Phi \in \mathcal{M}_+(X)$, *and* $\mu = e(\Phi)$. *If* $\mu \in I_0$, *then* $\Phi_{\mathrm{ac}} = 0$.

PROOF. Let $L^1(X) = L^1(X, m_X)$ be the space of all functions integrable on the group X with respect to the measure m_X. For functions $f_1, f_2 \in L^1(X)$ their convolution $f = f_1 * f_2$ is defined by the relation

$$f(x) = (f_1 * f_2)(x) = \int_X f_1(x - t) f_2(t) \, dm_X(t). \tag{1}$$

It is well known that the integral on the right-hand part of (1) exists for almost all x (with respect to m_X) and defines a function from $L^1(X)$. If at of the functions f_j is bounded, then the convolution f is continuous. The convolution of n copies of the function f is denoted by f^{*n} (see [HR1, §20]).

Let $\Phi_{\mathrm{ac}} \neq 0$. Then $\Phi_{\mathrm{ac}} = f(x) m_X$ for some $f \in L^1(X)$ and there exist a closed subset $A \subset X$ and $\delta > 0$ such that $m_X(A) > 0$ and $f(x) \geq \delta$,

$x \in A$. One may assume A to be the support of the measure with the density $g(x) = \chi_A(x)$ (see Lemma 4.13). Denote by P the measure on X with the density $\delta g(x)$ and set $\nu = e(P)$.

By Lemma 6.2 the distribution ν is concentrated on the set $B = \{0\} \cup M^+(A)$. Let us verify that B is a closed subgroup of X. Since $m_X(A) > 0$, there exists an open set U such that $U \subset (2)A$. Let $U \supset x_0 + V$, where V is a symmetric neighborhood of zero in X. Since the group X is compact, one can find a natural k_0 such that $k_0 x_0 \in V$. Hence there is a symmetric neighborhood of zero $W \subset X$ such that

$$(2k_0)A \supset (k_0)U \supset k_0 x_0 + V \supset W.$$

Let $H = M^+(W)$. Then H is an open subgroup of X and $H \subset B$. Since the group X is compact and the subgroup H is open, the factor-group X/H is finite. Consider the decomposition of the group X by the subgroup H:

$$X = H \cup (x_1 + H) \cup \cdots \cup (x_{n-1} + H),$$

and note that if $(x_j + H) \cap B \neq \varnothing$, then $x_j + H \subset B$. Therefore $B = \bigcup_{x_j \in B}(x_j + H)$ is a closed set. Let $a = x_j + h \in B$, $h \in H$. Then $na = nx_j + nh = -h_1 \in H$. Hence $a + (n-1)a + h_1 = 0$, i.e., $-a = (n-1)a + h_1 \in B$. Since B is a semigroup, we have obtained that B is a group. Obviously, $B \not\approx \mathbb{Z}(2)$.

Let us now verify that $g^{*4k_0}(0) > 0$. Indeed, set $\varphi(x) = g^{*2k_0}(x)$. This function is continuous and the support of the measure with the density $\varphi(x)$ is $(2k_0)A \supset W$. Let

$$0 g^{*4k_0}(0) = \int_X \varphi(-x)\varphi(x)\,dm_X(x).$$

Then

$$0 = \int_B \varphi(-x)\varphi(x)\,dm_B(x) \geq \int_W \varphi(-x)\varphi(x)\,dm_B(x). \qquad (2)$$

Since the neighborhood W is contained in the support of the measure with the density $\varphi(x)$, $\varphi(\zeta) > 0$ at some point $\zeta \in W$. Then $\varphi(x) > 0$ for all x from some neighborhood $W_\zeta \subset W$. Now (2) implies $\varphi(-x) = 0$ for all $x \in W_\zeta$, but that is impossible because $-W_\zeta \subset W \subset (2k_0)A$.

Since $g^{*4k_0}(0) > 0$, the inequality $g^{*(4k_0+n)}(x) > 0$ hold for any $x \in (n)A$. Therefore the continuous function

$$\psi(x) = \sum_{l=2}^{\infty} \frac{g^{*l}(x)}{l!}$$

is positive at every point $x \in (n)A$, and, hence, at every point $x \in B$. Since the group B is compact, $\psi(x) \geq \varepsilon > 0$ for some $\varepsilon > 0$ and any $x \in B$. This implies that the distribution ν satisfies condition 4.16(i). By Theorem 4.17 we have $\nu \notin I_0$. Since ν is a divisor of μ, we have $\mu \notin I_0$ as well. $\square$

6.22. PROPOSITION. *Let X be a nondiscrete group that consists of compact elements, $\Phi \in \mathscr{M}_+(X)$, and $\mu = e(\Phi)$. If $\mu \in I_0$, then $\Phi_{ac} = 0$.*

PROOF. By Theorem 1.14 we have $X \approx \mathbb{R}^n + G$, where $n \geq 0$ and the group G contains an open compact subgroup K. By the hypothesis, $X = X_0$. Therefore $n = 0$ and the factor group G/K is periodic. We assume that $X = G$ and K is an open compact subgroup of X. Since K is open, the factor group X/K is discrete. Let P be the restriction of the measure Φ_{ac} to an arbitrary conjugacy class $x + K$. Since the factor group X/K does not contain elements of infinite order, the class $x + K$ generates a compact subgroup H of X and as can easily be seen, the measure P, when considered on H, is absolutely continuous with respect to m_H. Since the distribution $e(P)$ is a divisor of μ, we have $e(P) \in I_0$. By Lemma 6.21 we get $P = 0$ and therefore $\Phi_{ac} = 0$. $\square$

6.23. PROPOSITION. *Let X be a nondiscrete group containing a noncompact element. Then on X there exists a measure $\Phi \in \mathscr{M}_+(X)$ such that $\Phi_{ac} > 0$, $\Phi_s > 0$, $\Phi_d > 0$, and $e(\Phi) \in I_0$.*

PROOF. By Theorem 1.14 one may assume $X = \mathbb{R}^n + G$, where $n \geq 0$ and the group G contains an open compact subgroup K. The following two cases are possible.

1. $n > 0$. Then $X = \mathbb{R} + H$. Let $\Phi \in \mathscr{M}_+(X)$ and $\sigma(\Phi) \subset [a, b] + H$, where $0 < b < 2a$. By Corollary 6.19 $\mu = e(\Phi) \in I_0$. Clearly, the measure Φ can be chosen in such a way that $\Phi_{ac} > 0$, $\Phi_s > 0$, $\Phi_d > 0$.

2. $n = 0$. In this case the factor group X/K is discrete, since K is open and contains an element of infinite order. Let the conjugacy class $x_0 + K$ be an element of infinite order in the factor group X/K. For any $A \in \mathscr{B}(X)$ we have $A \cap (x_0 + K) = x_0 + E_A$, where $E_A \in \mathscr{B}(K)$. Let $P \in \mathscr{M}_+(K)$ be an arbitrary measure on K. Define a measure $\Phi \in \mathscr{M}_+(X)$ by setting $\Phi(A) = P(E_A)$. By the construction $\sigma(\Phi) \subset x_0 + K$. Therefore the measures Φ^{*n} and Φ^{*m} are pairwise singular for any natural n, m, $n \neq m$. By Theorem 6.13, $\mu = e(\Phi) \in I_0$. If $P_{ac} > 0$, $P_s > 0$, and $P_d > 0$, then measure Φ possesses the same property. $\square$

Let a measure $\Phi \in \mathscr{M}_+(X)$ be concentrated on an independent finite set, and let ζ be an element of order 2 in the group X. By Theorem 6.5 $\mu = e(\Phi + bE_\zeta) \in I_0$ for any $b \geq 0$. It turns out that the condition on the measure Φ in this statement cannot be weakened to requiring that the measure Φ have pairwise singular powers.

6.24. PROPOSITION. *Let $X = \mathbb{Z}(2) + \mathbb{Z}(3) + \mathbb{Z}$, and let ζ be an element of order two in the group X. Then there exist a measure $\Phi \in \mathscr{M}_+(X)$ and a sufficiently small number $b > 0$ such that the measures Φ^{*n} and Φ^{*m} are pairwise singular for all natural n, m, $n \neq m$, and $e(\Phi + bE_\zeta) \notin I_0$.*

PROOF. Let η be an element of order 3 in the group X and h a generator of the subgroup $\mathbb{Z}$. Denote

$$P = k(E_\eta + E_{-n}) * E_h, \qquad k > 0.$$

Consider the charge

$$\Lambda = \varepsilon k(E_\eta - E_h) * E_\eta, \qquad \varepsilon > 0,$$

and verify that for $\varepsilon > 0$ small enough the charges $e(P \pm \Lambda * E_\zeta + cE_\zeta)$, $c > 0$, are distributions. Obviously, we can confine ourself to proving this for the charge $e(P + \Lambda * E_\zeta + cE_\zeta)$. Let ν be an arbitrary charge on the group X and A a subset of X. Denote by $\nu|_A$ the restriction of ν to A. Set $P(X) = a$, $B_n = \mathbb{Z}(3) + nh$, $C_n = \zeta + \mathbb{Z}(3) + nh$, $n = 0, 1, 2, \ldots$. We have

$$\begin{aligned} e(P + \Lambda * E_\zeta)|_{B_n} &= k_n[(P + \Lambda)^{*n} + (P - \Lambda)^{*n}], \\ e(P + \Lambda * E_\zeta)|_{C_n} &= k_n[(P + \Lambda)^{*n} - (P - \Lambda)^{*n}] * E_\zeta, \end{aligned} \tag{1}$$

where $k_n = e^{-a/2n!}$.

Note that

$$e(cE_\zeta) = tE_0 + (1 - t)E_\zeta, \qquad \frac{1}{2} < t < 1, \tag{2}$$

and consider the restrictions

$$\mu_n = e(P + \Lambda * E_\zeta + cE_\zeta)|_{B_n}, \qquad \nu_n = e(P + \Lambda * E_\zeta + cE_\zeta)|_{C_n}.$$

Using (1) and (2) one can easily obtain

$$\begin{aligned} \mu_n &= k_n[(P + \Lambda)^{*n} + d(P - \Lambda)^{*n}], \\ \nu_n &= k_n[(P + \Lambda)^{*n} - d(P - \Lambda)^{*n}] * E_\zeta, \end{aligned}$$

where $d = 2t - 1$, $0 < d < 1$.

In order that the charge $e(P + \Lambda * E_\zeta + cE_\zeta)$ be a measure, it is necessary and sufficient that the charges μ_n and ν_n be measures for any natural n. We have

$$P + \Lambda = k(uE_\eta + vE_{-\eta}) * E_h, \qquad P - \Lambda = k(vE_\eta + uE_{-\eta}) * E_h,$$

where $u = 1 + \varepsilon$ and $v = 1 - \varepsilon$. Set

$$\lambda_n(u, v) = (uE_\eta + vE_{-\eta})^{*n} \in \mathcal{M}_+(\mathbb{Z}(3)).$$

Then

$$\begin{aligned} \mu_n &= k_n k^n (\lambda_n(u, v) + d\lambda_n(u, v)) * E_{nh}, \\ \nu_n &= k_n k^n (\lambda_n(u, v) - d\lambda_n(u, v)) * E_{nh} * E_\zeta. \end{aligned} \tag{3}$$

Obviously, $\mu_n \in \mathcal{M}_+(X)$. The proof of the fact that $\nu_n \in \mathcal{M}_+(X)$ will be carried out in the case $n = 3l$. The cases $n = 3l \pm 1$ are quite similar. Let $\eta_0 = \exp\{2\pi i/3\}$. We have

$$\lambda_n(u, v) = C_0^n u^n E_{n\eta} + C_1^n u^{n-1} v E_{(n-2)\eta} + \cdots + C_n^n v^n E_{-n\eta}.$$

For $n = 3l$ we obtain

$$\lambda_n(u, v)(\{0\}) = \frac{1}{3}[(u + v)^n + (u + \eta_0 v)^n + (u + \eta_0^2 v)^n], \qquad (4)$$

$$\lambda_n(u, v)(\{\eta\}) = \frac{1}{3}[(u + v)^n + \eta_0^2(u + \eta_0 v)^n + \eta_0(u + \eta_0^2 v)^n], \qquad (5)$$

$$\lambda_n(u, v)(\{-\eta\}) = \frac{1}{3}[(u + v)^n + \eta_0(u + \eta_0)v)^n + \eta_0^2(u + \eta_0^2 v)^n]. \qquad (6)$$

Relations (4)–(6) imply

$$\lambda_n(u, v)(\{0\}) = \lambda_n(v, u)(\{0\}), \qquad \lambda_n(u, v)(\{\eta\}) = \lambda_n(v, u)(\{-\eta\}).$$

Taking this into account we get

$$(\lambda_n(u, v) - d\lambda_n(v, u))(\{0\}) = (1 - d)\lambda_n(u, v)(\{0\}) > 0,$$

hence $\nu_n(\{\zeta + nh\}) > 0$, $n = 0, 1, 2, \ldots$. Further we have

$$(\lambda_n(u, v) - d\lambda_n(v, u))(\{\eta\}) = \lambda_n(u, v)(\{\eta\}) - d\lambda_n(u, v)(\{-\eta\})$$

$$= \frac{1}{3}(u + v)^n \left(1 - d + (\eta_0^2 - d\eta_0) \left(\frac{u + \eta_0 v}{u + v}\right)^n \right.$$

$$\left. + (\eta_0 - d\eta_0^2) \left(\frac{u + \eta_0^2 v}{u + v}\right)^n \right). \qquad (7)$$

Since $1 - d > 0$, for any $\varepsilon_0 > 0$ there exists number n_0 such that the right-hand part of equality (7) is positive for all $0 < \varepsilon < \varepsilon_0$ and $n > n_0$. Since the left-hand part of equality (7) is positive for any fixed n and $\varepsilon = 0$, it remains positive for $\varepsilon > 0$ small enough and for all $n = 0, 1, \ldots, n_0$. Returning to (3) we obtain that $\nu_n(\{\zeta + \eta + nh\}) > 0$ for all $n = 0, 1, 2, \ldots$. Similarly, $\nu_n(\{\zeta - \eta + nh\}) > 0$, $n = 0, 1, 2, \ldots$. So $\nu_n \in \mathcal{M}_+(X)$. Let $\Phi = 2P$. The measures Φ^{*n} and Φ^{*m} are pairwise singular for all natural n, m, $n \neq m$. If $\varepsilon > 0$ is small enough, we have the decomposition of μ as a convolution of distributions

$$\mu = e(\Phi + bE_\zeta) = \alpha * \beta = e\left(P + \Lambda * E_\zeta + \frac{b}{2}E_\zeta\right) * e\left(P - \Lambda * E_\zeta + \frac{b}{2}E_\zeta\right).$$

Since $\|P \pm \Lambda * E_\zeta + \frac{b}{2}E_\zeta\| \to 0$ as $k \to 0$ and $b \to 0$, by Lemma 4.14 the distributions α and β are not infinitely divisible when k and b are sufficiently small. Applying Theorem 4.3 we get $\mu \notin I_0$. $\square$

§7. Group analogs of Linnik's theorems

In this section we prove a group analog of the Linnik theorem which states that the convolution of Gaussian and Poisson distributions belongs to the class I_0 and of the Linnik theorem on the structure of the Lévy measure of distributions from the class I_0 with nontrivial Gaussian divisors.

7.1. Let us recall the formulation of the Linnik theorem on decomposition of the convolution of Gaussian and Poisson distributions [LinO, Chapter VI, §3].

LINNIK'S THEOREM. *Let* $\gamma \in \Gamma(\mathbb{R})$ *and* $\pi = e(\Phi)$, *where* $\Phi = aE_t$, $a > 0$, *and* $t \in \mathbb{R}$. *Then* $\mu = \gamma * \gamma\pi \in I_0$.

In the general case an abelian locally compact group X there may exist both Gaussian and Poisson distributions that do not belong to the class I_0. Therefore, for their convolution to belong to the class I_0, it is necessary that each of them belong to the class I_0.

In what follows we describe a class of groups, for which this necessary condition is also sufficient. Let $x_0 \in X$ and $a > 0$. Denote by π the Poisson distribution $\pi = e(aE_{x_0})$. By Proposition 6.6, $\pi \in I_0$ if and only if the order of x_0 is either infinity or two.

Let $\gamma \in \Gamma(X)$. When studying the decomposition of the convolution $\mu = \gamma * \pi$, one may assume, without loss of generality, that $\gamma \in \Gamma^s(X)$. Then, by Proposition 5.5, $\sigma(\gamma)$ is a connected subgroup of the group X.

7.2. THEOREM. *Let* $\gamma \in \Gamma^s(X)$, *let* $\sigma(\gamma)$ *be a connected finite-dimensional subgroup of the group* X, *and let* π *be a Poisson distribution on* X. *If both* γ *and* π *belong to* I_0, *then their convolution* $\mu_0 = \gamma * \pi$ *belongs to the class* I_0.

PROOF. Set $\sigma(\gamma) = G$. Since $\gamma \in I_0$, then by Proposition 5.28, there exist a continuous monomorphism $p: \mathbb{R}^n \to G$ and a distribution $\nu \in \Gamma^s(\mathbb{R}^n)$ such that $\gamma = p(\nu)$. Let $\pi = e(\Phi)$, where $\Phi = aE_{x_0}$, $a > 0$, and the order of x_0 is either infinity or two.

Consider the group $\mathbb{Z}(m)$. Set $\zeta = \exp\{2\pi i/m\}$. Then $\mathbb{Z}(m) = \{\zeta^k : k = 0, 1, \ldots, m-1\}$.

Then following cases are possible.

1. $x_0 \in p(\mathbb{R}^n)$. Let $x_0 = p(t_0)$, $t_0 \in \mathbb{R}^n$. Then $\pi = p(\lambda)$, where $\lambda = e(aE_{t_0})$ is a Poisson distribution on $\mathbb{R}^n$ and $\mu_0 = p(\nu * \lambda)$.

2. $x_0 \notin p(\mathbb{R}^n)$.

a) The element x_0 is of order two. In this case, setting $p_1(t, \zeta^k) = p(t) + kx_0$, for $t \in \mathbb{R}^n$ and $k = 0, 1$, we extend the monomorphism p to a continuous monomorphism

$$p_1 : \mathbb{R}^n + \mathbb{Z}(2) \to X.$$

Then $\pi = p_1(\lambda)$, where $\lambda = e(aE_{(0,\zeta)})$ is a Poisson distribution on $\mathbb{R}^n + \mathbb{Z}(2)$ and $\mu_0 = p_1(\nu * \lambda)$.

b) The element x_0 is of infinite order and $qx_0 \notin p(\mathbb{R}^n)$ for any natural q. In this case, setting $p_1(t, k) = p(t) + kx_0$, where $t \in \mathbb{R}^n$ and $k \in \mathbb{Z}$, we extend the monomorphism p to a continuous monomorphism $p_1 : \mathbb{R}^n \to \mathbb{Z} \to X$. Then $\pi = p_1(\lambda)$, where $\lambda = e(aE_{(0,\zeta)})$, is a Poisson distribution on $\mathbb{R}^n + \mathbb{Z}$, and $\mu_0 = p_1(\nu * \lambda)$.

c) The element x_0 is of infinite order and $qx_0 \in p(\mathbb{R}^n)$ for some natural q. Let q be such that $x_0, 2x_0, \ldots, (q-1)x_0 \notin p(\mathbb{R}^n)$, and $qx_0 = p(t_0) \in p(\mathbb{R}^n)$, $t_0 \in \mathbb{R}^n$. Set $x_1 = x_0 - p(t_0/q)$. Then x_1 is an element of order q. Setting $p_1(t, \zeta^k) = p(t) + kx_1$, where $t \in R^n$ and $k = 0, 1, \ldots, q-1$, we extend the monomorphism p to a continuous monomorphism

$$p_1 : \mathbb{R}^n + \mathbb{Z}(q) \to X.$$

Then $\pi = p_1(\lambda)$ where $\lambda = e(aE_{t_0/q, \zeta})$ is a Poisson distribution on $\mathbb{R}^n + \mathbb{Z}(q)$ and $\mu_0 = p_1(\nu * \lambda)$.

Thus in all cases the distribution μ_0 is the continuous monomorphic image of the convolution of Gaussian and Poisson distributions that belong to the class I_0 on one of the following groups: $\mathbb{R}^n$, $\mathbb{R}^n + \mathbb{Z}$, or $\mathbb{R}^n + \mathbb{Z}(q)$. The group $\mathbb{R}^n + \mathbb{Z}$ is a subgroup of the $\mathbb{R}^{n+1}$. Using Proposition 4.18 one may therefore obtain Theorem 7.2 in the cases 1 and 2b) as a direct consequence of a multivariate analog of the Linnik theorem proved by Ostrovskii and Cuppens [LinO, Chapter VI, §3]. It also follows from Proposition 4.18 that in the cases 2a), c) Theorem 7.2 is a direct consequence of the corresponding result for the group $X = \mathbb{R}^n + \mathbb{Z}(q)$. So we proceed to this case. Since the proof is rather awkward, we split it into several parts.

1. Assume that $\pi = e(aE_{x_0})$, $x_0 = (\delta, \zeta)$, with $q \geq 2$ if $\delta \neq 0$, and $q = 2$ if $\delta = 0$. For simplicity the elements of the group $\mathbb{Z}(q)$ are denoted by k instead of ζ^k, $k = 0, 1, \ldots, q-1$.

At first consider the case $n = 1$. Denote the elements of the group $\mathbb{R} + \mathbb{Z}(q)$ by (t, k), where $t \in \mathbb{R}$, $k \in \mathbb{Z}(q)$, and the elements of the group $(\mathbb{R} + \mathbb{Z}(q))^* \approx \mathbb{R} + \mathbb{Z}(q)$ by (s, l), where $s \in \mathbb{R}$, $l \in \mathbb{Z}(q)$.

Let $\mu_0 = \mu_1 * \mu_2$, where $\mu_j \in \mathcal{M}^1(\mathbb{R} + \mathbb{Z}(q))$. Set $\alpha_j(s, l) = \hat{\mu}_j(s, l)$, $j = 0, 1, 2$. Then

$$\ln \alpha_0(s, l) = -\sigma s^2 + a \left(\exp\left\{ i\left(\delta s + \frac{2\pi l}{q} \right) \right\} - 1 \right), \qquad \sigma \geq 0, \ a \geq 0. \quad (1)$$

It follows from §2.10(d) that

$$\alpha_0(s, l) = \alpha_1(s, l)\alpha_2(s, l). \quad (2)$$

To complete the proof it is sufficient to verify that

$$\ln \alpha_1(s, l) = -\sigma_1 s^2 + a_1 \left(\exp\left\{ i\left(\delta s + \frac{2\pi l}{q} \right) \right\} - 1 \right) + i\beta s + \ln(k, l), \quad (3)$$

where $0 \leq \sigma_1 \leq \sigma$, $0 \leq a_1 \leq a$, $\beta \in \mathbb{R}$, and (k, l) is a character of the group $\mathbb{Z}(q)$, and that there is a similar representation for $\ln \alpha_2(s, l)$.

Setting $l = 0$ in (2) and using the Lévy theorem 3.13, we conclude that the functions $\alpha_j(s, 0)$, $j = 1, 2$, are entire. Then by Lemma 6.9 the functions

$\alpha_j(s, l)$, $j = 1, 2$, are entire with respect to s for any fixed $l \in \mathbb{Z}(q)$. In addition, the equality

$$\alpha_j(s, l) = \int_{R+Z(q)} \exp\left\{i\left(ts + \frac{2\pi kl}{q}\right)\right\} d\mu_j(t, k), \qquad j = 0, 1, 2, \quad (4)$$

and equality (2) are fulfilled for any $s \in \mathbb{C}$ and any $l \in \mathbb{Z}(q)$.

It follows from (4) that the function $\alpha_j(-iy + x, l)/\alpha_j(-iy, 0)$ is a characteristic function of the variable $(x, l) \in \mathbb{R} + \mathbb{Z}(q)$ for any fixed $y \in \mathbb{R}$. Hence we have

$$\left|\frac{\alpha_j(-iy + x, l)}{\alpha_j(-iy, 0)}\right| \le 1, \qquad j = 0, 1, 2,$$

for any $x, y \in \mathbb{R}$, $l \in \mathbb{Z}(q)$. With the account of (2) this yields

$$\left|\frac{\alpha_0(-iy + x, l)}{\alpha_0(-iy, 0)}\right| \le \left|\frac{\alpha_j(-iy + x, l)}{\alpha_j(-iy, 0)}\right| \le 1, \qquad j = 1, 2. \quad (5)$$

It follows from equality (2) that for any $l \in \mathbb{Z}(q)$ the entire function $\alpha_1(s, l)$ does not vanish everywhere in the complex s-plane, and, thus, has the representation $\alpha_1(s, l) = \exp\{f(s, l)\}$, where the branch $f(s, l)$ of the function $\ln \alpha_1(s, l)$ is chosen in such a way that $f(s, l) = \overline{f(-s, -l)}$.

2. Let us prove that the function $f(s, l)$ coincides with the right-hand side of equality (3). Using (1) we may reduce inequalities (5) to the form

$$0 \le \mathrm{Re}[f(-iy, 0) - f(-iy + x, l)] \le \sigma x^2 + 2ae^{\delta y} \sin^2\left(\frac{\delta x + 2\pi l/q}{2}\right), \quad (6)$$

$x, y \in \mathbb{R}$, $l \in \mathbb{Z}(q)$. Since

$$f(-iy + x, l) = \overline{f(-iy - x, -l)}, \qquad x, y \in \mathbb{R}, \ l \in \mathbb{Z}(q), \quad (7)$$

one may rewrite inequality (6) as

$$0 \le f(-iy, 0) - \frac{1}{2}[f(-iy + x, l) + f(-iy - x, -l)]$$
$$\le \sigma x^2 + 2ae^{\delta y} \sin^2\left(\frac{\delta x + 2\pi l/q}{2}\right), \qquad x, y \in \mathbb{R}, \ l \in \mathbb{Z}(q). \quad (8)$$

Below we distinguish between the two cases: $\delta \ne 0$ and $\delta = 0$.

3. Consider the case $\delta \ne 0$ and $q \ge 2$.

Without loss of generality one may assume $\delta = 1$. Setting $x = -2\pi l/q$ in (8) we obtain

$$f(-iy, 0) - \frac{1}{2}\left[f\left(-iy - \frac{2\pi l}{q}, l\right) + f\left(-iy + \frac{2\pi l}{q}, -l\right)\right] = O(1),$$
$$|y| \to \infty, \quad (9)$$

for every $l \in \mathbb{Z}(q)$.

Let us note that by the Linnik theorem the assertion we are in the process of proving is true for the group $\mathbb{R}$. Relations (1) and (2) with $l = 0$ imply

$$f(s, 0) = -\sigma_1 s^2 + a_1(e^{is} - 1) + i\beta s, \qquad s \in \mathbb{R}, \tag{10}$$

for $0 \le \sigma_1 \le \sigma$, $0 \le a_1 \le a$, and $\beta \in \mathbb{R}$. Without loss of generality one may assume $\beta = 0$. From (6) and (10) we obtain

$$|\operatorname{Re} f(-iy + x, l)| = O(\exp(\operatorname{Re} y) + |y|^2), \qquad |y| \to \infty, \tag{11}$$

for any fixed $x \in \mathbb{R}$ and $l \in \mathbb{Z}(q)$. Let us apply the Schwarz formula 3.6 to the function $f(-iy - i\xi + x, l)$ in the disk $|\xi| < 1$. We obtain

$$f(-iy - i\xi + x, l) = \frac{1}{2\pi} \int_0^{2\pi} \operatorname{Re} f(-iy - ie^{i\varphi} + x, l) \frac{e^{i\varphi}\xi}{e^{i\varphi} - \xi} \, d\varphi + iC.$$

Differentiating this equality with respect to ξ and setting $\xi = 0$, we get

$$-if'(-iy + x, l) = \frac{1}{\pi} \int_0^{2\pi} \operatorname{Re} f(-iy - ie^{i\varphi} + x, l) e^{i\varphi} \, d\varphi.$$

Accounting for (11) we obtain

$$|f'(-iy + x, l)| = O(\exp(\operatorname{Re} y) + |y|^2), \qquad |y| \to \infty. \tag{12}$$

This yields

$$|f(-iy + x, l)| = O(|y| \exp(\operatorname{Re} y) + |y|^3), \qquad |y| \to \infty. \tag{13}$$

From (13) for $\operatorname{Re} y = 0$ we obtain the estimate

$$f(-iy, 0) - \frac{1}{2}\left[f\left(-iy - \frac{2\pi l}{q}, l\right) + f\left(-iy + \frac{2\pi l}{q}, -l\right) \right] = O(|y|^3),$$
$$|y| \to \infty. \tag{14}$$

Applying the Phragmén-Lindelöf theorem to the left-hand side of (14) in each of the quadrants of the complex y-plane, we extend this estimate to the whole y-plane. Therefore the left-hand side of (14) is a polynomial of order not exceeding 3, which is possible, by (9), only if this polynomial is a constant. As a result we have

$$f(-iy, 0) - \frac{1}{2}\left[f\left(-iy - \frac{2\pi l}{q}, l\right) + f\left(-iy + \frac{2\pi l}{q}, -l\right) \right] = C_l, \tag{15}$$

where C_l are some real constants. From here, using (10) we obtain

$$\frac{1}{2}\left[f\left(-iy - \frac{2\pi l}{q}, l\right) + f\left(-iy + \frac{2\pi l}{q}, -l\right) \right] = -C_l + \sigma_1 y^2 + a_1(e^y - 1). \tag{16}$$

Note that if $x = -2\pi l/q$, $y \in \mathbb{R}$ and $l \in \mathbb{Z}(q)$, then equality (7) implies

$$f\left(-iy - \frac{2\pi l}{q}, l\right) = \overline{f\left(-iy + \frac{2\pi l}{q}, -l\right)}.$$

Set

$$h\left(-iy - \frac{2\pi l}{q}, l\right) = -C_l + \sigma_1 y^2 + a_1(e^y - 1).　\quad (17)$$

Then the entire function g, determined by the equality

$$ig\left(-iy - \frac{2\pi l}{q}, l\right) = f\left(-iy - \frac{2\pi l}{q}, l\right) - h\left(-iy - \frac{2\pi l}{q}, l\right),　\quad (18)$$

is real for $y \in \mathbb{R}$ because

$$h\left(-iy - \frac{2\pi l}{q}, l\right) = \operatorname{Re} f\left(-iy - \frac{2\pi l}{q}, l\right)$$

for $y \in \mathbb{R}$. Replacing x by $x - 2\pi l/q$ in inequality (8) and using (10), (17), and (18) we obtain

$$0 \le \sigma_1 x^2 + 2a_1 e^y \sin^2 \frac{x}{2} + C_l + \operatorname{Im} g\left(-iy - \frac{2\pi l}{q} + x, l\right)$$

$$\le \sigma\left(x - \frac{2\pi l}{q}\right)^2 + 2ae^y \sin^2 \frac{x}{2}, \quad x, y \in \mathbb{R}, \ l \in \mathbb{Z}(q).　\quad (19)$$

Since the function g is real for real y, we have

$$\operatorname{Im} g\left(-iy - \frac{2\pi l}{q} + x, l\right) + \frac{1}{2i}\left[g\left(-iy - \frac{2\pi l}{q} + x, l\right)\right.$$

$$\left. - g\left(-iy - \frac{2\pi l}{q} - x, l\right)\right].$$

Substituting this expression into (19), we obtain

$$0 \le \sigma_1 x^2 + 2a_1 e^y \sin^2 \frac{x}{2}$$

$$+ C_l + \frac{1}{2i}\left[g\left(-iy - \frac{2\pi l}{q} + x, l\right) - g\left(-iy - \frac{2\pi l}{q} - x, l\right)\right]$$

$$\le \sigma\left(x - \frac{2\pi l}{q}\right)^2 + 2ae^y \sin^2 \frac{x}{2},　\quad (20)$$

$$x, y \in \mathbb{R}, \qquad l \in \mathbb{Z}(q).$$

It follows from (18) and (13) that

$$\left|g\left(-iy - \frac{2\pi l}{q} + x, l\right)\right| = O(|y| \exp(\operatorname{Re} y) + |y|^3), \qquad |y| \to \infty,　\quad (21)$$

for all fixed $x \in \mathbb{R}$, $l \in \mathbb{Z}(q)$. Setting $x = 2\pi$ in inequality (20), we find

$$0 \le 4\pi^2 \sigma_1 + C_l + \frac{1}{2i}\left[g\left(-iy - \frac{2\pi l}{q} + 2\pi, l\right) - g\left(-iy - \frac{2\pi l}{q} - 2\pi, l\right)\right]$$

$$\le \sigma\left(2\pi - \frac{2\pi l}{q}\right)^2, \qquad y \in \mathbb{R}, \ l \in \mathbb{Z}(q).　\quad (22)$$

Taking into account inequalities (21) and (22), one may apply the Phragmén-Lindelöf principle 3.5 to the function $\frac{1}{2i}[g(-iy - \frac{2\pi l}{q} + 2\pi, l) - g(-iy - \frac{2\pi l}{q} - 2\pi, l)]$ in each of the quadrants of the complex y-plane. Then

$$\frac{1}{2i}\left[g\left(-iy - \frac{2\pi l}{q} + 2\pi, l\right) - g\left(-iy - \frac{2\pi l}{q} - 2\pi, l\right)\right] = A_l, \qquad (23)$$

where A_l are some real constants.

4. Let us solve each of the difference equations (23). Set

$$h_l(y) = g\left(-iy - \frac{2\pi l}{q}, l\right).$$

It follows from (20) and (23) that the function $h_l'(y)$ is a 4π-periodic entire function of exponential type $\sigma \le 1$. Applying Theorem 3.3 we obtain

$$h_l'(y) = \sum_{k=-2}^{2} b_k^{(l)} e^{iky/2}.$$

Hence

$$h_l(y) = \sum_{k=-2}^{2} c_k^{(l)} e^{iky/2} + b_0^{(l)} y.$$

Using (23), we finally obtain

$$g\left(-iy - \frac{2\pi l}{q}, l\right) = \sum_{k=-2}^{2} a_k^{(l)} e^{ky/2} + \frac{A_l y}{2\pi}, \qquad (24)$$

where $a_k^{(l)}$ are some real constants. Let us verify that $a_k^{(l)} = 0$ when $k \ne 0$.

Accounting of (24) we reduce inequality (20) to the form

$$0 \le \sigma_1 x^2 + 2a_1 e^y \sin^2 \frac{x}{2} + C_l + \sum_{k=-2}^{2} a_k^{(l)} e^{ky/2} \sin \frac{kx}{2} + \frac{A_l x}{2\pi}$$

$$\le \sigma\left(x - \frac{2\pi l}{q}\right)^2 + 2ae^y \sin^2 \frac{x}{2}, \qquad x, y \in \mathbb{R}, \ l \in \mathbb{Z}(q). \qquad (25)$$

Dividing the latter inequality by e^{-y} and then passing to the limit as $y \to -\infty$, we get $a_{-2}^{(l)} = 0$. After this we similarly obtain $a_{-1}^{(l)} = 0$. Substitute these values of $a_{-1}^{(l)}$ and $a_{-2}^{(l)}$ into (25) and divide the resulting inequality by e^y. Letting y to $+\infty$, we have

$$0 \le 2a_1 \sin^2 \frac{x}{2} + a_2^{(l)} \sin x \le 2a \sin^2 \frac{x}{s}.$$

Hence $a_2^{(l)} = 0$.

Return now to inequality (25) and pass to the limit as $y \to -\infty$. We have

$$0 \le \sigma_1 x^2 + C_l + \frac{A_1 x}{2\pi} \le \sigma\left(x - \frac{2\pi l}{q}\right)^2.$$

Hence

$$\sigma_1 x^2 + C_l + \frac{A_l x}{2\pi} = \sigma \left(x - \frac{2\pi l}{q} \right)^2. \tag{26}$$

From here, taking into consideration that $a_k^{(l)} = 0$, $k = -2, -1, 2$, we reduce inequality (25) to the form

$$0 \leq \sigma_1 \left(x - \frac{2\pi l}{q} \right)^2 + 2a_1 e^y \sin^2 \frac{x}{2} + a_1^{(l)} e^{y/2} \sin \frac{x}{2}$$
$$\leq \sigma \left(x - \frac{2\pi l}{q} \right)^2 + 2ae^y \sin^2 \frac{x}{2}. \tag{27}$$

Setting $x = -2\pi l/q$ in (27) we obtain

$$0 \leq 2a_1 e^y \sin^2 \frac{\pi l}{q} + a_1^{(l)} e^{y/2} \sin \frac{\pi l}{q} \leq 2ae^y \sin^2 \frac{\pi l}{q}. \tag{28}$$

Dividing inequality (28) by $e^{y/2}$ and letting y to $-\infty$, we obtain

$$0 \leq a_1 \sin \frac{\pi l}{q} \leq 0.$$

Since $\sin(\pi l/q) \neq 0$ for $0 < l \leq q - 1$, $a_1^{(l)} = 0$. From (24) we find

$$g \left(-iy - \frac{2\pi l}{q}, l \right) = q_0^{(l)} + A_l \frac{y}{2\pi}. \tag{29}$$

It follows from (17), (18), and (29) that

$$f \left(-iy - \frac{2\pi l}{q}, l \right) = h \left(-iy - \frac{2\pi l}{q} l \right) + ig \left(-iy - \frac{2\pi l}{q}, l \right)$$
$$= -C_l + \sigma_1 y^2 + a_1 (e^y - 1) + ia_0^{(l)} + iA_l \frac{y}{2\pi}.$$

From here and from (26) we have

$$f(s, l) = -\sigma_1 s^2 + a_1 \left(\exp \left\{ i \left(s + \frac{2\pi l}{q} \right) \right\} - 1 \right) + ia_0^{(l)}. \tag{30}$$

5. To complete our proof we shall show that the function $\exp\{ia_0^{(l)}\}$ is a character of the group $\mathbb{Z}(q)$. In fact $\exp\{f(-iy + x, l) - f(-iy, 0)\}$ is a characteristic function of the variable $(x, l) \in \mathbb{R} + \mathbb{Z}(q)$ for any fixed $y \in \mathbb{R}$.

Assume that $\sigma_1 \neq 0$ in (30) and set $y_k = -k\pi/\sigma_1$. Define a function β_k on the group $\mathbb{Z}$ by the equality

$$\beta_k(l) = \exp \left\{ f \left(-iy_k - \frac{l}{k}, l \ (\text{mod } q) \right) - f(-iy_k, 0) \right\}$$
$$= \exp \left\{ -\sigma_1 \left(\frac{1}{k} \right)^2 + a_1 \exp \left\{ -\frac{k\pi}{\sigma_1} + \frac{i2\pi l}{q} \right\} \right.$$
$$\left. \times \left(\exp \left\{ \frac{il}{k} \right\} - 1 \right) + ia_0^{(l \ (\text{mod } q))} \right\}, \qquad l \in \mathbb{Z}.$$

By 2.10(h) the function $\beta_k(l)$ is characteristic for any $k \in \mathbb{Z}$. Hence the function

$$\exp\{ia_0^{(l \pmod q)}\} = \lim_{k \to +\infty} \beta_k(l)$$

is characteristic as well. Therefore the function $\exp\{ia_0^{(l)}\}$, $l \in \mathbb{Z}(q)$, is characteristic too. It follows from inequality (5) that

$$|\beta_k(l)| \geq \left| \exp\left\{ -\sigma\left(\frac{1}{k}\right)^2 - \frac{2i\sigma l}{k\sigma_1} \right.\right.$$
$$\left.\left. + a\exp\left\{\frac{-k\pi}{\sigma_1} + 2i\frac{\pi l}{q}\right\}\left(\exp\left\{\frac{il}{k}\right\} - 1\right)\right\}\right|.$$

Passing to the limit as $k \to +\infty$, we obtain $|\exp\{ia_0^{(l)}\}| \geq 1$. So $|\exp\{ia_0^{(l)}\}| \equiv 1$, i.e., $\exp\{ia_0^{(l)}\}$ is a character of the group $\mathbb{Z}(q)$.

If $\sigma_1 = 0$ in (30), then set $y_k = -k\pi$ and use the same reasoning as in the case $\sigma_1 \neq 0$.

6. Now let us consider the case $\delta = 0$, $q = 2$.

Reduce inequality (8) to the form

$$0 \leq f(-iy, 0) - \frac{1}{2}[f(-iy + x, l) + f(-iy - x, -l)]$$
$$\leq \sigma x^2 + 2a\sin^2\frac{\pi l}{2}, \qquad x, y \in \mathbb{R}, \ l \in \mathbb{Z}(2). \tag{31}$$

The function $\exp\{f(s, 0)\}$ is a divisor of the characteristic function of a Gaussian distribution. Therefore, by the Cramér theorem

$$f(s, 0) = -\sigma_1 s^2 + i\beta s, \qquad s \in \mathbb{R}, \tag{32}$$

for $0 \leq \sigma_1 \leq \sigma$, $\beta \in \mathbb{R}$. Without loss of generality one may assume $\beta = 0$. From (31) with $l = 1$ we find

$$0 \leq \operatorname{Re}[f(-iy, 0) - f(-iy + x, 1)] \leq \sigma x^2 + 2a, \qquad x, y \in \mathbb{R}. \tag{33}$$

It follows from (33) and (32) that

$$\operatorname{Re} f(-iy, 1) = O(|y|^2)$$

and applying the Carathéodory inequality 3.4, we obtain

$$|f(-iy, 1)| = O(|y|^2).$$

Therefore $f(-iy, 1)$ is a polynomial of degree not exceeding 2. Since $l = -l$ for $l \in \mathbb{Z}(2)$, the function $f(-iy, 1)$ is real for $y \in \mathbb{R}$. Therefore,

$$f(-iy, 1) = A_1 y^2 + B_1 y + C_1, \tag{34}$$

where $A_1, B_1, C_1 \in \mathbb{R}$. By substituting (32) and (34) into (33) we obtain

$$0 \leq \sigma_1 y^2 - A_1 y^2 + A_1 x^2 - B_1 y - C_1 \leq \sigma x^2 + 2a, \qquad x, y \in \mathbb{R}.$$

Setting $x = 0$ here, we obtain $A_1 = \sigma_1$, $B_1 = 0$, and setting $x = y = 0$, we obtain $0 \le -C_1 \le 2a$. Let $a_1 = -C_1/2$. Then $0 \le a_1 \le a$ and

$$f(s, l) = -\sigma_1 s^2 + a_1 \left(\exp \left\{ \frac{2i\pi l}{2} \right\} - 1 \right), \qquad s \in \mathbb{R}, \ l \in \mathbb{Z}(2).$$

The proof for $n = 1$ is complete. Let us consider now the case of an arbitrary n.

7. Denote elements of the group $\mathbb{R}^n + \mathbb{Z}(q)$ by (t, k), where $t = (t_1, \ldots, t_n)$ $\in \mathbb{R}^n$ and $k \in \mathbb{Z}(q)$, and elements of the group $(\mathbb{R}^n + \mathbb{Z}(q))^* \approx \mathbb{R}^n + \mathbb{Z}(q)$ by (s, l), where $s = (s_1, \ldots, s_n) \in \mathbb{R}^n$ and $l \in \mathbb{Z}(q)$. Let $\mu_0 = \mu_1 * \mu_2$, where $\mu_j \in \mathcal{M}^1(\mathbb{R}^n + \mathbb{Z}(q))$. Set $\alpha_j(s, l) = \hat{\mu}_j(s, l)$, $j = 0, 1, 2$. Then

$$\ln \alpha_0(s, l) = -\langle As, s \rangle + a \left(\exp \left\{ i \left(\langle \delta, s \rangle + \frac{2\pi l}{q} \right) \right\} - 1 \right), \qquad (35)$$

where A is a symmetric positive semidefinite matrix and $a \ge 0$, $\delta \in \mathbb{R}^n$. It follows from 2.10(d) that

$$\alpha_0(s, l) = \alpha_1(s, l)\alpha_2(s, l). \qquad (36)$$

To prove the theorem it suffices to verify that

$$\ln \alpha_1(s, l) = -\langle A_1 s, s \rangle$$
$$+ a_1 \left(\exp \left\{ i \left(\langle \delta, s \rangle + \frac{2\pi l}{q} \right) \right\} - 1 \right) + i\langle \beta, s \rangle + \ln(k, l),$$
$$(37)$$

where A_1 is a symmetric positive semidefinite matrix such that $\langle A_1 s, s \rangle \le \langle As, s \rangle$, $s \in \mathbb{R}^n$, $\beta \in \mathbb{R}^n$, and (k, l) is a character of the group $\mathbb{Z}(q)$, and that $\ln \alpha_2(s, l)$ has a similar representation.

Setting $l = 0$ in (36) and applying Theorem 3.15 we conclude that $\alpha_j(s, 0)$, $j = 1, 2$, entire functions. Then by Lemma 6.9 the functions $\alpha_j(s, l)$, $j = 1, 2$, are entire with respect to s for any fixed $l \in \mathbb{Z}(q)$. Besides, the equalities

$$\alpha_j(s, l) = \int_{R^n + Z(q)} \exp \left\{ i \left(\langle t, s \rangle + \frac{2\pi k l}{q} \right) \right\} d\mu_j(t, k), \qquad j = 0, 1, 2,$$
$$(38)$$

and, consequently also (36), hold for all $s \in \mathbb{C}^n$ and $l \in \mathbb{Z}(q)$.

It follows from (38) that the function $\alpha_j(-iy + x, l)/\alpha_j(-iy, 0)$ is a characteristic function of the variable $(x, l) \in \mathbb{R}^n + \mathbb{Z}(q)$ for any fixed $y \in \mathbb{R}^n$. Hence from (36) we obtain

$$\left| \frac{\alpha_0(-iy + x, l)}{\alpha_0(-iy, 0)} \right| \le \left| \frac{\alpha_j(-iy + x, l)}{\alpha_j(-iy, 0)} \right| \le 1, \qquad j = 1, 2,$$

for any $x, y \in \mathbb{R}^n$, $l \in \mathbb{Z}(q)$. Equality (36) implies that the entire functions $\alpha_j(s, l)$ do not vanish in $\mathbb{C}^n$. Therefore one may reduce the last inequality to the form

$$0 \le \mathrm{Re}[f_1(-iy, 0) - f_1(-iy+x, l)] \le \mathrm{Re}[f_0(-iy, 0) - f_0(-iy+x, l)], \quad (39)$$

where the branches $f_j(s, l)$ of the functions $\ln \alpha_j(s, l)$ are chosen in such a way that $f(s, l) = \overline{f(-s, -l)}$.

The proof is carried out by induction with respect to n. Let $n \ge 2$, and let a point $x = (x_1, \ldots, x_n) \in \mathbb{R}^n$ be fixed. Set $x = x^{(1)} + x^{(2)}$, where $x^{(1)} = (x_1, 0, \ldots, 0)$ and $x^{(2)} = (0, x_2, \ldots, x_n)$. Identify $x^{(1)}$ and $x^{(2)}$ with the corresponding elements of $\mathbb{R}$ and $\mathbb{R}^{n-1}$. Without loss of generality we assume that if $\delta \ne 0$, then $\delta = (1, \ldots, 1)$. The function $\alpha_1(-iy^{(1)} + x^{(2)}, l)/\alpha_1(-iy^{(1)}, 0)$ is a characteristic function of the variable $(x^{(2)}, l) \in \mathbb{R}^{n-1} + \mathbb{Z}(q)$. By (36), the induction hypothesis yields

$$
\begin{aligned}
f_1(-iy^{(1)} &+ x^{(2)}, l) - f_1(-iy^{(1)}, 0) \\
&= -\langle B_1(y^{(1)})x^{(2)}, x^{(2)} \rangle \\
&\quad + \beta_1(y^{(1)}) \left(\exp \left\{ i \left(\langle \delta, x^{(2)} \rangle + \frac{2\pi l}{q} \right) \right\} - 1 \right) \\
&\quad + i \langle C_1(y^{(1)}), x^{(2)} \rangle + \ln(k_1(y^{(1)}), l),
\end{aligned}
\quad (40)
$$

where $B_1(y^{(1)})$ is a symmetric positive semidefinite matrix in $\mathbb{R}^{n-1}$, $\beta_1(y^{(1)}) \ge 0$, $C_1(y^{(1)}) \in \mathbb{R}^{n-1}$, and $(k_1(y^{(1)}), l)$ is a character of the group $\mathbb{Z}(q)$. Similarly we obtain

$$
\begin{aligned}
f_1(-iy^{(2)} &+ x^{(1)}, l) - f_1(-iy^{(2)}, 0) \\
&= -B_2(y^{(2)})(x^{(1)})^2 + \beta_2(y^{(2)}) \left(\exp \left\{ i \left(\langle \delta, x^{(1)} \rangle + \frac{2\pi l}{q} \right) \right\} - 1 \right) \\
&\quad + iC_2(y^{(2)}), x^{(1)} + \ln(k_2(y^{(2)}), l),
\end{aligned}
\quad (41)
$$

where $B_2(y^{(2)}) \ge 0$, $\beta_2(y^{(2)}) \ge 0$, $C_2(y^{(2)}) \in \mathbb{R}$, and $(k_2(y^{(2)}), l)$ is a character of the group $\mathbb{Z}(q)$.

Note that equalities (40) and (41) are also fulfilled for complex values of the variable x, because their left-hand sides are analytic with respect to x. Setting $x^{(2)} = -iy^{(2)}$ in (40) we obtain

$$
\begin{aligned}
f_1(-iy, l) &- f_1(-iy^{(1)}, 0) \\
&= -\langle B_1(y^{(1)})y^{(2)}, y^{(2)} \rangle + \beta_1(y^{(1)}) \left(\exp \left\{ \langle \delta, y^{(2)} \rangle + \frac{2\pi i l}{q} \right\} - 1 \right) \\
&\quad + \langle C_1(y^{(1)}), y^{(2)} \rangle + \ln(k_1(y^{(1)}), l).
\end{aligned}
\quad (42)
$$

Setting $l = 0$ here we find

$$
\begin{aligned}
f_1(-iy, 0) &- f_1(-iy^{(1)}, 0) \\
&= -\langle B_1(y^{(1)})y^{(2)}, y^{(2)} \rangle \\
&\quad + \beta_1(y^{(1)})(\exp\{\langle \delta, y^{(2)} \rangle\} - 1) + \langle C_1(y^{(1)}), y^{(2)} \rangle.
\end{aligned}
\quad (43)
$$

From (42) and (43) we get

$$f_1(-iy, l) - f_1(-iy, 0) = \beta_1(y^{(1)}) \exp\{\langle \delta, y^{(2)} \rangle\}$$
$$\times \left(\exp\left\{ \frac{2\pi i l}{q} \right\} - 1 \right) + \ln(k_1(y^{(1)}), l). \quad (44)$$

Substituting $x^{(1)} = -iy^{(1)}$ into (41) we have

$$f_1(-iy, l) - f_1(-iy, 0) = \beta_2(y^{(2)}) \exp\{\langle \delta, y^{(1)} \rangle\}$$
$$\times \left(\exp\left\{ \frac{2\pi i l}{q} \right\} - 1 \right) + \ln(k_2(y^{(2)}), l). \quad (45)$$

It follows from (44) and (45) that

$$\beta_1(y^{(1)}) \exp\{\langle \delta, y^{(2)} \rangle\} \left(\exp\left\{ \frac{2\pi i l}{q} \right\} - 1 \right) + \ln(k_1(y^{(1)}), l)$$

$$= \beta_2(y^{(2)}) \exp\{\langle \delta, y^{(1)} \rangle\} \left(\exp\left\{ \frac{2\pi i l}{q} \right\} - 1 \right) + \ln(k_2(y^{(2)}), l). \quad (46)$$

Setting $y^{(1)} = 0$ here, we find that, for $l \neq 0$,

$$\beta_1(0) \exp\{\langle \delta, y^{(2)} \rangle\} = \beta_2(y^{(2)}) + \frac{\ln(k_2(y^{(2)}), l) - \ln(k_1(0), l)}{\exp\{2\pi i l/q\} - 1}. \quad (47)$$

Since the left-hand side of equality (47) is real for any $y \in \mathbb{R}^n$, it follows from (47) that

$$k_2(y^{(2)}) = k_1(0). \quad (48)$$

Likewise, setting $y^{(2)} = 0$ in (46) we find that, for $l \neq 0$,

$$\beta_2(0) \exp\{\langle \delta, y^{(1)} \rangle\} = \beta_1(y^{(1)}) + \frac{\ln(k_1(y^{(1)}), l) - \ln(k_2(0), l)}{\exp\{2\pi i l/q\} - 1}. \quad (49)$$

Therefore,

$$k_1(y^{(1)}) = k_2(0). \quad (50)$$

Combining (48) and (50) we obtain

$$k_1(y^{(1)}) = k_2(y^{(2)}) = k \quad (51)$$

for any $y \in \mathbb{R}^n$. Now (49) implies

$$\beta_1(y^{(1)}) = \beta_2(0) \exp\{\langle \delta, y^{(1)} \rangle\}. \quad (52)$$

Substituting (51) and (52) into (44) we get

$$f_1(-iy, l) - f_1(-iy, 0)$$
$$= \beta_2(0) \exp\{\langle \delta, y \rangle\} \left(\exp\left\{ \frac{2\pi i l}{q} \right\} - 1 \right) + \ln(k, l). \quad (53)$$

We now use the multivariate analog of the Linnik theorem on decomposition of a convolution of Gaussian and Poisson distributions [LinO, Chapter VI, §3]. Applying this result to the characteristic function $\alpha_1(s, 0)$ we have

$$f_1(s, 0) = -\langle A_1 s, s \rangle + a_1(\exp\{i\langle \delta, s \rangle\} - 1) + i\langle \beta_1, s \rangle, \quad (54)$$

where A_1 is a symmetric positive semidefinite matrix such that $\langle A_1 s , s \rangle \leq \langle A s , s \rangle$, $s \in \mathbb{R}^n$, $\beta_1 \in \mathbb{R}^n$, and $0 \leq a_1 \leq a$. With this in hand, from (53) we obtain

$$f_1(-iy , l) = \langle A_1 y , y \rangle + a_1 (\exp\{\langle \delta , y \rangle\} - 1) + \langle \beta_1 , y \rangle$$
$$+ \beta_2(0) \exp\{\langle \delta , y \rangle\} \left(\exp\left\{ \frac{2\pi i l}{q} \right\} - 1 \right) + \ln(k , l). \quad (55)$$

This implies the representation (37) for $\delta = 0$ and $q = 2$, because as it follows from (36), $\beta_2(0) \leq a$ if $s = 0$. So the theorem is proved for this case.

Let $\delta \neq 0$. To verify that $\beta_2(0) = a_1$, let us set $y = y^{(1)} = (u, 0, \dots, 0)$, $x = x^{(1)} = (v, 0, \dots, 0)$ in inequality (39). Taking account of (55) and (35) we obtain

$$0 \leq \sigma_1 v^2 + 2(a_1 - \beta_2(0)) e^u \sin^2 \frac{v}{2} + 2\beta_2(0) e^u \sin^2 \frac{v + 2\pi l/q}{2}$$
$$\leq \sigma v^2 + 2a e^u \sin^2 \frac{v + 2\pi l/q}{2} , \quad (56)$$

where σ_1 and σ are determined from the equalities $\sigma_1 v^2 = \langle A_1 x , x \rangle$, $\sigma v^2 = \langle A x , x \rangle$. Dividing the inequality (56) by $2e^u$ and then passing to the limit as $u \to +\infty$, we find

$$0 \leq (a_1 - \beta_2(0)) \sin^2 \frac{v}{2} + \beta_2(0) \sin^2 \frac{v + 2\pi l/q}{2} \leq a \sin^2 \frac{v + 2\pi l/q}{2} .$$

Setting $v = -2\pi l/q$ here and taking into account that $\sin(\pi l/q) \neq 0$, we obtain $a_1 = \beta_2(0)$. Now representation (37) follows from (55). $\square$

7.3. REMARK. The definition of a Gaussian distribution on the group $\mathbb{R}^\infty$, given in §5.8, may be directly extended to the group $\mathbb{R}^\infty + \mathbb{Z}(q)$. Let $(t, k) \in \mathbb{R}^\infty + \mathbb{Z}(q)$ and $a > 0$. Consider the measure $P = aE_{(t,k)}$ and the distribution λ generated by this measure by the formula

$$\lambda = e(P) = \exp\{-a\} \left(E_0 + P + \frac{P^{*2}}{2!} + \cdots + \frac{P^{*n}}{n!} + \cdots \right).$$

By a Poisson distribution on the group $\mathbb{R}^\infty + \mathbb{Z}(q)$ we mean a shift of such a distribution λ. The theorem on decomposition of convolutions of Gaussian and Poisson distributions for the group $\mathbb{R}^n + \mathbb{Z}(q)$, proved in §7.2, immediately implies the corresponding result for the group $\mathbb{R}^\infty + \mathbb{Z}(q)$.

7.4. PROPOSITION. *Let ν be a Gaussian distribution and $\lambda \in I_0$ a Poisson distribution on the group $\mathbb{R}^\infty + \mathbb{Z}(q)$. Then their convolution $\nu * \lambda \in I_0$.*

This proposition allows us to strengthen Theorem 7.3 in the following way.

7.5. PROPOSITION. *Let $\gamma \in \Gamma^s(X)$, let π be a Poisson distribution on X, and let $\sigma(\gamma)$ be a connected subgroup of X, not containing a subgroup* *topologically isomorphic to* $\mathbb{T}^\infty$. *If* γ, $\pi \in I_0$, *then their convolution* $\gamma * \pi$ *also belongs to* I_0.

PROOF. Denote $\sigma(\gamma)$ by G. If $\dim G < \infty$, then the assertion follows from Theorem 7.2. If $\dim G = \infty$, then by 5.29 and 5.21 there exist a subspace $L \subset \mathbb{R}^\infty$ and a distribution $\nu \in \Gamma^s(\mathbb{R}^\infty)$ such that $\nu(L) = 1$ and $\gamma = p(\nu)$, where $p : \mathbb{R}^\infty \to X$ is a continuous homomorphism and, besides, the restriction of p to L is a monomorphism. Now the proof of Proposition 7.5 is reduced to the proof of Proposition 7.4 in the same way as the proof of Theorem 7.2 for an arbitrary group X was reduced to the proof of the same theorem for the group $\mathbb{R}^n + \mathbb{Z}(q)$. $\square$

7.6. Recall the Linnik theorem on the structure of the Lévy measure of an infinitely divisible distribution from the class I_0, having a nontrivial Gaussian divisor [LinO, Chapter 4].

LINNIK'S THEOREM. *Let* μ *be an infinitely divisible distribution from the class* I_0 *on the group* $X = \mathbb{R}$, *and let the characteristic function of* μ *have the representation* (x, Φ, φ), *where* $\varphi \not\equiv 0$. *Then the measure* Φ *is discrete and is concentrated on the set* $\{a_{k,1}\}_{k=-\infty}^{\infty} \cup \{a_{k,2}\}_{k=-\infty}^{\infty}$, *where* $a_{k,1} > 0$, $a_{k,2} < 0$, *and* $a_{k+1,j}/a_{k,j}$ *are natural numbers greater than* 1, $j = 1, 2$, *and* $k \in \mathbb{Z}$.

Let us describe the class of all those groups X for which even a weak analog of this theorem does not exist (see §7.13). To this end let us prove several lemmas.

7.7. LEMMA. *Let* $\mu \in \mathscr{M}^1(\mathbb{R}^{n+1})$, *and let the characteristic function* $\hat{\mu}(s, u)$ *have the form*

$$\hat{\mu}(s, u) = \exp\left\{-\langle As, s\rangle + \int_a^b [e^{iut} - 1] d\Phi(t)\right\}, \tag{1}$$

where $s = (s_1, \ldots, s_n) \in \mathbb{R}^n$, $u \in \mathbb{R}$, $A = (a_{kj})_{k,j=1}^n$ *is a symmetric positive semidefinite matrix*, $\Phi \in \mathscr{M}_+(\mathbb{R})$, $\sigma(\Phi) \subset [a, b]$, *and* $0 < b < 2a$. *Then* $\mu \in I_0$.

PROOF. First we note that if the characteristic function $\hat{\nu}(s, u)$ of some distribution $\nu \in \mathscr{M}^1(\mathbb{R}^{n+1})$ is entire, then, for any $u \in \mathbb{R}$ and $s \in \mathbb{R}^n$, the functions

$$f_u(s, \nu) = \frac{\hat{\nu}(s, iu)}{\hat{\nu}(0, iu)}, \qquad g_s(u, \nu) = \frac{\hat{\nu}(is, u)}{\hat{\nu}(is, 0)}, \tag{2}$$

are characteristic functions of some distributions of variables $s \in \mathbb{R}^n$ and $u \in \mathbb{R}$, respectively.

Let $\mu = \mu_1 * \mu_2$. It follows from 2.10(d) that

$$\hat{\mu}(s, u) = \hat{\mu}_1(s, u)\hat{\mu}_2(s, u). \tag{3}$$

Since $\hat{\mu}(s, u)$ is an entire function on $\mathbb{C}^{n+1}$, Theorem 3.15 implies that $\hat{\mu}_j(s, u)$, $j = 1, 2$, also are entire functions and by Theorem 3.14 equality (3) is true in $\mathbb{C}^{n+1}$. We find from (1) and (2) that

$$f_u(s, \mu) = \exp\{-\langle As, s\rangle\}, \qquad g_s(u, \mu) = \exp\left\{\int_a^b [e^{iut} - 1]\, d\Phi(t)\right\}. \quad (4)$$

It follows from (2) and (3) that

$$f_u(s, \mu) = f_u(s, \mu_1) \cdot f_u(s, \mu_2), \qquad g_s(u, \mu) = g_s(u, \mu_1) g_s(u, \mu_2). \quad (5)$$

Applying the Cramér theorem to the characteristic function $f_u(s, \mu)$ and Ostrovskiĭ theorem (see §6.16, p. 78) to the characteristic function $g_s(u, \mu)$ we obtain from (4) and (5) that

$$f_u(s, \mu_1) = \exp\{i\langle C(u), s\rangle - \langle A(u)s, s\rangle\},$$

where $C(u) = (C_k(u))_{k=1}^n \in \mathbb{R}^n$ and $A(u) = (\alpha_{kj}(u))_{k,j=1}^n$ is a symmetric positive semidefinite matrix such that $\langle A(u)s, s\rangle \leq \langle As, s\rangle$ for $s \in \mathbb{R}^n$, and

$$g_s(u, \mu_1) = \exp\left\{id(s)u + \int_a^b [e^{iut} - 1]\, d\Phi^s(t)\right\},$$

where $d(s) \in \mathbb{R}$, $\Phi^s \in \mathscr{M}_+(\mathbb{R})$, and $\Phi^s \leq \Phi$.

On the other hand by the Cramér and Ostrovskiĭ theorems we find from (3)

$$\mu_1(s, 0) = \exp\{i\langle C, s\rangle - \langle A^{(1)}s, s\rangle\},$$

where $C = (C_k)_{k=1}^n \in \mathbb{R}^n$ and $A^{(1)} = (\alpha_{kj}^{(1)})_{k,j=1}^n$ is a symmetric positive semidefinite matrix such that $\langle A^{(1)}s, s\rangle \leq \langle As, s\rangle$ for $s \in \mathbb{R}^n$, and

$$\hat{\mu}_1(0, u) = \exp\left\{idu + \int_a^b [e^{iut} - 1]\, d\Phi_1(t)\right\},$$

where $d \in \mathbb{R}$, $\Phi_1 \in \mathscr{M}_+(\mathbb{R})$, and $\Phi_1 \leq \Phi$. Now we obtain from (2) that

$$\mu_1(s, iu) = f_u(s, \mu_1)\hat{\mu}_1(0, iu)$$

$$= \exp\left\{i\langle C(u), s\rangle - \langle A(u)s, s\rangle - du + \int_a^b [e^{-ut} - 1]\, d\Phi_1(t)\right\}, \quad (6)$$

$$\hat{\mu}_1(is, u) = g_s(u, \mu_1)\hat{\mu}_1(is, 0)$$

$$= \exp\left\{id(s)u + \int_a^b [e^{iut} - 1]\, d\Phi^s(t) - \langle C, s\rangle + \langle A^{(1)}s, s\rangle\right\}. \quad (7)$$

Replacing s by is in (6) and u by iu in (7) we conclude that

$$-\langle C(u), s\rangle + \langle A(u)s, s\rangle - du + \int_a^b [e^{-ut} - 1]\, d\Phi_1(t)$$

$$= -d(s)u + \int_a^b [e^{-ut} - 1]\, d\Phi^s(t) - \langle C, s\rangle + \langle A^{(1)}s, s\rangle \quad (8)$$

for any $s \in \mathbb{R}^n$ and $u \in \mathbb{R}$. Setting $s = e_k$, where $\{e_k\}_{k=1}^n$ is the standard basis in $\mathbb{R}^n$, in (8) we have

$$- C_k(u) + \alpha_{kk}(u) - du + \int_a^b [e^{-ut} - 1] d\Phi_1(t)$$

$$= -d(e_k)u + \int_a^b [e^{-ut} - 1] d\Phi^{e_k}(t) - C_k - \alpha_{kk}^{(1)}. \qquad (9)$$

Since

$$\int_a^b [e^{-ut} - 1] d\Phi_1(t) = O(1), \qquad \int_a^b [e^{-ut} - 1] d\Phi^{e_k}(t) = O(1)$$

for $u \to +\infty$ and $0 \le \alpha_{kk}(u) \le \alpha_{kk}$, it follows from (9) that

$$C_k(u) = O(u), \qquad u \to +\infty. \qquad (10)$$

Setting $u = s_k s_j$ in (8) we get

$$d(s)s_k s_j = \langle C(s_k s_j), s \rangle - \langle A(s_k s_j)s, s \rangle + ds_k s_j - \int_a^b [e^{-s_k s_j t} - 1] d\Phi_1(t)$$

$$+ \int_a^b [e^{-s_k s_j t} - 1] d\Phi^s(t) - \langle C, s \rangle + \langle A^{(1)}s, s \rangle.$$

After dividing both the sides of this equality by $(s_k s_j)^2$ and accounting for (10) we pass to the limit as $s_k \to +\infty$, $s_j \to +\infty$

$$\lim_{s_k \to +\infty, \ s_j \to +\infty} \frac{d(s)}{s_k s_j} = 0. \qquad (11)$$

Taking (11) into account, we divide both sides of equality (8) by $s_k s_j$ and pass to the limit as $s_k \to +\infty$, $s_j \to +\infty$. We obtain $\alpha_{kj}(u) = \alpha_{kj}^{(1)}$, i.e.,

$$A(u) = A^{(1)}. \qquad (12)$$

Setting $b_k = d(e_k) - d$ and $P_k = \Phi_1 - \Phi^{e_k}$ for $k = 1, \dots, n$ and accounting for (12) we find from (9) that

$$C_k(u) = b_k u + C_k + \int_a^b [e^{-ut} - 1] dP_k(t). \qquad (13)$$

Hence $C_k(u)$, $k = 1, \dots, n$ are entire functions. We conclude from (6) and (12) that the function $\hat{\mu}_1(s, iu)$ has the form

$$\hat{\mu}_1(s, iu) = \exp\left\{ i\langle C(u)s, \rangle - \langle A^{(1)}s, s \rangle - du + \int_a^b [e^{-ut} - 1] d\Phi_1(t) \right\}$$

for any $s \in \mathbb{R}^n$ and $u \in \mathbb{R}$. Since the functions on both sides of this equality are entire, this representation for $\hat{\mu}_1(s, iu)$ is valid for any complex $s \in \mathbb{C}^n$ and $u \in \mathbb{C}$. Therefore

$$\hat{\mu}_1(s, u) = \exp\left\{ i\langle C(u), s \rangle - \langle A^{(1)}s, s \rangle + idu + \int_a^b [e^{iut} - 1] d\Phi_1(t) \right\} \qquad (14)$$

for $s \in \mathbb{R}^n$ and $u \in \mathbb{R}$. So from (13) and (14) we obtain

$$\hat{\mu}_1(s, iu) = \exp\left\{\left(\sum_{k=1}^n b_k s_k\right)u + i\langle C, s\rangle + i\sum_{k=1}^n\left(\int_a^b [e^{iut} - 1]dP_k(t)\right)s_k \right. $$
$$\left. - \langle A^{(1)}s, s\rangle + idu + \int_a^b [e^{iut} - 1]d\Phi_1(t)\right\}.$$

(15)

Since, for any fixed $s \in \mathbb{R}^n$, the function $\hat{\mu}(s, u)$ is bounded as a function of the variable $u \in \mathbb{R}$, we conclude from (15) that $b_k = 0$ for $k = 1, \ldots, n$. Hence

$$\hat{\mu}_1(s, u) = \exp\left\{i\langle C, s\rangle + i\sum_{k=1}^n\left(\int_a^b [e^{iut} - 1]dP_k(t)\right)s_k \right. $$
$$\left. - \langle A^{(1)}s, s\rangle + idu + \int_a^b [e^{iut} - 1]d\Phi_1(t)\right\}.$$

(16)

Let $w, v \in \mathbb{R}^n$ and $\eta, \zeta \in \mathbb{R}$. Then we find from (16)

$$\left|\frac{\hat{\mu}_1(w + iv, \eta + i\zeta)}{\hat{\mu}_1(iv, i\zeta)}\right| = \exp\left\{2\sum_{k=1}^n\left(\int_a^b e^{-\zeta t}\sin^2\frac{\eta t}{2}dP_k(t)\right)v_k \right. $$
$$- \sum_{k=1}^n\left(\int_a^b e^{-\zeta t}\sin\eta t\, dP_k(t)\right)w_k - \langle A^{(1)}w, w\rangle$$
$$\left. - 2\int_a^b e^{-\zeta t}\sin^2\frac{\eta t}{2}d\Phi_1(t)\right\}.$$

(17)

Now let us use the fact that whatever the entire characteristic function $\hat{\nu}(z)$, the function $\hat{\nu}(x + iy)/\hat{\nu}(iy)$ is a characteristic function of the variable $x \in \mathbb{R}^m$ for any fixed $y \in \mathbb{R}^m$. Therefore

$$\left|\frac{\hat{\mu}_1(w + iv, \eta + i\zeta)}{\hat{\mu}_1(iv, i\zeta)}\right| \le 1.$$

Denote the expression in the braces in (17) by $H(w, v, \eta, \zeta)$. Since $H(w, v, \eta, \zeta) \le 0$, we have, in particular,

$$H(0, v, \eta, \zeta) = 2\sum_{k=1}^n\left(\int_a^b e^{-\zeta t}\sin^2\frac{\eta t}{2}dP_k(t)\right)v_k$$
$$- 2\int_a^b e^{-\zeta t}\sin^2\frac{\eta t}{2}d\Phi_1(t) \le 0.$$

Therefore

$$\int_a^b e^{-\zeta t}\sin^2\frac{\eta t}{2}dP_k(t) = 0, \qquad k = 1, \ldots, n,$$

for any $\zeta \in \mathbb{R}$, and hence for any $\zeta \in \mathbb{C}$. Setting $\zeta = -i\psi$, $\psi \in \mathbb{R}$, we obtain

$$\int_a^b e^{i\psi t} \sin^2 \frac{\eta t}{2} \, dP_k(t) = 0, \qquad k = 1, \ldots, n.$$

By 2.10(c) the charges $(\sin^2(\eta t/2))P_k$, $k = 1, \ldots, n$, are equal to zero for any $\eta \in \mathbb{R}$. Hence $P_k = 0$, $k = 1, \ldots, n$. Now from (16) we finally find

$$\hat{\mu}_1(s, u) = \exp\left\{ i\langle C, s\rangle - \langle A^{(1)}s, s\rangle + idu + \int_a^b [e^{iut} - 1]d\Phi_1(t) \right\}.$$

This representation for the characteristic function of an arbitrary divisor μ_1 of the distribution μ shows that $\mu \in I_0$. $\square$

7.8. LEMMA. *Let* $\Phi \in \mathcal{M}_+(X)$, $\mu = e(\Phi)$, *and the measure* Φ *satisfy the conditions*:

  (i) $\Phi(X) < \frac{1}{2}\ln 2$.

  (ii) *The measures* Φ^{*n} *and* Φ^{*m} *are mutually singular for any natural* n, m, $n \neq m$.

Then any divisor of μ *has the form* $\mu_1 = e(\Phi_1) * E_{x_1}$, *where* $\Phi_1 \in \mathcal{M}_+(X)$ *and* $\Phi_1 \leq \Phi$.

PROOF. Let $\mu = \mu_1 * \mu_2$, where $\mu_j \in \mathcal{M}^1(X)$, $j = 1, 2$. Then, as proved in Theorem 6.13, $\mu_j = e(\Phi_j) * E_{x_j}$ with $\Phi_j \in \mathcal{M}_+(X)$. We have

$$\mu = e(\Phi) = e(\Phi_1 + \Phi_2) * E_{x_0}, \tag{1}$$

where $x_0 = x_1 + x_2$. By Lemma 4.11 relation (1) implies

$$\Phi(X) = \Phi_1(X) + \Phi_2(X). \tag{2}$$

Observe that $\mu(\{0\}) \geq \exp\{-\Phi(X)\} > \frac{1}{2}$. On the other hand, it follows from (1) and (2) that $\mu(\{x_0\}) > \frac{1}{2}$. But this is possible only if $x_0 = 0$. Hence $e(\Phi) = e(\Phi_1 + \Phi_2)$. By Lemma 4.12 $\Phi = \Phi_1 + \Phi_2$ and thus $\Phi_1 \leq \Phi$. $\square$

7.9. LEMMA. *Let* $X = \mathbb{R}^n + K$, *where* $n \geq 0$ *and* K *is a nondiscrete compact periodic group, and let* $\mu = \gamma * e(\Phi)$, *where* $\gamma \in \Gamma^s(\mathbb{R}^n)$, $\Phi \in \mathcal{M}_+(K)$ *and the measure* Φ *satisfies the conditions* (i) *and* (ii) *of Lemma* 7.8. *Then* $\mu \in I_0$.

PROOF. Let $L = K^*$. Elements of the group $Y \approx \mathbb{R}^n + L$ will be denoted by (s, l), where $s \in \mathbb{R}^n$ and $l \in L$. The characteristic function of the distribution μ has the form

$$\hat{\mu}(s, l) = \exp\left\{ -\langle As, s\rangle + \int_K [(k, l) - 1]d\Phi(k) \right\}, \tag{1}$$

where A is a symmetric positive semidefinite matrix. Let $\mu = \mu_1 * \mu_2$ where $\mu_j \in \mathcal{M}^1(X)$, $j = 1, 2$. Then, by 2.10(d),

$$\hat{\mu}(s, l) = \hat{\mu}_1(s, l)\hat{\mu}_2(s, l). \tag{2}$$

It follows from (1) that $\hat{\mu}(s, 0)$ is an entire function on $\mathbb{C}^n$. Setting $l = 0$ in (2) and applying Theorem 3.15 we conclude that $\hat{\mu}_1(s, 0)$ is an entire function as well. Therefore, by Lemma 6.9, $\hat{\mu}_1(s, l)$ is an entire function with respect to s for any fixed $l \in L$, and equality (2) is true for any $s \in \mathbb{C}^n$ and $l \in L$. Hence $\hat{\mu}_1(s, l) \neq 0$ for any $s \in \mathbb{C}^n$ and $l \in L$. By Theorem 3.8 we have

$$\hat{\mu}_1(s, l) = \exp\{f(s, l)\},$$

where the branch $f(s, l)$ of the entire function $\ln \hat{\mu}_1(s, l)$ is chosen in such a way that $f(s, l) = \overline{f(-s, -l)}$. Setting $l = 0$ in (1) and (2) and applying the Cramér theorem, we obtain

$$\hat{\mu}_1(s, 0) = \exp\{i\langle B_1, s\rangle - \langle A_1 s, s\rangle\}, \tag{3}$$

where $B_1 \in \mathbb{R}^n$ and A_1 is a symmetric positive semidefinite matrix such that

$$\langle A_1 s, s\rangle \leq \langle As, s\rangle$$

for $s \in \mathbb{R}^n$. It follows from (6.9) that

$$\max_{s \in B_r, l \in L} |\hat{\mu}_1(s, l)| = \max_{s \in B_r} |\hat{\mu}_1(s, 0)| = C \exp\{ar^2\}$$

for some $C > 0$ and $a > 0$. From here we find

$$\max_{s \in B_r} \operatorname{Re} f(s, l) = O(r^2).$$

By Theorem 3.9, $f(s, l)$ is a polynomial of degree not exceeding 2. Hence

$$\hat{\mu}_1(s, l) = \exp\{\langle A(l)s, s\rangle + \langle B(l), s\rangle + C(l)\}, \tag{4}$$

where $A(l)$ is some matrix and $B(l) \in \mathbb{C}^n$, $C(l) \in \mathbb{C}$.

Observe now that if the characteristic function $\hat{\nu}(s, l)$ of some distribution $\nu \in \mathscr{M}^1(\mathbb{R}^n + K)$ is entire with respect to s for any fixed $l \in L$, then, for any fixed $s \in \mathbb{R}^n$, the function

$$g_s(l, \nu) = \frac{\hat{\nu}(is, l)}{\hat{\nu}(is, 0)} \tag{5}$$

is the characteristic function of some distribution on K. It follows from (2) that

$$g_s(l, \mu) = g_s(l, \mu_1)g_s(l, \mu_2). \tag{6}$$

But (1) implies

$$g_s(l, \mu) = \exp\left\{\int_K [(k, l) - 1]\,d\Phi(k)\right\}.$$

Therefore, by Lemma 7.8, from (6) we get

$$g_s(l, \mu_1) = (k_s, l)\exp\left\{\int_K [(k, l) - 1]\,d\Phi^s(k)\right\}, \tag{7}$$

where $k_s \in K$, $\Phi^s \in \mathscr{M}_+(K)$, and $\Phi^s \le \Phi$. Now, according to (3) and (7), we find from (5) that

$$\hat{\mu}_1(is, l) = g_s(l, \mu_1)\hat{\mu}_1(is, 0)$$
$$= (k_s, l) \exp \left\{ \int_K [(k, l) - 1] d\Phi^s(k) - \langle B_1, s \rangle + \langle A_1 s, s \rangle \right\}. \quad (8)$$

Let us verify that the element k_s depends continuously on s. Set $\mu[s] = e(\Phi^s) * E_{k_s}$, $\mu[s] \in \mathscr{M}^1(K)$. Since $\Phi^s(K) \le \Phi(K) < \frac{1}{2}\ln 2$, we have

$$\mu[s](\{k_s\}) > \frac{1}{\sqrt{2}}. \quad (9)$$

The function $\hat{\mu}_1(is, l)$ is continuous on the group $\mathbb{R}^n + L$. By 2.10(f), it follows from (8) that $\mu[s] \Rightarrow \mu[s_0]$ as $s \to s_0$. If the element k_s as a function of s has a discontinuity at the point s_0, then we can find a sequence $s_n \to s_0$ and a neighborhood U of the element k_{s_0} such that $k_{s_n} \notin U$ for all $n = 1, 2, 3, \ldots$. Let $f(k)$ be a continuous function on K such that $0 \le f(k) \le 1$ for any $k \in K$ and $f(k) = 0$ for $k \notin U$ and $f(k_{s_0}) = 1$. Taking (9) into account, we have

$$\int_K f(k) \, d\mu[s_n](k) = \int_U f(k) \, d\mu[s_n](k) \le \mu[s_n](U) \le \mu[s_n](K \backslash \{k_{s_n}\})$$
$$\le 1 - \frac{1}{\sqrt{2}};$$

$$\int_K f(k) \, d\mu[s_0](k) \ge \mu[s_0](\{k_{s_0}\}) > \frac{1}{\sqrt{2}}.$$

But these inequalities contradict the fact that

$$\lim_{n \to \infty} \int_K f(k) \, d\mu[s_n](k) = \int_K f(k) \, d\mu[s_0](k).$$

So the element k_s depends continuously on s. Since the function (k_s, l) takes only a finite number of values, it follows from Theorem 1.18 that $(k_s, l) = Q(l) = \exp\{q(l)\}$, where $q(l)$ is some function on L. This allows one to rewrite (8) in the form

$$\hat{\mu}_1(is, l) = \exp \left\{ \int_K [(k, l) - 1] d\Phi^s(k) - \langle B_1, s \rangle + \langle A_1 s, s \rangle + q(l) \right\}. \quad (10)$$

Replacing s by is in (4) and comparing the expression obtained with (10) we get

$$- \langle A(l)s, s \rangle + i \langle B(l), s \rangle + C(l)$$
$$= \langle A_1 s, s \rangle - \langle B_1, s \rangle$$
$$+ \int_K [(k, l) - 1] d\Phi^s(k) + q(l) + 2\pi i \, n(s, l), \quad (11)$$

where $n(s, l) \in \mathbb{Z}$. Since the function $n(s, l)$ is continuous with respect to s for any fixed $l \in L$ and all its values are integers, we have $n(s, l) = n(l)$. Since

$$\int_K [(k, l) - 1] d\Phi^s(k) = O(1)$$

as $|s_j| \to \infty$, $j = 1, \ldots, n$, (11) implies

$$A(l) = -A_1, \qquad B(l) = iB_1 \tag{12}$$

for any $l \in L$.

Setting $s = 0$ in (2) and accounting for (1), by Lemma 7.8 we obtain

$$\hat{\mu}_1(0, l) = (k_0, l) \exp \left\{ \int_K [(k, l) - 1] d\Phi_1(k) \right\}, \tag{13}$$

where $k_0 \in K$, $\Phi_1 \in \mathcal{M}_+(K)$, and $\Phi_1 \leq \Phi$. Combining (4) with $s = 0$ and (13) we conclude that

$$\exp\{C(l)\} = (k_0, l) \exp \left\{ \int_K [(k, l) - 1] d\Phi_1(k) \right\}. \tag{14}$$

From (4), (12), and (14) we finally find

$$\hat{\mu}_1(s, l) = (k_0, l) \exp \left\{ -\langle A_1 s, s \rangle + i \langle B_1, s \rangle + \int_K [(k, l) - 1] d\Phi_1(k) \right\},$$

whence $\mu \in I_0$. $\square$

Let a group X be such that any neighborhood of zero in X contains an element of infinite order. In [Rud] W. Rudin has constructed a perfect independent subset of X. Below we generalize this construction.

7.10. LEMMA. *Let G be a connected compact group and $p: \mathbb{R}^n \to G$ a continuous homomorphism such that $p(\mathbb{R}^n) \neq G$. Denote $E_l = \{t \in \mathbb{R}^n : |t_j| \leq l, \ j = 1, \ldots, n\}$ and $A_l = p(E_l)$. Let $n_1, \ldots, n_k$ be given integers, not all zero, and let $E = \{(x_1, \ldots, x_k) \in G^k : n_1 x_1 + \cdots + n_k x_k \notin A_l\}$. Then E is an open dense subset of G^k.*

PROOF. Let a mapping $f: G^k \to G$ be defined by the formula

$$f(x_1, \ldots, x_k) = n_1 x_1 + \cdots + n_k x_k.$$

Suppose that $f^{-1}(A_l)$ contains a nonempty open subset $V = V_1 \times \cdots \times V_k$, $V_j \subset G$, $j = 1, \ldots, k$. Fix a j such that $n_j \neq 0$, and fix elements $x_i \in V_i$, $i \neq j$. Set

$$x_0 = \sum_{i \neq j} n_i x_i.$$

For any $x \in V_j$ we then have $n_j x = -x_0 + p(t_x)$ with some $t_x \in E_l$. Set $W = \{w \in G : w = x - y, \ x, y \in V_j\}$. For any $w \in W$ we have

$$n_j w = n_j x - n_j y = -x_0 + p(t_x) + x_0 - p(t_y) = p(t_x - t_y) \in p(\mathbb{R}^n).$$

G is a connected compact group. Therefore, by 1.20, the mapping $f_n\colon G \to G$ $(f_n(x) = nx)$ is an epimorphism for any $n = 2, 3, 4, \ldots$. Hence f_n is an open mapping [HR1, §5] and so $f_{n_j}(W)$ is an open set. But then $m_G(p(\mathbb{R}^n)) \geq m_G(f_{n_j}(W)) > 0$. This is impossible because the Haar measure of any proper Borel subgroup of the connected group G is equal to zero. Thus $f^{-1}(A_i)$ does not contain a nonempty open set. Hence E is a dense set.

Since the mapping f is continuous and the set A_l is closed, the set $f^{-1}(A_l)$ is closed as well. Therefore E is open. $\square$

7.11. LEMMA. *Let the group* G *and the homomorphism* p *be the same as in Lemma 7.10. Then there exists a perfect independent subset* B *of the group* G *such that*

$$(\text{i}) \quad M(B) \cap p(\mathbb{R}^n) = \{0\}.$$

PROOF. By a compact neighborhood we mean a compact set that is the closure of an open set. Let us construct by induction a sequence of compact neighborhoods $B_l \subset G$, $l = 1, 2, 3, \ldots$. Let B_1 be an arbitrary compact neighborhood, and $B_l = B_l^{(1)} \cup \cdots \cup B_l^{(s)}$, where $s = 2^{l-1}$ and $B_l^{(i)}$ are pairwise disjoint compact neighborhoods. Consider the set F of all points $(x_1, \ldots, x_{2s}) \in G^{2s}$ such that if

$$|n_1| + \cdots + |n_{2s}| > 0, \qquad |n_i| \leq l, \ i = 1, \ldots, 2s, \tag{1}$$

then $n_1 x_1 + \cdots + n_{2s} x_{2s} \notin A_l$. Applying Lemma 7.10 a finite number of times we see that F is an open dense subset of G^{2s}. Hence the set $B_l^{(1)} \times B_l^{(1)} \times B_l^{(2)} \times B_l^{(2)} \times \cdots \times B_l^{(s)} \times B_l^{(s)}$ contains the open set $V_1 \times \cdots \times V_{2s} \subset F$. Let us choose pairwise disjoint compact neighborhoods $B_{l+1}^{(i)} \subset V_i$, $i = 1, \ldots, 2s$, whose diameters are less than $1/l$. Set $B_{l+1} = B_{l+1}^{(1)} \cup \cdots \cup B_{l+1}^{(2s)}$. Then $B_{l+1} \subset B_l$. If $x_i \in B_{l+1}^{(i)}$, $i = 1, \ldots, 2s$, and the numbers n_i satisfy condition (1), then $n_1 x_1 + \cdots + n_{2s} x_{2s} \notin A_l$. Thus if $x_1, \ldots, x_j$ are distinct points of B_{l+1}, no two of which belong to one and the same set $B_{l+1}^{(i)}$, and $|n_1| + \cdots + |n_j| > 0$, $|n_i| \leq l$, $i = 1, \ldots, j$, then

$$n_1 x_1 + \cdots + n_j x_j \notin A_l. \tag{2}$$

Set $B = \bigcap_{l=1}^{\infty} B_l$. Then B evidently is a perfect set. Let us verify that (i) is fulfilled. Assume the opposite to be true. Then there exist $\zeta \in p(\mathbb{R}^n)$, $\zeta \neq 0$, $x_1, \ldots, x_j \in B$, and $n_1, \ldots, n_j \in \mathbb{Z}$ such that

$$\zeta = n_1 x_1 + \cdots + n_j x_j. \tag{3}$$

Choose l so large that $\zeta \in A_l$, $|n_i| \leq l$, $i = 1, \ldots, j$, and no set $B_{l+1}^{(i)}$ contains more than one of the points $x_1, \ldots, x_j$. Then (2) is fulfilled (as was mentioned before), which contradicts (3).

If we assume $\zeta = 0$ and $|n_1| + \cdots + |n_j| > 0$ in (3), then this argument shows that B is an independent set. $\square$

7.12. LEMMA. *Let groups X and X' be given, and let $A \subset X$, $A' \subset X'$ be independent Borel F_σ-subsets. Let $h \colon A \to A'$ be a bijection that preserves Borel sets in both directions). Let $\tilde{h} \colon M(A) \to M(A')$ be the algebraic isomorphism of the groups $M(A)$ and $M(A')$ generated by h. Then the $\tilde{h}$ images and preimages of Borel sets are again Borel sets.*

PROOF. Set $A_{k_1,k_2} = (k_1)A + (k_2)(-A)$, where k_1, k_2 are nonnegative integers $((0)A = \{0\})$ and $A_n = \bigcup_{k_1-k_2=n} A_{k_1,k_2}$. Since the set A is independent, the sets A_n, $n \in \mathbb{Z}$ are mutually disjoint.

Let us prove that the image of any Borel set $E \subset M(A)$ under the mapping $\tilde{h}$ is a Borel set. The reasoning for preimages is similar. Clearly, it suffices to verify that if $E \in \mathscr{B}(X)$ and $E \subset A_n$, then $\tilde{h}(E) \in \mathscr{B}(X')$. We shall confine ourselves to the case $n \geq 0$ (the case $n < 0$ is similar).

Note that $A_n = \bigcup_{l=0}^{\infty} A_{n+l,l}$ and $A_{n+l,l} \subset A_{n+k,k}$ for $l < k$. Therefore, it is enough to prove that if $E \subset A_{n+k,k} \backslash A_{n+k-1,k-1}$, then $\tilde{h}(E) \in \mathscr{B}(X')$. Let $A_{n+k,k} = A_{k_1,k_2}$. Consider the Cartesian products

$$\tilde{A}_{k_1,k_2} = A \times \cdots \times A, \qquad \tilde{A}'_{k_1,k_2} = A' \times \cdots \times A',$$

where the number of multipliers in every product is k_1+k_2. Let $h_{k_1,k_2} \colon \tilde{A}_{k_1,k_2} \to \tilde{A}'_{k_1,k_2}$ be the mapping generated by h, that is,

$$h_{k_1,k_2}(x_1, \ldots, x_{k_1}; y_1, \ldots, y_{k_2}) = (h(x_1), \ldots, h(x_{k_1}); h(y_1), \ldots, h(y_{k_2})).$$

By standard topological arguments (see [Kur, §31, VI]), if the images and preimages of Borel sets under the mapping h are again Borel sets, then the mapping h_{k_1,k_2} has the same property. Denote by π_{k_1,k_2} the mapping of $\tilde{A}_{k_1,k_2}$ onto A_{k_1,k_2} defined by the formula

$$\pi_{k_1,k_2}(x_1, \ldots, x_{k_1}; y_1, \ldots, y_{k_2}) = x_1 + \cdots + x_{k_1} - y_1 - \cdots - y_{k_2}.$$

This mapping is continuous and its restriction to the set $A_{n+k,k} \backslash A_{n+k-1,k-1}$ has the following property: the preimage of every one-point set is finite. In the same way we define a mapping π'_{k_1,k_2} of the set $\tilde{A}'_{k_1,k_2}$ onto A'_{k_1,k_2}.

Let $E \in \mathscr{B}(X)$, $E \subset A_{n+k,k} \backslash A_{n+k-1,k-1}$. Obviously

$$\tilde{h}(E) = \pi'_{k_1,k_2}(h_{k_1,k_2}(\pi^{-1}_{k_1,k_2}(E))).$$

Since the mapping π_{k_1,k_2} is continuous, the set $\pi^{-1}_{k_1,k_2}(E)$ is a Borel set and its image $h_{k_1,k_2}(\pi^{-1}_{k_1,k_2}(E))$ also is a Borel set, because the mapping h_{k_1,k_2} preserves Borel sets. It remains to check whether the mapping π_{k_1,k_2} transforms Borel sets into Borel ones. This is a consequence of the following theorem of Lusin (see [Kur, §39, VII]). Let E be a Borel set in a complete separable space. If a continuous mapping f of E is such that the preimage of any one-point set is at most countable, then $f(E)$ is a Borel set. $\square$

7.13. THEOREM. *Let the component of zero C_X of the group X differ from zero and have finite dimension, and let the group X be not topologically isomorphic to the group $\mathbb{R}^n + D$, $n \geq 0$, where D is a discrete group. Let γ be an arbitrary symmetric Gaussian distribution of the class I_0 on X. Then there exists a continuous measure $\Phi \in \mathcal{M}_+(X)$ such that the distribution $\mu = \gamma * e(\Phi)$ belongs to the class I_0.*

PROOF. Since $\gamma \in \Gamma^s(X)$, we have $\sigma(\gamma) \subset C_X$ by Proposition 5.5. Since $\gamma \in I_0$, it follows from Propositions 5.21 and 5.28 that there exist a continuous monomorphism $p \colon \mathbb{R}^n \to C_X$ and a distribution $\nu \in \Gamma^s(\mathbb{R}^n)$ such that $\gamma = p(\nu)$. Two cases are possible.

1. $C_X \not\approx \mathbb{R}^n$. In this case it is obvious that $p(\mathbb{R}^n) \neq C_X$. By Lemma 7.11 there exists a perfect independent subset B of C_X such that condition 7.11(i) is fulfilled. Let us consider the distribution $\delta = \nu * e(P)$ on the group $\mathbb{R}^{n+1}$, where $P \in \mathcal{M}_+(\mathbb{R}^{n+1})$ and $\sigma(P) = E \subset \{t \in \mathbb{R}^{n+1} : t = (0, \dots, 0, t_{n+1})$, $t_{n+1} \in [a, b]$, $0 < b < 2a\}$. Assume also that E is a perfect independent set and P is a nonzero continuous measure. The characteristic function of the distribution δ has the same form as in 7.7(1) and by Lemma 7.7 we have $\delta \in I_0$. Note that the distribution δ is concentrated on the Borel subgroup $G = \mathbb{R}^n + M(E) \subset \mathbb{R}^{n+1}$. It is known (see [Kur, §37, II]) that there exists a one-to-one mapping h of the set E onto B for which images and preimages of Borel sets are Borel sets. Let us extend h to an algebraic isomorphism $\tilde{h}$ of the groups $M(E)$ and $M(B)$. By Lemma 7.12 the images and preimages of Borel sets under the mapping $\tilde{h}$ are also Borel sets. Let G' be the subgroup of C_X generated (algebraically) by $p(\mathbb{R}^n)$ and $M(B)$, and let ψ be the mapping of G into G', defined by the formula

$$\psi(t, t_{n+1}) = p(t) + \tilde{h}(t_{n+1}), \qquad t \in \mathbb{R}^n, \ t_{n+1} \in M(E).$$

Obviously, ψ is an algebraic isomorphism of the groups G and G'. Let us verify that images and preimages of Borel sets under this mapping are Borel sets. To do this we consider the mapping $\psi_1 \colon G \to C_X^2$, defined by the formula

$$\psi_1(t, t_{n+1}) = (p(t), \tilde{h}(t_{n+1})),$$

where $t \in \mathbb{R}^n$ and $t_{n+1} \in M(E)$. Since any one-to-one continuous image of a Borel set is a Borel set (see [Kur, §39, IV]), images and preimages of Borel sets under the mapping p also are Borel sets. The mapping $\tilde{h}$ possesses the same property. By standard topological reasoning (see [Kur, §31, VI]) images and preimages of Borel sets under the mapping ψ_1 are also Borel sets. Define a mapping $\psi_2 \colon \psi_1(G) \to G'$ by the formula

$$\psi_2(p(t), \tilde{h}(t_{n+1})) = p(t) + \tilde{h}(t_{n+1}).$$

It is a continuous one-to-one mapping. As mentioned above, images and preimages of Borel sets under the mapping ψ_2 are also Borel sets, and the mapping $\psi = \psi_2 \circ \psi_1$ possesses the same property.

By Proposition 2.6 we have

$$\psi(\delta) = \psi(\nu * e(P)) = \psi(\nu) * \psi(e(P)) = p(\nu) * \tilde{h}(e(P)) = \gamma * e(\tilde{h}(P)).$$

Obviously, the measure $\Phi = \tilde{h}(P) \in \mathscr{M}_+(X)$ is continuous. It follows from Remark 4.19 that $\mu = \gamma * e(\Phi) \in I_0$.

2. $C_X \approx \mathbb{R}^n$. In this case, as it follows from Theorem 1.14, $X \approx \mathbb{R}^n + X_1$, where $n \geq 1$ and the group X_1 contains a compact open subgroup. By the hypothesis the group X_1 is nondiscrete. There are two further possibilities.

a) Any neighborhood of zero in the group X_1 contains an element of infinite order.

In this case let us construct a perfect independent subset B (see [Rud]) of the group X_1. Assume $C_X = \mathbb{R}^n$, and let $p: \mathbb{R}^n \to C_X$ be the identity mapping. Condition 7.17(i) for the group X will be fulfilled automatically. The reasoning from here on is similar to one in the case 1.

b) There exists a neighborhood of zero in the group X_1 consisting only of elements of finite order.

Let K be a compact open subgroup of X_1. Then, in K, there exists a neighborhood of zero U, consisting of elements of finite order. By Theorem 1.19 there exists a subgroup K_1 of K such that $K_1 \subset U$ and $K/K_1 \approx T^m + F$, where $m \geq 0$ and the group F is finite. Since the group K is totally disconnected, $m = 0$. Therefore the group K is periodic. It follows from the hypothesis of the theorem that the group K is nondiscrete. Let $\Phi \in \mathscr{M}_+(K)$; let the measure Φ be continuous and satisfy the conditions of Lemma 7.8. The existence of such a measure Φ is ensured by Theorem 6.15. One can also assume that $C_X = \mathbb{R}^n$. Let $\mu = \gamma * e(\Phi)$. By Lemma 7.9 we obtain $\mu \in I_0$. $\square$

7.14. REMARK. Let the component C_X of zero of the group X differ from zero, and let it be not topologically isomorphic to the group $\mathbb{R} + D$, where D is a discrete group. Then there exist $\gamma \in \Gamma(X)$ and a continuous measure $\Phi \in \mathscr{M}_+(X)$ such that the measure $\mu = \gamma * e(\Phi)$ belongs to the class I_0. Indeed if $C_X \not\approx \mathbb{R}$, then accounting for Remark 1.21 we may construct a continuous monomorphism $p: \mathbb{R} \to C_X$ such that $p(\mathbb{R}) \neq C_X$. The further reasoning is the same as in the proof of Theorem 7.13 in the case 1, where $n = 1$ is to be assumed, and in the role of ν an arbitrary Gaussian distribution on $\mathbb{R}$ is to be considered.

But if $C_X \approx \mathbb{R}$, then the proof is the same as the proof of Theorem 7.13 in the case 2.

§8. General theorems on distributions of class I_0

It is well known (see [LinO, Chapter VI, §4]) that for the group $X = \mathbb{R}$ the class I_0 is dense in the class of all infinitely divisible distributions and that any infinitely divisible distribution may be represented as a finite or infinite convolution of distributions from the class I_0 (see [O1]). In this section we give a complete description of groups X for which the analogs of these theorems are true.

First, consider groups on which the class I_0 is dense in the class of all infinitely divisible distributions. The description of such groups depends on whether the group X is discrete or not. To study the case when the group X is discrete we need the following lemmas.

8.1. LEMMA. *Let K be a compact subgroup of the group X. Then the distribution m_K may be approximated by generalized Poisson distributions.*

PROOF. Let n be a natural number. Denote $\Phi_n = nm_K$. By 2.14(i) we have

$$e(\hat{\Phi}_n)(y) = \exp\left\{\int_K [(x, y) - 1]\, d\Phi_n(x)\right\}$$
$$= \begin{cases} 1, & y \in A(Y, K), \\ \exp\{-n\}, & y \notin A(Y, K). \end{cases}$$

By 2.10(f) this implies that $e(\Phi)_n \Rightarrow m_K$. □

8.2. LEMMA. *In order that the group X contain a countable dense independent set, it is necessary and sufficient that any neighborhood of zero in X contain an element of infinite order.*

PROOF. The necessity is obvious. Let us prove the sufficiency. By Theorem 1.14, we may assume $X = \mathbb{R}^n + G$, where $n \geq 0$ and the group G contains an open compact subgroup K. Two cases are possible.

1. $n > 0$. Clearly, it is enough to construct a countable dense independent set on $\mathbb{R}$. For this, in turn, it suffices to find, for any independent set $\{x_1, \ldots, x_n\} \subset \mathbb{R}$ and any interval $(a, b) \subset \mathbb{R}$, a point $x_{m+1} \in (a, b)$ such that the set $\{x_1, \ldots, x_m, x_{m+1}\} \subset \mathbb{R}$ remains independent. Such a point evidently exists since the set of those $x \in \mathbb{R}$, for which the collection $\{x_1, \ldots, x_m, x\}$ is dependent, is countable.

2. $n = 0$. Let m be a natural number. Consider the subgroup $E_m(K) = \{x \in K: mx = 0\}$ of K, and verify that $E_m(K)$ is nowhere dense. Let the opposite be true. Since the subgroup $E_m(K)$ is closed, $E_m(K) \supset V$ for some open set V. Hence the subgroup $E_m(K)$ is open. Then the factor group $K/E_m(K)$ is discrete, and, since K is a compact group, $K/E_m(K)$ is finite. Therefore, there exists a natural number q such that $K^{(q)} \subset E_m(K)$. Hence $K = E_{mq}(K)$. We have obtained a compact periodic open subgroup of X. We arrived at a contradiction. Thus the subgroup $E_m(K)$ is nowhere dense.

Let us now prove that, for fixed m and z, the solutions of the equation

$$mx = z \tag{1}$$

form a set of the first category in X. To do this consider the factor group X/K. It is discrete and, since X is separable, X/K is countable. Therefore it suffices to prove that the set of all solutions of equation (1) belonging to one conjugacy class is a set of the first category. Let x_0 be a fixed solution of equation (1) and x a solution such that $x - x_0 \in K$. Then $m(x - x_0) = 0$, i.e., $x - x_0 \in E_m(K)$ and $x \in x_0 + E_m(K)$. Since we already know that the subgroup $E_m(K)$ is nowhere dense, the set of solutions of equation (1) belonging to one coset is also nowhere dense.

Let $\{x_1, \ldots, x_k\}$ be an independent set of elements of the group X. The linearly dependence of elements $x_1, \ldots, x_k, x$ means that x is a solution of the equation

$$m_1 x_1 + \cdots + m_k x_k + mx = 0 \tag{2}$$

for some integers $m_1, \ldots, m_k$ and natural m. The set of solutions of equation (2) is a set of the first category for any fixed set of integers $m_1, \ldots, m_k$, m and hence for all integers $m_1, \ldots, m_k, m$.

Consider an arbitrary neighborhood U in the group X. Since U is a set of the second category, one may always choose an element $x_{k+1} \in U$ such that the set $\{x_1, \ldots, x_k, x_{k+1}\}$ remains linearly independent.

By a standard construction one can easily obtain a countable independent set dense in X. $\square$

8.3. PROPOSITION. *Let the group X be nondiscrete. Then the class I_0 is dense in the class of all infinitely divisible distributions.*

PROOF. By Theorem 2.21 any infinitely divisible distribution μ may be represented as a convolution $\mu = m_K * \nu$, where K is some compact subgroup of X and the distribution ν is infinitely divisible and does not have non-degenerate idempotent divisors. By Lemma 8.1 the distribution m_K may be approximated by generalized Poisson distributions. Furthermore, as proved in [PRV1] (see also [He2, Chapter V]), the distribution ν may also be approximated by generalized Poisson distributions. The proposition will be proved if we verify that any generalized Poisson distribution $e(\Phi)$, $\Phi \in \mathcal{M}_+(X)$, may be approximated by a generalized Poisson distribution from the class I_0.

Since the measure Φ is regular [Gr, Chapter II, §2.1], it may be approximated by a measure $\Lambda \in \mathcal{M}_+(X)$ such that $\sigma(\Lambda)$ is a compact set. In its turn, the measure Λ may be approximated by a measure P that is concentrated at a finite set of points

$$P = \sum_{i=1}^{n} a_i E_{x_i}, \qquad a_i \geq 0, \ x_i \in X. \tag{1}$$

Let us check that a measure P of the form (1) may be approximated by a measure $\Delta \in \mathcal{M}_+(X)$ such that $e(\Delta) \in I_0$. Two cases are possible.

1. Every neighborhood of zero in the group X contains an element of infinite order.

By Lemma 8.2 the measure P may be approximated by measures of the form $\Delta = \sum_{i=1}^{n} a_i E_{z_i}$, where $\{z_1, \ldots, z_n\}$ is an independent set of points in X. By Theorem 6.5 $e(\Delta) \in I_0$.

2. There exists a neighborhood of zero in X that does not contain elements of infinite order.

Let us prove that the measure P may be approximated by a measure $\Delta \in \mathscr{M}_+(X)$ such that the measures Δ^{*p} and Δ^{*q} are mutually singular for any natural p, q, $p \neq q$. Then by Theorem 6.1 the distribution $e(\Delta)$ would belong to the class I_0. Let us proceed to construct such a measure Δ.

As was noted when proving Theorem 7.13, case 2b), the group X contains a compact periodic open subgroup K. Let us use Theorem 1.18 and represent the group K as a direct product

$$K = \mathsf{P}_{i=1}^{n} K_i,$$

where K_i are nondiscrete groups, $i = 1, \ldots, n$. Let U_i be a neighborhood of zero in K_i. By Theorem 6.15 for every $i = 1, \ldots, n$ there exists a distribution $\Phi_i \in \mathscr{M}^1(K_i)$ such that the distribution Φ_i^{*p} and $\Phi_i^{*q} * E_\zeta$ are mutually singular for any natural p, q, $p \neq q$, and any $\zeta \in K_i$, and $\sigma(\Phi_i) \subset U_i$. Set

$$\Delta = \sum_{i=1}^{n} a_i E_{x_i} * \Phi_i.$$

Clearly, P may be approximated by measures of the form. Let us check that Δ^{*p} and Δ^{*q} are mutually singular for any natural p, q, $p \neq q$. Since

$$\Delta^{*p} = \sum_{m_1 + \cdots + m_n = p} c_{m_1, \ldots, m_n} E_{m_1 x_1 + \cdots + m_n x_n} * \Phi_1^{*m_1} * \cdots * \Phi_n^{*m_n},$$

it suffices to verify that for $p \neq q$ and any $\eta \in X$ the measures $\Phi_1^{*m_1} * \cdots * \Phi_n^{*m_n}$ and $\Phi_1^{*l_1} * \cdots * Q_n^{*l_n} * E_\eta$ are mutually singular when $m_1 + \cdots + m_n = p$ and $l_1 + \cdots + l_n = q$. Two cases are possible.

a) $\eta \in K$. Then $\eta = \eta_1 + \cdots + \eta_n$, $\eta_i \in K_i$, $i = 1, \ldots, n$, and $\phi_1^{*l_1} * \cdots * \Phi_n^{*l_n} * E_\eta = (\Phi_1^{*l_1} * E_{\eta_1}) * \cdots * (\Phi_n^{*l_n} * E_{\eta_n})$. Since $p \neq q$, we have $m_{i_0} \neq l_{i_0}$ for some $i = i_0$ and then the measures $\Phi_{i_0}^{*m_{i_0}}$ and $\Phi_{i_0}^{*l_{i_0}} * E_{\eta_{i_0}}$ are mutually singular. But if μ_i are arbitrary finite measures on K_i, $i = 1, \ldots, n$, then

$$*_{i=1}^{n} \mu_i = \bigotimes_{i=1}^{n} \mu_i.$$

Since it is possible to choose disjoint sets on which the measures $\Phi_{i_0}^{*m_{i_0}}$ and $\Phi_{i_0}^{*l_{i_0}} * E_{\eta_{i_0}}$ are concentrated, the measures $\Phi_1^{*m_1} * \cdots * \Phi_n^{*m_n}$ and $\Phi_1^{*l_1} * \cdots * \Phi_n^{*l_n} * E_\eta$ will also be concentrated on disjoint sets and, therefore, mutually singular.

b) $\eta \notin K$. Then $\sigma(\Phi_1^{*m_1} * \cdots * \Phi_n^{*m_n}) \subset K$, while $K \cap \sigma(\Phi_1^{*l_1} * \cdots * \Phi_n^{*l_n} * E_\eta) = \varnothing$. Therefore the measures $\Phi_1^{*m_1} * \cdots * \Phi_n^{*m_n}$ and $\Phi_1^{*l_1} * \cdots * \Phi_n^{*l_n} * E_\eta$ are mutually singular as well. $\square$

Consider the case of a discrete group X. We shall need the following lemmas.

8.4. LEMMA. *Let the group* X *be discrete, and let* $\Phi, \Phi_n \in \mathcal{M}_+(X)$, $\Phi(\{0\}) = \Phi_n(\{0\}) = 0$, $n = 1, 2, 3, \ldots$, *and*

$$\text{(i)} \quad e(\Phi_n) * E_{x_n} \Rightarrow e(\Phi),$$

where x_n *is some sequence of elements of the group* X. *Then the sequence of the measures* $\{\Phi_n\}$ *is conditionally compact.*

PROOF. We first establish that $\sup_n \Phi_n(X) < \infty$. It it were not so, one could choose a subsequence $\{\Phi_{n_k}\}$ such that $\Phi_{n_k}(X) \geq 2k$, $k = 1, 2, 3, \ldots$. Let Δ_k, $k = 1, 2, 3, \ldots$, be distributions on X such that

$$\Delta_k(A) \leq \frac{1}{k} \Phi_{n_k}(A) \tag{1}$$

for any $A \in \mathcal{B}(X)$.

Since the distribution $\lambda_k = e(\Delta_k)$ is a divisor of $e(\Phi_{n_k})$, by Theorem 2.2 it follows from (i) that the sequence $\{\lambda_k\}$ is shift-compact. Let λ be a limit of shifts of the λ_k. It follows from (1) that any power of λ is a divisor of $e(\Phi)$. Therefore, the sequence $\{\lambda^n\}$ is shift-compact as well and any limit of shifts of the λ^n is a divisor of $e(\Phi)$. Any such limit is idempotent, and, since $e(\Phi)$ has no nondegenerate idempotent divisors, any such limit has to be degenerate. Hence the distribution λ itself is degenerate. Therefore $\lambda_k * \bar{\lambda}_k \Rightarrow E_0$. Denoting $X' = X \backslash \{0\}$ we have

$$e(\Delta_k + \bar{\Delta}_k)(X') \to 0$$

as $k \to \infty$. But on the other hand

$$e(\Delta_k + \bar{\Delta}_k)(X') = \exp\{-(\Delta_k + \bar{\Delta}_k(X))\}$$
$$\times \sum_{p=0}^\infty \frac{(\Delta_k + \bar{\Delta}_k)^{*p}}{p!}(X') \geq e^{-2} \Delta_k(X') = e^{-2},$$

which leads to a contradiction. Thus $\sup_n \Phi_n(X) = L < \infty$.

Set $P_n = \Phi_n + \bar{\Phi}_n$. Relation (i) implies $e(P_n) \Rightarrow e(\Phi + \bar{\Phi})$. Hence, for any $\varepsilon > 0$, there exists a compact set $C \subset X$ such that $e(P_n)(X \backslash C) < \varepsilon$, $n = 1, 2, \ldots$. Then

$$\varepsilon > e(P_n)(X \backslash C) \geq \exp\{-P_n(X)\} P_n(X \backslash C)$$
$$\geq \exp\{-2L\} P_n(X \backslash C) \geq \exp\{-2L\} \Phi_n(X \backslash C)$$

for all $n = 1, 2, 3, \ldots$. Using the Prokhorov theorem [He2, Chapter 1.1] we conclude that the sequence of measures $\{\Phi_n\}$ is conditionally compact. $\square$

8.5. **LEMMA.** *Let* $X = \mathbb{Z}$ *and* $\mu = e(E_1 + \frac{3}{2}E_3 + \frac{3}{2}E_4)$. *Then* $\mu \notin I_0$.

PROOF. Consider the polynomial

$$g(z) = 1 + 2z - z^2 + 3z^3 + 3z^4.$$

This polynomial has the following property: the polynomials $g^n(z)$, $n = 2, 3, 4, \ldots$, have nonnegative coefficients. For $n = 2$ and $n = 3$ this is checked directly:

$$g^2(z) = 1 + 4z + 2z^2 + 2z^3 + 19z^4 + 6z^5 + 3z^6 + 18z^7 + 9z^8,$$
$$g^3(z) = 1 + 6z + 9z^2 + 5z^3 + 36z^4 + 60z^5 + 8z^6 + 81z^7$$
$$+ 117z^8 + 27z^9 + 54z^{10} + 81z^{11} + 27z^{12}.$$

For $n > 3$ this follows from the fact than any $n > 3$ may be represented either as $n = 2q$ or as $n = 2q + 3$, where q is a natural number.

Let $a > 0$. Set

$$f_a(z) = \exp\{a(g(z) - 8)\} = e^{-8a} \sum_{n=0}^{\infty} \sum_{n=0}^{\infty} \frac{1}{n!} a^n g^n(z) = \sum_{k=0}^{\infty} c_k(a) z^k.$$

Clearly, all the coefficients $c_k(a)$, except maybe $c_2(a)$, are nonnegative and

$$\sum_{k=0}^{\infty} c_k(a) = f_a(1) = 1.$$

Since

$$c_2(a) = \frac{1}{2} f''_a(0) = \frac{1}{2} e^{-7a} \{a^2 g'^2(0) + a g''(0)\} = a e^{-7a}(2a - 1),$$

when $a \geq \frac{1}{2}$ we have $c_2(a) \geq 0$.

Therefore the following decomposition is valid:

$$\hat{\mu}(e^{it}) = \exp\left\{\frac{1}{2}(2e^{it} + 3e^{3it} + 3e^{4it} - 8)\right\}$$

$$= \exp\left\{\frac{1}{2}(e^{2it} - 1)\right\} \times \exp\left\{\frac{1}{2}(2e^{it} - e^{2it} + 3^{3it} + 3e^{4it} - 7)\right\}$$

$$= f_1(e^{it})f_2(e^{it}),$$

and the above-said shows that $f_2(e^{it})$ is a characteristic function. Hence $\mu = \mu_1 * \mu_2$, where $\mu_j \in \mathcal{M}^1(\mathbb{Z})$ and $\hat{\mu}_j(e^{it})$, $j = 1, 2$. Since the representation of the characteristic function of an infinitely divisible distribution on the group $X = \mathbb{Z}$ is unique (see Remark 2.22), we get that the distribution μ_2 is not infinitely divisible. By Theorem 4.3 $\mu_2 \notin I_0$ and hence $\mu \notin I_0$. $\square$

We can now answer the question of whether the class I_0 is dense in the class of all infinitely divisible distributions for discrete groups.

8.6. PROPOSITION. *Let X be a discrete group. The class I_0 is dense in the class of all infinitely divisible distributions if and only if $X \approx \mathbb{Z}(2)$.*

PROOF. Let $X \approx \mathbb{Z}(2)$, and let $\zeta \in X$ be an element of order two. Clearly, $\mathscr{M}^1(X) = \{m_X, e(aE_\zeta), e(aE_\zeta) * E_\zeta, \ a \geq 0\}$ and the assertion is obviously true in this case.

Assume that the class I_0 is dense in the class of all infinitely divisible distributions; we verify that the group X has no elements of infinite order. If an infinitely divisible distribution on a discrete group has no nondegenerate idempotent divisors, then it is a generalized Poisson distribution. This fact follows from §2.20, Proposition 5.5 and Theorem 2.21. It follows from Theorem 4.3 that, in particular, any distribution from the class I_0 is a degenerated Poisson distribution. Suppose

$$E(\Phi_n) * E_{x_n} \Rightarrow e(\Phi), \qquad n \to \infty, \tag{1}$$

where $\Phi_n, \Phi \in \mathscr{M}_+(X)$ and $\Phi_n(\{0\}) = \Phi(\{0\}) = 0$. By Lemma 8.4 the sequence $\{\Phi_n\}$ is conditionally compact. Therefore, there exists a measure $P \in \mathscr{M}_+(X)$ such that

$$\Phi_{n_k} \Rightarrow P, \qquad k \to \infty. \tag{2}$$

Hence

$$e(\Phi_{n_k}) \Rightarrow e(P), \qquad k \to \infty. \tag{3}$$

It follows from (1) and (3) that $e(P) = e(\Phi) * E_{x_0}$, for some element $x_0 \in X$. Remark 2.22 yields $\Phi(\{x\}) = P(\{x\})$ for every element $x \in X$ of infinite order.

Let ζ be an element of infinite order in X. Set $\Phi_0 = 6(E_\zeta + E_{3\zeta} + E_{4\zeta})$, and note that if $\Omega \in \mathscr{M}_+(X)$ and $\Omega(\{x\}) \geq \frac{1}{2}\Phi_0(\{x\})$ for $x \in \{\zeta, 3\zeta, 4\zeta\}$, then by Lemma 8.5 $e(\Omega) \notin I_0$.

Since the class I_0 is dense in the class of all infinitely divisible distributions, the distribution $e(\Phi_0)$, in particular, may be approximated by distributions from the class I_0. Hence there exist measures $\Phi_n \in \mathscr{M}_+(X)$, $\Phi_n(\{0\}) = 0$, and elements $x_n \in X$, $n = 1, 2, 3, \ldots$, such that $e(\Phi_n) \in I_0$ and $e(\Phi_n) * E_{x_n} \Rightarrow e(\Phi_0)$ as $n \to \infty$. As mentioned above, this implies relation (2) and the equality $P(\{x\}) = \Phi_0(\{x\})$ for $x \in \{\zeta, 3\zeta, 4\zeta\}$. On the other hand, (2) yields $\Phi_{n_k}(\{x\}) \to P(\{x\})$ as $k \to \infty$ for any $x \in X$. In other words, starting from some index k, we have $\Phi_{n_k}(\{x\}) \geq \frac{1}{2}\Phi_0(\{x\})$ when $x \in \{\zeta, 3\zeta, 4\zeta\}$ and, hence, $(\Phi_{n_k}) \notin I_0$. Thus X has no element, of infinite order. It follows from Propositions 6.6 and 6.11 that X has no elements of order $p > 2$ and no more than one element of second order. Hence $H \approx \mathbb{Z}(2)$. $\square$

Combining Proposition 8.3 and 8.6 we obtain the following result.

8.7. THEOREM. *On a group X the class I_0 is dense in the class of all infinitely divisible distributions if and only if the group X is either nondiscrete or $X \approx \mathbb{Z}(2)$.*

Let us describe those groups X on which the class I_0 forms some sort of a basis in the class of all infinitely divisible distributions.

8.8. THEOREM. *In order that any infinitely divisible distribution on a group X may be represented as a finite or infinite convolution of distributions of the class I_0, it is necessary and sufficient that X be isomorphic to the group $\mathbb{R}^n + D$, where $n \geq 0$ and D is a discrete group without elements of finite order $p > 2$.*

PROOF. *Necessity.* By Theorem 1.14 we may assume $X = \mathbb{R}^n + G$, where $n \geq 0$ and G contains a compact open subgroup K. Consider two cases.

1. The group K is discrete. Then so is the group G. Suppose that there is an element $\zeta \in G$ of finite order $p > 2$. Let $F = M(\zeta)$ be the subgroup generated by ζ. Clearly, $F \approx \mathbb{Z}(p)$. As was shown when proving Proposition 8.6, any distribution μ from the class I_0 on a discrete group is a generalized Poisson distribution. According to Proposition 6.6 this yields that the class I_0 on the group F consists either of the distributions of the form $e(aE_\eta) * E_x$, where $a \geq 0$ and $x \in F$ and η is the unique element of order two in F, if $p = 2n$, or of the distributions of the form E_x, where $x \in F$, if $p = 2n + 1$.

Let μ be an infinitely divisible distribution on F such that

$$\mu = *_{n=1}^{\infty} \mu_n, \tag{1}$$

where $\mu_n \in I_0(X)$. By Proposition 2.5 we have $\sigma(\mu_n * E_{x_n}) \subset F$ for some $x_n \in X$ and, hence, $\mu_n * E_{x_n} \in I_0(F)$, i.e., the distribution $\mu_n * E_{x_n}$ has the desired form. Clearly, not every infinitely divisible distribution on the group F is representable in the form (1). Therefore the group G has no elements of order $p > 2$, and the proof of the necessity in this case is complete.

2. The group K is nondiscrete. Let us prove that this case cannot occur. Choose a number a such that $0 < a < \ln \frac{3}{2}$ and consider the distribution $\mu = e(am_K)$ on the group K. Let (1) hold, where the distributions μ_n belong to $I_0(X)$, $n = 1, 2, \ldots$. By Proposition 2.5 for every n there exists an element $x_n \in X$ such that $\sigma(\mu_n * E_{x_n}) \subset K$ and, hence, $\mu_n * E_{x_n} \in I_0(K)$. According to Theorems 2.21 and 4.3 the characteristic function of the distribution μ_n admits the representation (z_n, Φ_n, φ_n), where $z_n \in X$ and Φ_n is a measure on $K \backslash \{0\}$. Since the function φ in representation 2.22(i) is unique, (1) implies $\varphi_n = 0$ for $n = 1, 2, 3, \ldots$.

Observe now that $\mu(\{0\}) > \frac{2}{3}$ and that the infinitely divisible distribution ν_n having the characteristic function $(0, \Phi_1 + \cdots + \Phi_n, 0)$ is a divisor of μ. By Lemma 4.13 we have

$$\Phi_1(X) + \cdots + \Phi_n(X) < \ln 2.$$

Hence the series $\sum_{n=1}^{\infty} \Phi_n$ is norm convergent. Denote its sum by Φ. It then follows from (1) that the distributions $E_{z_1 + \cdots + z_n}$ converge to some distribution and that $\mu = e(\Phi) * E_z$. But $\Phi(X) \leq \ln 2$, hence $\mu(\{z\}) \geq \frac{1}{2}$, which is impossible if $z \neq 0$. So $z = 0$, and then $e(am_K) = e(\Phi)$. Since $\|am_K\| + \|\Phi\| < 2\pi$, by Lemma 4.12 we have

$$am_K = \Phi = \sum_{n=1}^{\infty} \Phi_n.$$

The latter is impossible because by Lemma 6.21 $(\Phi_n)_{\mathrm{ac}} = 0$ for $n = 1, 2,$ $3, \ldots$. The necessity is proved completely.

Sufficiency. By Theorem 2.21 any infinitely divisible distribution on the group X may be represented as a convolution $\lambda = m_K * \mu$, where K is a compact subgroup of X and μ is an infinitely divisible distribution without nondegenerate generate idempotent divisors. Let $X = \mathbb{R}^n + D$. By hypothesis any nonzero subgroup K of the group X is finite and consists of elements of order two. By the structural Theorem 1.24, the subgroup K may be represented as a direct product

$$K = \mathsf{P}_{i=1}^{m} K_i,$$

where $K_i \approx \mathbb{Z}(2)$. Then $m_K = *_{i=1}^{m} m_{K_i}$. But the distribution $m_{\mathbb{Z}(2)}$ is representable as an infinite convolution

$$m_{\mathbb{Z}(2)} = *_{j=1}^{\infty} e(P_j),$$

where $P_j = E_\eta$, $j = 1, 2, 3, \ldots$, and η is an element in $\mathbb{Z}(2)$ of second order. Since $e(E_\eta) \in I_0$, the theorem is proved for the distribution m_K.

To prove the corresponding assertion for the distribution μ, we use the representation 2.22(i). Let γ be a distribution with the characteristic function $\hat{\gamma}(y) = \exp\{\varphi(y)\}$. Since $C_X = \mathbb{R}^n$, the Cramér theorem yields $E_{x_0} * \gamma \in I_0$.

Represent $X \setminus \{0\}$ as the union of disjoint sets

$$X \setminus \{0\} = \bigcup_{j=1}^{\infty} A_j, \qquad A_j \in \mathscr{B}(X),$$

such that the restriction Φ_j of Φ to A_j is a finite measure. By the properties of the function $g(x, y)$ defined in Lemma 2.19 and from Theorem 1.3 it follows that

$$\exp\left\{\int_{X \setminus \{0\}} [(x, y) - 1 - ig(x, y)] d\Phi_j(x)\right\}$$

$$= (x_j, y) \exp\left\{\int_{X \setminus \{0\}} [(x, y) - 1] d\Phi_j(x)\right\}.$$

Hence it suffices to prove the theorem for distributions of the form $\Delta = e(P)$, where $P \in \mathscr{M}_+(X)$. The proof is carried out by induction on n. Let $n = 0$, i.e., $X = D$. Then

$$P = \sum_{k=1}^{\infty} a_k E_{x_k},$$

where $a_k \geq 0$ and the elements x_k are either of infinite order or of order two. By virtue of Proposition 6.6 we have $e(a_k E_{x_k}) \in I_0$, which proves the assertion for $n = 0$.

Let $n > 0$. Then $X = \mathbb{R} + X_1$, where $X_1 = \mathbb{R}^{n-1} + D$. Represent the group $\mathbb{R}$ as the union

$$R = \bigcup_{k=-\infty}^{\infty} B_k,$$

where $B_0 = \{0\}$, $B_k = [c_k, d_k)$, $k > 0$; $B_k = (-d_{-k}, -c_{-k}]$, $k < 0$; $0 < d_k < 2c_k$, $k = 1, 2, 3, \ldots$. Denote by P_k the restriction of the measure P to $B_k + X_1$. Then

$$\Delta = *_{k=-\infty}^{\infty} \Delta_k,$$

where $\Delta_k = e(P_k)$. By Corollary 6.19, $\Delta_k \in I_0$ for $k \neq 0$ and by the inductive assumption the distribution Δ_0 may be represented as a finite or infinite convolution of distributions of the class I_0. The theorem is proved. $\square$

CHAPTER III

Characterization Problems

Throughout this chapter, unless stated otherwise, we assume that X is a locally compact abelian group satisfying the second axiom of countability.

§9. Bernstein's characterization of Gaussian distribution

The following theorem was proved by S. N. Bernstein. Let ξ and η be independent random variables with equal variance. For the random variables $\xi + \eta$ and $\xi - \eta$ to be independent as well, it is necessary and sufficient that both ξ and η be Gaussian. In this section we give a complete description of those groups X to which this characterization theorem may be extended. Some related subjects are considered as well.

9.1. DEFINITION. A distribution $\mu \in \mathcal{M}^1(X)$ is called Gaussian in the sense of Bernstein (B-Gaussian) if for any independent identically distributed random variables ξ and η with values in X and with distribution μ, the random variables $\xi + \eta$ and $\xi - \eta$ are independent.

The set of B-Gaussian distributions on the group X will be denoted $\Gamma_B(X)$ and the set of all symmetric distributions from $\Gamma_B(X)$ by $\Gamma_B^s(X)$.

9.2. PROPOSITION. *A distribution μ belongs to $\Gamma_B(X)$ if and only if its characteristic function satisfies the equation*

$$\text{(i)} \quad \hat{\mu}(y_1 + y_2)\hat{\mu}(y_1 - y_2) = \hat{\mu}(y_1)^2|\hat{\mu}(y_2)|^2$$

for any $y_1, y_2 \in Y$.

PROOF. As usual, by E we denote the mathematical expectation. By the definition $\mathsf{E}(\xi, y) = \mathsf{E}(\eta, y) = \hat{\mu}(y)$. Clearly, random variables $\xi + \eta$ and $\xi - \eta$ are independent if and only if

$$\mathsf{E}(\xi + \eta, y_1)(\xi - \eta, y_2) = \mathsf{E}(\xi + \eta, y_1)\mathsf{E}(\xi - \eta, y_2) \tag{1}$$

for any $y_1, y_2 \in Y$.

Taking into account the independence of the random variables ξ and η we transform the left- and right-hand sides of (1) as follows:

$$\mathsf{E}(\xi + \eta, y_1)(\xi - \eta, y_2) = \mathsf{E}(\xi, y_1 + y_2)(\eta, y_1 - y_2)$$
$$= \mathsf{E}(\xi, y_1 + y_2)\mathsf{E}(\eta, y_1 - y_2) = \hat{\mu}(y_1 + y_2)\hat{\mu}(y_1 - y_2);$$
$$\mathsf{E}(\xi + \eta, y_1)\mathsf{E}(\xi - \eta, y_2) = \mathsf{E}(\xi, y_1)(\eta, y_1)\mathsf{E}(\xi, y_2)(\eta, -y_2)$$
$$= \mathsf{E}(\xi, y_1)\mathsf{E}(\eta, y_1)\mathsf{E}(\xi, y_2)\mathsf{E}(\eta, -y_2) = \hat{\mu}(y_1)^2 |\hat{\mu}y_2)|^2.$$

Therefore (i) and (1) are equivalent. $\square$

It should be noted that (i) implies $\Gamma(X) \subset \Gamma_B(X)$; $\Gamma_B(X)$ may contain also idempotent distributions.

Denote $I_B(X) = I(X) \cap \Gamma_B(X)$. To describe the class $I_B(X)$ the following lemma is required.

9.3. LEMMA. *Let n be a natural number and G a closed subgroup of X. The following statements are equivalent:*

(i) $\overline{G(n)} = G$.
(ii) *If $ny \in A(Y, G)$, then $y \in A(Y, G)$.*

PROOF. (i) $\Rightarrow$ (ii). Let $ny \in A(Y, G)$, i.e., $(x, ny) = 1$ for all $x \in G$. Hence $(nx, y) = 1$ for all $x \in G$, i.e., $y \in A(Y, \overline{G^{(n)}})$. Since $\overline{G^{(n)}} = G$ by hypothesis, we obtain $y \in A(Y, G)$.

(ii) $\Rightarrow$ (i). If $y \in A(Y, \overline{G^{(n)}})$, then $(nx, y) = 1$ for all $x \in G$. Hence $(x, ny) = 1$ for all $x \in G$, i.e., $ny \in A(Y, G)$ and, by the hypothesis, $y \in A(Y, G)$. So, $A(Y, \overline{G^{(n)}}) \subset A(Y, G)$. Hence $A(Y, \overline{G^{(n)}}) = A(Y, G)$ and $\overline{G^{(n)}} = G$. $\square$

In what follows a special class of groups will be significant.

9.4. DEFINITION. A group X is called a Corwin group if $X^{(2)} = X$.

Here are some examples. The groups $X = \mathbb{T}$, $X = \mathbb{Z}(2k + 1)$, and $X = \mathbb{Z}(p^\infty)$ are Corwin groups, the groups $X = \mathbb{Z}$ and $X = \mathbb{Z}(2k)$ are not.

9.5. PROPOSITION. *Let K be a compact subgroup of a group X. The following statements are equivalent:*

(i) *K is a Corwin group.*
(ii) *If $2y \in A(Y, K)$, then $y \in A(Y, K)$.*
(iii) *$m_K \in I_B(X)$.*

PROOF. The equivalence of (i) and (ii) follows from Lemma 9.3 since $K^{(2)} = \overline{K^{(2)}}$.

(ii) $\Rightarrow$ (iii). According to Proposition 9.2 it suffices to verify that the characteristic function $\hat{m}_K(y)$ satisfies equation 9.2(i). Let us use representation 2.14(i). If $y_1, y_2 \in A(Y, K)$, then both sides of 9.2(i) are equal to one. If either $y_1 \in A(Y, K)$, $y_2 \notin A(Y, K)$ or $y_1 \notin A(Y, K)$, $y_2 \in A(Y, K)$, then $y_1 + y_2 \notin A(Y, K)$ and both sides of 9.2(i) vanish. Let $y_1, y_2 \notin A(Y, K)$. In this case the right-hand side of 9.2(i) vanishes.

Let the left-hand side of 9.2(i) differ from zero. By 2.14(i) this would mean that $y_1 + y_2$, $y_1 - y_2 \in A(Y, K)$. But then $2y_1 \in A(Y, K)$ and by (ii) $y_1 \in A(Y, K)$. We obtain a contradiction. Hence the left-hand side of 9.2(i) also vanishes.

(iii) $\Leftrightarrow$ (ii). By Proposition 9.2 the characteristic function $\hat{m}_k(y)$ satisfies equation 9.2(i). Let $2y \in A(Y, K)$ and set $y_1 = y_2 = y$ in 9.2(i). The left-hand side of 9.2(i) is then equal to one. Hence the right-hand side is also equal to one, i.e., $y \in A(Y, K)$. $\square$

In view of 2.10(d), Proposition 9.2 implies that $\Gamma_B(X)$ is a semigroup in $\mathscr{M}^1(X)$. Therefore,

$$I_B(X) * \Gamma(X) \subset \Gamma_B(X). \tag{1}$$

Our first task is to describe those groups X for which

$$I_B(X) * \Gamma(X) = \Gamma_B(X); \tag{2}$$

the description is given in Theorem 9.10. Observe that in view of 2.14 and Proposition 9.5 the equality (2) for a group X means the following: every distribution $\mu \in \Gamma_B(X)$ is invariant with respect to some compact Corwin subgroup $K \subset X$, and under the natural homomorphism $X \to X/K$ the distribution μ induces a Gaussian distribution on the factor group X/K.

To prove Theorem 9.10 we need several lemmas.

9.6. LEMMA. *Let* $\mu \in \Gamma_B(x)$ *and* $\hat{\mu}(y) \neq 0$ *for all* $y \in Y$. *Then the characteristic function* $\hat{\mu}(y)$ *admits the representation*

$$\text{(i)} \quad \hat{\mu}(y) = l(y)\exp\{-\varphi(y)\},$$

where $l(y)$ *is a continuous function,* $l(-y) = \overline{l(y)}$, $|l(y)| \equiv 1$, $l(0) = 1$, *and*

$$\text{(ii)} \quad l(y_1 + y_2) \cdot l(y_1 - y_2) = l^2(y_1)$$

for all $y_1, y_2 \in Y$; *the function* $\varphi(y)$ *is the same as in Definition 5.1.*

PROOF. By Proposition 9.2 $\hat{\mu}$ satisfies 9.2(i). Let $\nu = \mu * \overline{\mu}$. According to 2.10(d),(e) we have $\hat{\nu}(y) = |\hat{\mu}(y)|^2$, and the characteristic function $\hat{\nu}(y)$ also satisfies equation 9.2(i). Since $\hat{\nu}(-y) = \nu(y)$, 9.2(i) yields

$$\hat{\nu}(y_1 + y_2)\hat{\nu}(y_1 - y_2) = \hat{\nu}^2(y_1)\hat{\nu}^2(y_2)$$

for all $y_1, y_2 \in Y$. Taking logarithms of both sides of this equality and setting $\ln \nu(y) = -2\varphi(y)$, we see that the function $\varphi(y)$ is the same as in Definition 5.1 and $|\hat{\mu}(y)| = \exp\{-\varphi(y)\}$. Set $l(y) = \hat{\mu}(y)/|\hat{\mu}(y)|$. This function clearly is continuous, $l(-y) = \overline{l(y)}$, $|l(y)| \equiv 1$, $l(0) = 1$, and, in addition, the function $l(y)$ satisfies equation 9.2(i) which turns into (ii). $\square$

9.7. LEMMA. *Let* K *be a connected compact infinite-dimensional group that does not contain a subgroup topologically isomorphic to* $\mathbb{T}$. *Let* $D = K^*$, *and let* $f: D \to \mathbb{R}_0^\infty$ *be the monomorphism constructed in Proposition 5.9.*

Then, for any elements a_1, $a_2 \in D$, there exists a subgroup $B \subset D$ of a finite rank l such that a_1, $a_2 \in B$ and $\overline{f(B)} \approx \mathbb{R}^l$.

PROOF. Let us observe first that for any n there exists a number p such that $\mathbb{R}^n \subset \overline{f(D)} \cap \mathbb{R}^p$. Indeed, let a set $E = \{x_1, \ldots, x_k\} \subset \mathbb{R}^n$ be such that the group $M(E)$ it generates is dense in $\mathbb{R}^n$. By Lemma 5.17 each x_j, $j = 1, \ldots, k$, is the limit of some sequence $u_i^{(j)} \in f(D)$. By the definition of the topology in $\mathbb{R}_0^\infty$ all $u_i^{(j)}$, $j = 1, \ldots, k$, $i = 1, 2, 3, \ldots$, belong to some $\mathbb{R}^p$ and converge there. Therefore $x_j \in \overline{f(D) \cap \mathbb{R}^p}$, $j = 1, \ldots, k$, and, hence, $\mathbb{R}^n \subset \overline{f(D) \cap \mathbb{R}^p}$.

By Theorem 1.23(a) we have $f(D) \cap \mathbb{R}^p = G + F$, where $G \approx \mathbb{R}^{t_p}$, $F \approx \mathbb{Z}^{s_p}$, $t_p + s_p = p$. Therefore $\mathbb{R}^n \subset G$.

Let $f(a_1)$, $f(a_2) \in f(D) \cap \mathbb{R}^n$. Set $A = f(D) \cap G$ and $B = f^{-1}(A)$. Since $r(B) = r(A) = t_p$, B is the desired subgroup. $\square$

9.8. LEMMA. *Let the group X not contain a subgroup isomorphic to $\mathbb{T}^2$, $\mu \in \Gamma_B(X)$ and $\hat{\mu}(y) \neq 0$ for all $y \in Y$. Then $\mu \in \Gamma(X)$.*

PROOF. According to 2.10(d),(e), Lemma 9.6 implies that $\nu = \mu * \overline{\mu} \in \Gamma^s(X)$. By Proposition 5.5 we have $\sigma(\nu) = X_1$, where X_1 is a connected subgroup of X. According to Proposition 2.5, the distribution μ, being a divisor of ν, up to a shift is concentrated on the subgroup X_1. Therefore from the very beginning one may assume X to be connected. Two cases are possible.

1. No subgroup of X is topologically isomorphic to $\mathbb{T}$. Then by Theorem 5.22, $\mu \in \Gamma(X)$.

2. The group X contains a subgroup F topologically isomorphic to $\mathbb{T}$. By Proposition 1.22 the subgroup F is a topological direct summand of X, i.e., $X = F + G$, where G is a connected group without a subgroup topologically isomorphic to $\mathbb{T}$, since, by the hypothesis, X has no subgroup topologically isomorphic to $\mathbb{T}^2$. The group Y admits the representation $Y = L + H$, where L is topologically isomorphic to $\mathbb{Z}$ and $H \approx G^*$. Combining Theorem 1.15 and 1.11 and Corollary 1.10 we get $H \approx \mathbb{R}^n + D$, where D is a discrete torsion-free group. Let us assume $H \approx D$ (the general case is quite similar) and also $\dim X = \infty$. The assumption $\dim X < \infty$ only simplifies the reasoning.

According to Lemma 9.6 the characteristic function $\hat{\mu}(y)$ admits representation 9.6(i). To prove the lemma we have to verify that

$$l(y_1 + y_2) = l(y_1)l(y_2) \tag{1}$$

for all y_1, $y_2 \in Y$, i.e., that $l(y)$ is a character of the group Y.

Denote the generator of the group L by ζ and by $(n\zeta, h)$, $n \in \mathbb{Z}$ and $h \in H$, the elements of the group Y.

Setting $y_1 = y_2$ in equation 9.6(ii) and according for $l(0) = 1$, we obtain

$$l(2y) = l^2(y). \tag{2}$$

Together with 9.6(ii) this yields by induction $l(ny) = l^n(y)$, $y \in Y$, $n \in \mathbb{Z}$. Hence the function $l(n\zeta, 0)$ satisfies equation (1). On the other hand, in view of Proposition 9.2 the function $\hat{\mu}(y)$ satisfies equation 9.2(i) on H. Since, by 2.10(h), $\hat{\nu}(0, h)$ is the characteristic function of some distribution on the group G, this distribution obviously belongs to $\Gamma^s(G)$ and the group G does not contain a subgroup topologically isomorphic to $\mathbb{T}$. Using Theorem 5.22, we obtain that $\hat{\mu}(0, h)$ also is the characteristic function of some distribution from $\Gamma(G)$, and hence $l(0, h)$ satisfies equation (1).

Consider the function

$$\psi(n\zeta, h) = l(n\zeta, 0) \cdot l(0, h)$$

on the group Y. By the aforementioned this function satisfies equation (1). Set

$$\alpha(n\zeta, h) = \frac{l(n\zeta, h)}{\psi(n\zeta, h)}.$$

Clearly, the function $\alpha(n\zeta, h)$ satisfies equation 9.6(ii). Let us check that $\alpha(n\zeta, h) \equiv 1$. By construction, $\alpha(n\zeta, 0) = \alpha(0, h) = 1$ for all $n \in \mathbb{Z}$, $h \in H$. Setting $y_1 = (n\zeta, 0)$ and $y_2 = (n\zeta, h)$ in equation 9.6(ii) we obtain

$$\alpha(2n\zeta, h)\alpha(0, -h) = \alpha^2(n\zeta, 0).$$

Therefore $\alpha(2n\zeta, h) = 1$ for all $n \in \mathbb{Z}$, $h \in H$. In particular, $\alpha(2nz, 2h) \equiv 1$. But by (2) we have $\alpha(2n\zeta, 2h) = \alpha^2(n\zeta, h)$. Hence we have either $\alpha(n\zeta, h) = 1$ or $\alpha(n\zeta, h) = -1$ for any $(n\zeta, h) \in Y$.

Suppose $\alpha(n_0\zeta, h_0) = -1$ for some $n_0 \in \mathbb{Z}$, $h_0 \in H$. Let $d_1, \ldots, d_l, \ldots$ be a maximal independent system in H, and let $f: H \to \mathbb{R}_0^\infty$ be the monomorphism constructed in Proposition 5.9. The group G satisfies the condition of Lemma 9.7. Therefore, we may choose a subgroup $B \subset H$ of finite rank l such that $h_0 \in B$ and $\overline{f(B)} \approx \mathbb{R}^l$. Let π be a topological isomorphism $\pi: \overline{f(B)} \to \mathbb{R}^l$, and let $f_1 = \pi \circ f$. Then $\overline{f_1(B)} = \mathbb{R}^l$.

Consider the point $s_0 = f_1(h_0)/2 \in \mathbb{R}^l$ and choose an element $h_1 \in B$ such that

$$\|s_0 - f_1(h_1)\| < \frac{\varepsilon}{2}$$

(by $\|\cdot\|$ we denote the norm of a vector from $\mathbb{R}^l$). Setting $y_1 = (n_0\zeta, h_0 - h_1)$, $y_2 = (0, h_1)$ in equation 9.6(ii) and recalling that $\alpha^2(n\zeta, h) \equiv 1$ we obtain

$$\alpha(n_0\zeta, h_0)\alpha(n_0\zeta, h_0 - 2h_1) = 1,$$

i.e., $\alpha(n_0\zeta, h_0 - 2h_1) = -1$ and also

$$\|f_1(h_0 - 2h_1)\| = \|f_1(h_0) - 2f_1(h_1)\| = 2\|s_0 - f_1(h_1)\| < \varepsilon.$$

So, for any $\varepsilon > 0$, there exists an element $(h_0\zeta, h_2) \in L + B$, $h_2 = h_0 - 2h_1$, such that $\alpha(n_0\zeta, h_2) = -1$ and $\|f_1(h_2)\| < \varepsilon$.

Observe that the function $\alpha(n\zeta, h)\exp\{-\varphi(n\zeta, h)\}$ is characteristic and denote by $g(n\zeta, h)$ its restriction to the subgroup $L + B$. According to 2.10(h), the function $g(n\zeta, h)$ also is characteristic. It follows from the proof of Proposition 5.6 that the function $\varphi(n\zeta, h)$ admits the following representation on $L + B$:

$$\varphi(n\zeta, h) = an^2 + 2n\langle B, f_1(h)\rangle + \langle Af_1(h), f_1(h)\rangle, \tag{3}$$

where $a \geq 0$, $B \in \mathbb{R}^l$, and $A = (\alpha_{ik})^l_{i, k=1}$ is a symmetric positive semidefinite matrix.

Applying the inequality from 2.10(c) to the characteristic function $g(n\zeta, h)$ and setting $y_1 = (n_0\zeta, h_2)$, $y_2 = (n_0\zeta, 0)$ we obtain

$$|g(n_0\zeta, h_2) - g(n_0\zeta, 0)|^2 \leq 2(1 - g(0, h_2))$$

or, taking account of (3),

$$|-\exp\{-(an_0^2 + 2n_0\langle B, f_1(h_2)\rangle + \langle Af_1(h_2), f_1(h_2)\rangle)\} - \exp\{-an_0^2\}|^2$$
$$\leq 2(1 - \exp\{-\langle Af_1(h_2), f_1(h_2)\rangle\}).$$

The latter is impossible if the norm $\|f_1(h_2)\|$ is small enough. So, $\alpha(n\zeta, h) \equiv 1$ and, hence, the function $l(n\zeta, h) \equiv \psi(n\zeta, h)$ satisfies equation (1). $\square$

9.9. LEMMA. *There exists a distribution μ_0 on the group $X = \mathbb{T}^2$ such that $\mu_0 \in \Gamma_B(X)$, $\mu_0 \notin \Gamma(X)$, and $\hat{\mu}_0(y) \neq 0$ for every $y \in Y$.*

PROOF. Denote by (n, m), $n, m \in \mathbb{Z}$, the elements of the group $Y \approx \mathbb{Z}^2$, and consider the function

$$l(n, m) = \begin{cases} 1 & \text{if } nm \in \mathbb{Z}^{(2)}, \\ -1 & \text{if } nm \notin \mathbb{Z}^{(2)} \end{cases}$$

on this group. Let us check that this function satisfies equation 9.2(i). Let $y_1 = (n_1, m_1)$, $y_2 = (n_2, m_2)$. Then $y_1 + y_2 = (n_1 + n_2, m_1 + m_2)$, $y_1 - y_2 = (n_1 - n_2, m_1 - m_2)$, and it suffices to see that the values $l(y_1 + y_2)$ and $l(y_1 - y_2)$ have the same sign. But, clearly, $(n_1 + n_2)(m_1 + m_2) = (n_1 - n_2)(m_1 - m_2) + 2(n_1m_2 + n_2m_1)$, which implies the desired statement.

Let a number $a > 0$ be chosen in such a way that

$$\sum_{(n, m)\in Y, \; (n, m)\neq(0, 0)} \exp\{-a(n^2 + m^2)\} < 1. \tag{1}$$

Consider the function

$$p(t, s) = \sum_{(n, m)\in Y} l(n, m)\exp\{-a(n^2 + m^2) - i(nt + ms)\}$$

on the group X. By construction, $\rho(t, s) > 0$ and, hence, the distribution $\mu_0 \in \mathcal{M}^1(X)$ with the density ρ with respect to m_X has the characteristic function

$$\hat{\mu}_0(n, m) = l(n, m)\exp\{-a(n^2 + m^2)\}.$$

This function, clearly, satisfies equation 9.2(i) and by Proposition 9.2 we get $\mu_0 \in \Gamma_B(X)$. By construction, $\mu_0 \notin \Gamma(X)$ (since the function l is not a character) and $\hat{\mu}_0(y) \neq 0$ for any $y \in Y$. $\square$

9.10. THEOREM. *The equality* (9.5.2) *holds on a group* X *if and only if*

(i) *for any compact Corwin subgroup* $K \subset X$ *the factor-group* X/K *does not contain a subgroup topologically isomorphic to* $\mathbb{T}^2$.

(Below in Proposition 9.11 we give an easily checked sufficient condition for (i) to be fulfilled.)

PROOF. *Necessity.* Suppose that there is a compact Corwin subgroup $K \subset X$ such that the factor-group X/K contains a subgroup topologically isomorphic to $\mathbb{T}^2$. This allows one to treat the distribution μ_0 constructed in Lemma 9.9 as a distribution on X/K. Then $\mu_0 \in \Gamma_B(X/K)$ and $\mu_0 \notin \Gamma(X/K)$ and also $\hat{\mu}_0(h) \neq 0$ for $h \in (X/K)^*$. By Theorem 1.6, $(X/K)^* \approx A(Y/K)$ and we may consider the function $\hat{\mu}_0(h)$ to be defined on $A(Y, K)$.

Consider the function

$$f(y) = \begin{cases} \hat{\mu}_0(y), & y \in A(Y, K), \\ 0, & y \notin A(Y, K), \end{cases} \tag{1}$$

on the group Y. According to Theorem 2.9 and Remark 2.9 this function is positive definite. Since the group K is compact, by Theorem 1.7 the subgroup $A(Y, K)$ is open and, hence, the function $f(y)$ is continuous. It follows from Theorem 2.8 that there exists a distribution $\gamma \in \mathcal{M}^1(X)$ such that $\hat{\gamma}(y) = f(y)$. Let us verify that $\gamma \in \Gamma_B(X)$. According to Proposition 9.2 it suffices to prove that $f(y)$ satisfies equation 9.2(i). If $y_1, y_2 \in A(Y, K)$, then this equation is satisfied since $f(y) = \hat{\mu}_0(y)$ on $A(Y, K)$ and $\mu_0 \in \Gamma_B(X/K)$. If either $y_1 \in A(Y, K)$ and $y_2 \notin A(Y, K)$ or $y_1 \notin A(Y, K)$ and $y_2 \in A(Y, K)$, then $y_1 + y_2 \notin A(Y, K)$ and both sides of 9.2(i) vanish. If $y_1, y_2 \notin A(Y, K)$, then either $y_1 + y_2 \notin A(Y, K)$ or $y_1 - y_2 \notin A(Y, K)$, since otherwise $2y_1 \in A(Y, K)$ and by Proposition 9.5 $y_1 \in A(Y, K)$. Thus both sides of equation 9.2(i) vanish again. We have obtained that this equation is valid for all $y_1, y_2 \in Y$, i.e., $\gamma \in \Gamma_B(X)$.

Let us now check that $\gamma \notin I_B(X) * \Gamma(X)$. Suppose the contrary is true, i.e.,

$$\gamma = m_{K_1} * \gamma_0, \tag{2}$$

where K_1 is a compact Corwin subgroup and $\gamma_0 \in \Gamma(X)$. Since $\hat{\gamma}_0(y) \neq 0$ for all $y \in Y$, relation (2) together with 2.14(i) yields $\{y \in Y : \hat{\gamma}(y) \neq 0\} = A(Y, K_1)$. On the other hand since $\hat{\mu}_0(y) \neq 0$ for $y \in A(Y, K)$, (1) implies $\{y \in Y : \hat{\gamma}(y) \neq 0\} = A(Y, K)$. Hence $A(Y, K_1) = A(Y, K)$ and 1.5(i) gives $K_1 = K$. According to (2) we then obtain that the restriction of $\hat{\gamma}(y)$ to $A(Y, K)$ is the characteristic function of some distribution $\lambda \in \Gamma(X/K)$, but this is impossible since $\lambda = \mu_0 \notin \Gamma(X/K)$.

Sufficiency. Let $\mu \in \Gamma_B(X)$. Consider the set $N = \{y \in Y : \hat{\mu}(y) \neq 0\}$. Since by Proposition 9.2 the characteristic function $\hat{\mu}(y)$ satisfies equation 9.2(i), N is an open subgroup of Y. Set $K = A(X, N)$. Then by Theorem 1.7 K is a compact group. By equation 9.2(i) if $2y \in N$, then $y \in N$ and Lemma 9.3 implies that K is a Corwin group. By hypothesis, the factor-group X/K does not contain a subgroup topologically isomorphic to $\mathbb{T}^2$. Then, according to §2.10(h), the restriction of the characteristic function $\hat{\mu}(y)$ to N is the characteristic function of some distribution $\nu_0 \in \mathscr{M}^1(X/K)$. By Proposition 9.2 $\nu_0 \in \Gamma_B(X/K)$ and by Lemma 9.8 $\nu_0 \in \Gamma(X/K)$. Hence

$$\hat{\nu}_0(y) = ([x], y)\exp\{-\varphi_0(y)\}, \qquad y \in N,$$

where $[x] \in X/K$ and the function $\varphi_0(y)$ is the same as in Definition 5.1. By Proposition 5.4 the function φ_0 may be extended with preservation of its properties to a function φ on Y. Consider now the distribution ν on X with the characteristic function

$$\hat{\nu}(y) = (x, y)\exp\{-\varphi(y)\}.$$

Then $\nu \in \Gamma(X)$ and $\mu = m_K * \nu$. $\quad \square$

We now present a simple sufficient condition for condition 9.10(i) to be fulfilled.

9.11. PROPOSITION. *For a group X to satisfy condition 9.10(i) it is sufficient that the following condition be fulfilled:*

(i) *Either $\overline{Y^{(2)}} = Y$ or $Y/\overline{Y^{(2)}} \approx \mathbb{Z}(2)$.*

PROOF. Let us check that 9.11(i) implies 9.10(i). Suppose that a factor-group X/K, where K is some compact Corwin subgroup, contains a subgroup F topologically isomorphic to $\mathbb{T}^2$. By Proposition 1.22 the subgroup F is a topological direct summand in X/K. Since, by Theorem 1.6, $(X/K)^* \approx A(Y, K)$, the group $A(Y, K)$ contains a subgroup G, $G \approx \mathbb{Z}^2$, as a topological direct summand, i.e., $A(Y, K) = H + G$. Denote the elements of the group $A(Y, K)$ by $(h; n, m)$, $h \in H$, $n, m \in \mathbb{Z}$. If $2y \in A(Y, K)$, then by Proposition 9.5 $y \in A(Y, K)$. Let $2y = (h; n, m)$. Then $y = (h_1; n/2, m/2)$. Therefore, if $(n, m) \notin G^{(2)}$, then $(h; n, m) \notin Y^{(2)}$. In particular, $(0; n, m) \notin Y^{(2)}$. Since the group K is compact, by Theorem 1.7 the subgroup $A(Y, K)$ is open and, hence, $(0; n, m) \notin \overline{Y^{(2)}}$. But this contradicts condition 9.11(i) since for the natural homomorphism $\tau: Y \to Y/\overline{Y^{(2)}}$ we have $\tau(0; 0, 1) \neq 0$, $\tau(0; 1, 0) \neq 0$, $\tau(0; 0, 1) \neq \tau(0; 1, 0)$. $\quad \square$

9.12. REMARK. Condition 9.11(i) is not necessary for 9.10(i) to be fulfilled. Indeed, let G be the subgroup of all finite order elements of $\mathbb{T}$. Endow this group with the discrete topology and set $X = G^2$, $H = G^*$. One can easily see that $H/\overline{H^{(2)}} \approx \mathbb{Z}(2)$, and hence $Y/\overline{Y^{(2)}} \approx (\mathbb{Z}(2))^2$, i.e., condition 9.11(i) is not fulfilled. On the other hand, any factor-group X/K of the group X consists of elements of finite order; hence, 9.10(i) is true for this group.

9.13. PROPOSITION. *In order that*

$$\text{(i)} \quad \Gamma(X) = \Gamma_B(X)$$

for a group X, *it is necessary and sufficient that the only compact Corwin subgroup of* X *be the subgroup* $K = \{0\}$.

PROOF. *Necessity* is obvious.

Sufficiency. Since $\mathbb{T}^2$ is a compact Corwin group, no factor-group X/K, where K is a compact Corwin group (unique in this case and coinciding with X), contains a subgroup topologically isomorphic to $\mathbb{T}^2$. By Theorem 9.10 equality 9.5(b) is valid. Hence (i) is valid as well. $\square$

9.14. REMARK. Let $\mu \in \Gamma_B(X)$ be an infinitely divisible distribution. Then $\mu \in I_B(X) * \Gamma(X)$. To prove this statement one may argue in the same way as when proving sufficiency in Theorem 9.10. The following fact is used instead of Lemma 9.8: if $\gamma \in \Gamma_B(X)$ is an infinitely divisible distribution and $\hat{\gamma}(y) \neq 0$ for all $y \in Y$, then $\gamma \in \Gamma(X)$. This fact is a simple consequence of Lemma 9.6 and Proposition 5.21.

Denote by $I_B^s(X)$ the set of all symmetric distributions from $I_B(X)$. We may extend Theorem 9.10 as follows.

9.15. PROPOSITION. *In order that*

$$\text{(i)} \quad I_B^s(X) * \Gamma^s(X) = \Gamma_B^s(X)$$

for a group X, *it is necessary and sufficient that* X *satisfy condition* 9.10(i).

PROOF. *Necessity.* Consider the distribution $\gamma \in \Gamma_B(X)$, $\gamma \notin I_B(X) * \Gamma(X)$ constructed when proving necessity in Theorem 9.10. By 2.10(c),(f) this distribution is symmetric since $\hat{\gamma}(y) = \overline{\hat{\gamma}(y)}$.

Sufficiency. Let $\mu \in \Gamma_B^s(X)$. Theorem 9.10 yields $\mu = m_K * \nu$ where $m_K \in I_B(X)$ and $\nu \in \Gamma(X)$. Then $\nu = E_x * \nu_1$, where $\nu_1 \in \Gamma^s(X)$. If $\mu = \overline{\mu}$, then, by 2.10(e), the restriction of the characteristic function $\hat{\nu}(y)$ to $A(Y, K)$ is a real-valued function. Hence the restriction of the character (x, y) to $A(Y, K)$ is real valued as well. Therefore,

$$\lambda = m_K * E_x \in I_B^s(X), \qquad \mu = \lambda * \nu_1 \in I_B^s(X) * \Gamma^s(X). \quad \square$$

It follows from Definition 5.1 that $D(X) * \Gamma^s(X) = \Gamma(X)$. It turns out that there is an analogous result for B-Gaussian distributions.

9.16. PROPOSITION. *The following equality takes place*:

$$\text{(i)} \quad D(X) * \Gamma_B^s(X) = \Gamma_B(X).$$

PROOF. By Proposition 9.2 the inclusion $\mu \in \Gamma_B(X)$ is equivalent to the fulfillment of equation 9.2(i) for the characteristic function $\hat{\mu}(y)$. Our assertion will be proved if an element $x \in X$ could be found such that the function $\hat{\mu}(y)/(x, y)$ is real valued. As in the proof of the sufficiency part in Theorem 9.10, consider the set $N = \{y \in Y : \hat{\mu}(y) \neq 0\}$. One can easily see that it is enough to consider the case $\hat{\mu}(y) \neq 0$ for all $y \in Y$. Then it follows from Lemma 9.6 that the distribution μ is a divisor of the distribution $\nu = \mu * \overline{\mu} \in \Gamma^s(X)$. According to Proposition 5.5 $\sigma(\nu) = X_1$, where X_1 is a connected subgroup of the group X. By Proposition 2.5 the distribution μ, being a divisor of ν, also is concentrated up to a shift on the subgroup X_1. Therefore, we may assume that X is a connected group.

By Lemma 9.6 the characteristic function $\hat{\mu}(y)$ admits representation 9.6(i). Replacing y_1 by y_2 in 9.6(ii) we obtain

$$l(y_1 + y_2)l(y_2 - y_1) = l^2(y_2) \tag{1}$$

for all $y_1, y_2 \in Y$. Then multiplying 9.6(ii) and (1) we obtain, with account of $l(-y) = \overline{l(y)}$, that

$$l^2(y_1 + y_2) = l^2(y_1)l^2(y_2) \tag{2}$$

for all $y_1, y_2 \in Y$, i.e., $l^2(y)$ is a character of the group Y. Together with 9.8(2) this implies that the restriction of $l(y)$ to the subgroup $Y^{(2)}$ satisfies equation 9.8(1). Since X is a connected group, Theorem 1.15 and Corollary 1.10 imply that $Y^{(2)}$ is a closed subgroup of Y. By Theorem 1.6 there exists an element $x \in X$ such that $l(y) = (x, y)$ for $y \in Y^{(2)}$. Set $\alpha(y) = l(y)/(x, y)$. Then $\alpha(2y) = 1$ for each $y \in Y$. Clearly, the function $\alpha(y)$ satisfies equation 9.8(2). Therefore, $1 = \alpha(2y) = \alpha^2(y)$. This implies $\alpha(y) = \pm 1$, i.e., $\alpha(y)$ is a real-valued function. Therefore, the function $\hat{\mu}(y)/(x, y)$ is real valued as well. $\square$

Consider now the following problem. Let ξ and η be X-valued independent random variables (not necessarily identically distributed, in contrast to Theorem 9.10), and such that the random variables $\xi + \eta$ and $\xi - \eta$ are also independent. For what group X does this fact imply that the distributions of the random variables ξ and η belong to $I_B(X) * \Gamma(X)$? A complete description of such groups is given in Theorem 9.19. Before passing to this theorem let us prove some auxiliary statements.

9.17. PROPOSITION. *Let ξ and η be X-valued independent random variables, X a group, with distributions μ and ν. For the random variables $\xi + \eta$ and $\xi - \eta$ to be independent, it is necessary and sufficient that the characteristic functions $\hat{\mu}(y)$ and $\hat{\nu}(y)$ satisfy the equation*

(i) $\hat{\mu}(y_1 + y_2)\hat{\nu}(y_1 - y_2) = \hat{\mu}(y_1)\hat{\nu}(y_1)\hat{\mu}(y_2)\hat{\nu}(-y_2)$

for all $y_1, y_2 \in Y$.

The proof of this proposition is similar to that of Proposition 9.2 and we omit it.

9.18. PROPOSITION. *Let ξ and η be X-valued random variables, X a group, with distributions μ and ν. If the random variables $\xi + \eta$ and $\xi - \eta$ are independent, then $\sigma(\mu) \subset x_\mu + X_1$, $\sigma(\nu) \subset x_\nu + X_1$, where X_1 is a subgroup of X and X_1 is topologically isomorphic to the group $\mathbb{R}^n + K$, where $n \geq 0$ and K is a compact Corwin group.*

PROOF. By Theorem 1.14 we may assume $X = \mathbb{R}^n + G$, where $n \geq 0$ and the group G contains a compact open subgroup. Then $Y = \mathbb{R}^n + H$, $H \approx G^*$. Denote by L a compact open subgroup of H.

Set $N_1 = \{y \in Y : \hat{\mu}(y) \neq 0\}$, $N_2 = \{y \in Y : \hat{\nu}(y) \neq 0\}$, and $N = N_1 \cap N_2$. Proposition 9.17 provides the fulfillment of 9.17(i). Therefore, N is a subgroup of Y; clearly N is open. Consider the intersection $B = N \cap L$. Since every open subgroup is closed, B is a compact open subgroup of H. Setting $y_1 = y_2 = y$ and $y_1 = -y_2 = y$ in 9.17(i) we obtain

$$\hat{\mu}(2y) = \hat{\mu}^2(y)|\hat{\nu}(y)|^2, \qquad \hat{\nu}(2y) = |\hat{\mu}(y)|^2 \hat{\nu}^2(y), \tag{1}$$

which implies

$$|\hat{\mu}(2^m y)| = |\hat{\mu}(y)\hat{\nu}(y)|^{2^{2m-1}}, \qquad y \in Y. \tag{2}$$

Let $y \in B$. Since B is compact, there exists a convergent subsequence $\{2^{m_i}y\}$. Let $2^{m_i}y \to y_0 \in B$. Equality (2) yields

$$|\hat{\mu}(y_0)| = \lim_{i \to \infty} |\hat{\mu}(2^{m_i}y)| = \lim_{i \to \infty} |\hat{\mu}(y)\hat{\nu}(y)|^{2^{2m_i-1}}. \tag{3}$$

If $|\hat{\mu}(y)\hat{\nu}(y)| < 1$, then the limit in the right-hand side of equality (3) is zero, but this is impossible since $y_0 \in B \subset N$. Hence $|\hat{\mu}(y)| = |\hat{\nu}(y)| = 1$ for every $y \in B$. According to 2.10(i) and Theorem 1.6, one may replace the distributions μ and ν by their shifts μ_1 and ν_1 such that $\hat{\mu}_1(y) = \hat{\nu}_2(y) = 1$ for all $y \in B$. Clearly, the characteristic functions $\hat{\mu}_1(y)$ and $\hat{\nu}_1(y)$ still satisfy equation 9.17(i). It follows from Proposition 2.13 that $\sigma(\mu_1)$, $\sigma(\nu_1) \subset A(X, B)$. Taking into account Theorem 1.6 we obtain $A(X, B) \approx (Y/B)^* = ((\mathbb{R}^n + H)/B)^* \approx \mathbb{R}^n + (H/B)^* = \mathbb{R}^n + F$, where $F = (H/B)^*$. Since the subgroup B is open in H, the factor-group H/B is discrete. By Theorem 1.7, the group F is compact.

So, let now $X = \mathbb{R}^n + F$, where F is a compact group, and let μ_1 and ν_1 be distributions on X whose characteristic functions satisfy equation 9.17(i). Let $D = F^*$. By Theorem 1.7 D is a discrete group. Denote by D_2 the subgroup of D consisting of elements whose orders are powers of two. For any $y \in D_2$ there exists a natural m such that $2^m y = 0$. Therefore, (2) yields $|\hat{\mu}_1(y)| = |\hat{\nu}_1(y)| = 1$. Using 2.10(i) and Proposition 2.13 we obtain that $\sigma(\mu_2)$, $\sigma(\nu_2) \subset A(X, D_2)$ for some shifts μ_2 and ν_2 of the distributions μ_1 and ν_1. Set $X_1 = A(X, D_2)$. One can easily see that $X_1 \approx \mathbb{R}^n + K$, where K is a compact group. By Theorem 1.6 $X_1^* \approx Y/D_2$ and, clearly, the group Y/D_2 is free of elements of order two. According to Lemma 9.3, K is a Corwin group. Returning to the distributions μ and ν we obtain the desired statement. $\square$

9.19. THEOREM. *Let ξ and η be X-valued independent random variables with distributions μ and ν. Suppose that the group X satisfies the condition:*

(i) *for any compact Corwin subgroup $K \subset X$ the factor-group X/K does not contain a subgroup topologically isomorphic to $\mathbb{T}$.*

Then for the random variables $\xi + \eta$ and $\xi - \eta$ to be independent, it is necessary and sufficient that $\mu, \nu \in I_B(X) * \Gamma(X)$ and $\mu = \nu * E_x$.

PROOF. If $\mu, \nu \in I_B(X) * \Gamma(X)$ and $\mu = \nu * E_x$, then it follows from Proposition 9.5 that the characteristic functions $\hat{\mu}(y)$ and $\hat{\nu}(y)$ satisfy equation 9.17(i). Then by Proposition 9.17 the random variables $\xi + \eta$ and $\xi - \eta$ are independent. Thus, this part of the theorem is valid even without X to satisfy condition (i).

Let the random variables $\xi + \eta$ and $\xi - \eta$ be independent, and let X satisfy condition (i). Then every closed subgroup X_1 of the group X satisfies this condition. By Proposition 9.18 we may confine ourselves to the case $X = \mathbb{R}^n + K$, where $n \geq 0$ and K is a compact Corwin group. Applying Lemma 9.3 in the case $n = 2$ and $G = X$ we obtain that the group Y has no elements of order two. Therefore, each element of finite order of Y has an odd order and, hence, belongs to $Y^{(2)}$. Let us verify that each element of infinite order of Y belongs to $Y^{(2)}$ as well. It suffices to prove this fact for an arbitrary element h of infinite order from $K^* = H$. Let $h \notin H^{(2)}$. Consider the subgroup $L = M(h)$ generated by h. Clearly, if $2y \in L$, then $y \in H$ and, hence, $2y \in H^{(2)}$. Let us prove that $2y \in L$ implies $y \in L$. If $2y = (2m + 1)h$, then $h = 2y - 2mh \in H^{(2)}$ despite the choice of h. Therefore, if $2y \in L$, then necessarily $2y = 2mh$. This yields $y = mh \in L$ since Y has no elements of order two. Thus the subgroup $E = \mathbb{R}^n + L$ possesses the following property: if $2y \in E$, then $y \in E$. By Lemma 9.3, $A(X, E)$ is a Corwin group. This group is compact, since E is open. By Theorem 1.6 and 1.5(i) we have $(X/A(X, E))^* \approx E \approx \mathbb{R}^n + \mathbb{Z}$. Therefore $X/A(X, E) \approx \mathbb{R}^n + \mathbb{T}$, contrary to condition (i) of the theorem. So each element of infinite order of Y belongs to $Y^{(2)}$ and, hence,

$$Y^{(2)} = Y. \tag{1}$$

By Proposition 9.17 the characteristic functions $\hat{\mu}(y)$ and $\hat{\nu}(y)$ satisfy equation 9.17(i). Combining (1) and 9.18(1) we obtain $\{y \in Y : \hat{\mu}(y) \neq 0\} = \{y \in Y : \hat{\nu}(y) \neq 0\} = N$. Now 9.17(i) implies that N is an open subgroup of Y with the following property: if $2y \in N$, then $y \in N$. According to Theorem 1.7 and Lemma 9.3 $K = A(X, N)$ is a compact Corwin subgroup.

Replacing y_2 by $-y_2$ in equation 9.17(i) we obtain

$$\hat{\mu}(y_1 - y_2)\hat{\nu}(y_1 + y_2) = \hat{\mu}(y_1)\hat{\nu}(y_1)\hat{\mu}(-y_2)\hat{\nu}(y_2) \tag{2}$$

for all $y_1, y_2 \in Y$. Let $\lambda = \mu * \nu$. Multiplying the latter equality and 9.17(i) and then using 2.10(d), we obtain that the characteristic function $\hat{\lambda}(y)$ satisfies equation 9.2(i). In view of Proposition 9.2, $\lambda \in \Gamma_B(X)$. Denote by

λ', μ', ν' the distributions on X/K whose characteristic functions coincide with restriction of the characteristic functions $\hat{\lambda}(y)$, $\hat{\mu}(y)$ and $\hat{\nu}(y)$ to N (see 2.10(h)). Since no subgroup of X/K is isomorphic to $\mathbb{T}$, applying Lemma 9.8 to the group X/K and the distribution $\lambda' \in \Gamma_B(X/K)$ we obtain $\lambda' \in \Gamma(X/K)$. Then by Theorem 5.22 μ', $\nu' \in \Gamma(X/K)$. Arguing as in the proof of the sufficiency part in Theorem 9.10, we obtain μ, $\nu \in I_B(X)*\Gamma(X)$. Then (1) and 9.18(1) also imply $|\hat{\mu}(y)| = |\hat{\nu}(y)|$ and, hence, $\mu = \nu * E_x$. □
Condition 9.19(i) may be reformulated as follows.

9.20. PROPOSITION. *Let X be an arbitrary group. The following statements are equivalent:*

(i) *X satisfies condition 9.19(i).*
(ii) *If K is a compact Corwin subgroup of X, then K^* also is a Corwin group.*
(iii) *The group C_X does not contain elements of order two.*

PROOF. (i) $\Rightarrow$ (ii). This implication was established in the proof of Theorem 9.19.

(ii) $\Rightarrow$ (i). Let us verify that if G is a compact Corwin subgroup of X, then the factor-group X/G also satisfies condition (ii). Indeed, denote by $p: X \to X/G$ the natural homomorphism and suppose that K is a compact Corwin subgroup of X/G. Set $K_1 = p^{-1}(K)$. Then $K \approx K_1/G$ and K_1 is compact, since both K and G are compact. Clearly, K_1 is a Corwin group and, by condition (ii), K_1^* also is a Corwin group. Theorem 1.6 yields $K^* \approx A(K_1^*, G)$. Let $y \in A(K_1^*, G)$. Then $y = 2y'$ for some $y' \in K_1^*$. Since G is a Corwin group, Lemma 9.3 implies $y' \in A(K_1^*, G)$, i.e., K^* is a Corwin group. This, in particular, implies (i) since $\mathbb{T}$ is a Corwin group.

The relation (ii) $\Leftrightarrow$ (iii) is a consequence of Proposition 11.15. □

Theorem 9.19 is sharp.

9.21. THEOREM. *Let the group X not satisfy condition 9.19(i). Then there exist X-valued random variables ξ and η with distribution μ and ν such that μ, $\nu \notin I_B(X) * \Gamma(X)$ and the random variables $\xi + \eta$ and $\xi - \eta$ are independent.*

PROOF. Consider the functions

$$f_0(n) = \begin{cases} \exp\{-an^2\} & \text{if } n \in \mathbb{Z}^{(2)}, \\ (1+\varepsilon)\exp\{-an^2\} & \text{if } n \notin \mathbb{Z}^{(2)}; \end{cases}$$

$$g_0(n) = \begin{cases} \exp\{-an^2\} & \text{if } n \in \mathbb{Z}^{(2)}, \\ (1+\varepsilon)^{-1}\exp\{-an^2\} & \text{if } n \notin \mathbb{Z}^{(2)}, \end{cases}$$

on the group $\mathbb{Z}$. If $a > 0$ is large enough and $\varepsilon > 0$, then these functions are the characteristic functions of some distributions μ_0 and ν_0 on $\mathbb{T}$. By construction, the functions $\hat{\mu}_0(n)$ and $\hat{\nu}_0(n)$ are nonvanishing on $\mathbb{Z}$ and satisfy equation 9.17(i), and μ_0, $\nu_0 \notin \Gamma(\mathbb{T})$.

By the hypothesis, for some Corwin subgroup $K \subset X$ the factor-group X/K contains a subgroup topologically isomorphic to $\mathbb{T}$. We may treat the distributions μ_0 and ν_0 as distributions on X/K. By Theorem 1.6 $(X/K)^* \approx A(Y, K)$, so we may assume that the functions $\hat{\mu}_0(h)$ and $\hat{\nu}_0(h)$ are defined on $A(Y, K)$.

Consider the following functions on Y:

$$f(y) = \begin{cases} \hat{\mu}_0(y) & \text{if } y \in A(Y, K), \\ 0 & \text{if } y \notin A(Y, K); \end{cases}$$

$$g(y) = \begin{cases} \hat{\nu}_0(y) & \text{if } y \in A(Y, K), \\ 0 & \text{if } y \notin A(Y, K). \end{cases}$$

In view of Proposition 9.17, the theorem will be proven if, one, we verify that $f(y)$ and $g(y)$ are the characteristic functions of some distributions μ and ν on X and the functions $\hat{\mu}(y) = f(y)$ and $\hat{\nu}(y) = g(y)$ satisfy equation 9.17(i), and, two, we prove that $\mu, \nu \notin I_B(X) * \Gamma(X)$. All this may be done in exactly the same way as in the proof of the necessity part in Theorem 9.10. $\square$

9.22. REMARK. In the proof of Theorem 9.19 the following fact was essential. If condition 9.19(i) is fulfilled and the characteristic functions $\hat{\mu}(y)$ and $\hat{\nu}(y)$ satisfy equation 9.17(i), then

$$\{y \in Y : \hat{\mu}(g) \neq 0\} = \{y \in Y : \hat{\nu}(y) \neq 0\}. \tag{1}$$

Without condition 9.19(i) this equality might not hold. Indeed, let $X = \mathbb{T}$, $\mu = m_\mathbb{T}$, and let ν be an arbitrary distribution on $\mathbb{T}$ satisfying the condition $\hat{\nu}(2n) = 0$ for all $n \in \mathbb{Z}$, $n \neq 0$. Then the functions $\hat{\mu}(n)$ and $\hat{\nu}(n)$ clearly satisfy equation 9.17(i), but equality (1) does not hold if $\nu \neq m_\mathbb{T}$.

9.23. REMARK. Let ξ and η be independent infinitely divisible X-valued random variables, and let μ and ν be their distributions. The random variables $\xi + \eta$ and $\xi - \eta$ are independent if and only if $\mu, \nu \in I_B(X) * \Gamma(X)$ and $\mu = \nu * E_x$.

Indeed, according to Proposition 9.17 the independence of the random variables $\xi + \eta$ and $\xi - \eta$ implies that the characteristic functions $\hat{\mu}(y)$ and $\hat{\nu}(y)$ satisfy 9.17(i) and, hence, 9.18(1). By Corollary 2.18 the sets $N_1 = \{y \in Y : \hat{\mu}(y) \neq 0\}$ and $N_2 = \{y \in Y : \hat{\nu}(y) \neq 0\}$ are subgroups of Y and relations 9.18(1) imply $N_1 = N_2 = N$. The rest of the argument repeats that in the proof of Theorem 9.19 with Remark 9.14 used instead of Lemma 9.8 and Proposition 5.21 instead of Theorem 5.22.

9.24. REMARK. Let ξ and η be X-valued independent random variables with distributions μ and ν such that $\hat{\mu}(y)\hat{\nu}(y) \neq 0$ for any $y \in Y$. The following statement may be deduced from the proofs of Theorems 9.19 and 9.21. The independence of the random variables $\xi + \eta$ and $\xi - \eta$ implies $\mu, \nu \in \Gamma(X)$ if and only if the group X does not contain a subgroup topologically isomorphic to $\mathbb{T}$.

§10. Characterization of Gaussian distribution by independence of linear statistics

In this section we completely describe those groups X to which the following theorem of V. P. Skitovich and Darmois can be extended. This theorem characterizes a Gaussian distribution on the real line by independence of its linear statistics.

10.1. THEOREM. *Let ξ_j, $j = 1, \ldots, m$, $m \geq 2$, be independent random variables. If the linear forms $L_1 = a_1\xi_1 + \cdots + a_m\xi_m$ and $L_2 = b_1\xi_1 + \cdots + b_m\xi_m$, where all the coefficients differ from zero, are independent, then the random variables ξ_j are Gaussian* [KLR, §3.1].

10.2. DEFINITION. *A set of integers $\{a_j\}$ is called admissible for a group X if $X^{(a_j)} \neq \{0\}$ for all j.*

Let ξ_j, $j = 1, \ldots, m$, be X-valued random variables, X a group. The admissibility of the set $\{a_j\}$ when considering the linear form $a_1\xi_1 + \cdots + a_m\xi_m$ is the group analog of the condition $a_j \neq 0$, $j = 1, 2, \ldots, m$, in the case when $X = \mathbb{R}$.

10.3. PROPOSITION. *Let ξ_j, $j = 1, \ldots, m$, $m \geq 2$, be independent random variables with values in a group X, and let μ_j be their distributions. In order that the linear forms $L_1 = a_1\xi_1 + \cdots + a_m\xi_m$ and $L_2 = b_1\xi_1 + \cdots + b_m\xi_m$, $a_j, b_j \in \mathbb{Z}$ be independent, it is necessary and sufficient that the characteristic functions $\hat{\mu}_j(y)$ satisfy the equation*

$$(\text{i}) \quad \prod_{j=1}^{m} \hat{\mu}_j(a_j u + b_j v) = \prod_{j=1}^{m} \hat{\mu}_j(a_j u) \prod_{j=1}^{m} \hat{\mu}_j(b_j v)$$

for all $u, v \in Y$.

The proof of this proposition is entirely similar to the proof of Proposition 9.2 and is omitted.

The first result that can be considered as a group analog of Theorem 10.1, will be proved below (Theorem 10.5). First we shall prove the following lemma.

10.4. LEMMA. *Let X be a discrete torsion-free group, ξ_j, $j = 1, \ldots, m$, $m \geq 2$, independent random variables with values in X and with distributions μ_j, and let $\{a_j\}_1^m$, $\{b_j\}_1^m$ be sets of nonzero integers. If the linear forms $L_1 = a_1\xi_1 + \cdots + a_m\xi_m$, $L_2 = b_1\xi_1 + \cdots + b_m\xi_m$ are independent, then $\mu_j \in D(X)$, $j = 1, 2, \ldots, m$.*

PROOF. According to 1.10, Y is a connected compact group. The following two cases are possible.

1. $Y \not\approx \mathbb{T}$. By Remark 1.21 a continuous monomorphism $\psi: \mathbb{R} \to Y$ exists whose image $\psi(\mathbb{R})$ is dense in Y. Consider the restrictions of the

characteristic functions $\mu_j(y)$ to $\psi(\mathbb{R})$. It follows from 10.3 that the characteristic functions $\hat{\mu}_j(y)$ satisfy equation 10.3(i). Let $f_j(t) = \hat{\mu}_j(\psi(t))$, $t \in \mathbb{R}$. By Theorem 2.8 the functions $f_j(t)$ are characteristic functions on $\mathbb{R}$, which obviously satisfy equation 10.3(i). Proposition 10.3, applied to the group $\mathbb{R}$, and the Skitovich-Darmois Theorem 10.1 give us

$$f_j(t) = \hat{\mu}_j(\psi(t)) = \exp\{-\sigma_j t^2 + i\beta_j t\},$$

where $\sigma_j \geq 0$, $\beta_j \in \mathbb{R}$, $j = 1, \ldots, m$.

Let V be an arbitrary neighborhood of zero in Y. Since ψ is a monomorphism and $\overline{\psi(\mathbb{R})} = Y$, one can choose a sequence t_n such that $t_n \to \infty$ and $\psi(t_n) \in V$ for all n. If $\sigma_j > 0$ for some $j > 0$, then $|\hat{\mu}_j(\psi(t_n))| = \exp\{-\sigma_j t_n^2\} \to 0$ as $t_n \to \infty$, which contradicts the continuity of $\hat{\mu}_j(y)$ because $\hat{\mu}_j(0) = 1$. So $\sigma_j = 0$, $j = 1, \ldots, m$. Hence $|\hat{\mu}_j(\psi(t))| \equiv 1$ for $t \in \mathbb{R}$, and since $\psi(\mathbb{R})$ is dense in Y, we have $|\hat{\mu}_j(y)| \equiv 1$ for $y \in Y$, $j = 1, \ldots, m$. From 2.10(i) we obtain $\hat{\mu}_j(y) = (x_j, y)$ and, consequently, $\mu_j = E_{x_j}$, $j = 1, \ldots, m$. In this case the lemma is proved.

2. $Y \approx \mathbb{T}$. Then $X \approx \mathbb{Z}$. Without loss of generality we may assume that $X = \mathbb{Z}$ and that μ_j are distributions on $\mathbb{R}$ with 2π-periodic characteristic functions that satisfy equation 10.3(i). From Proposition 10.3 applied to the group $\mathbb{R}$ and Theorem 10.1 of Skitovich and Darmois it follows that $\hat{\mu}_j(t) = \exp\{-\sigma_j t^2 + i\beta_j t\}$, $\sigma_j \geq 0$, $\beta_j \in \mathbb{R}$, $j = 1, \ldots, m$. Since the functions $\hat{\mu}_j(t)$ must be 2π-periodic, we have $\sigma_j = 0$, $\beta_j \in \mathbb{Z}$, that is, $\mu_j \in D(X)$, $j = 1, \ldots, m$. $\square$

10.5. THEOREM. *Let ξ_j, $j = 1, \ldots, m$, $m \geq 2$, be independent random variables, with values in a group X and distributions μ_j, and let $\{a_j\}_{j=1}^m$, $\{b_j\}_{j=1}^m$ be sets of integers admissible for the group X. Assume that the linear forms $L_1 = a_1\xi_1 + \cdots + a_m\xi_m$ and $L_2 = b_1\xi_1 + \cdots + b_m\xi_m$ are independent. Then*

a) *if*

$$\text{(i)} \quad X \approx \mathbb{R}^n + D,$$

where $n \geq 0$ and D is a torsion-free discrete group, then $\mu_j \in \Gamma(X)$, $j = 1, \ldots, m$.

b) *If $X^{(2)} = \{0\}$, then $\mu_j \in D(X)$, $j = 1, \ldots, m$.* (*)

c) *If $X \approx \mathbb{Z}(3)$, then either $\mu_j \in D(X)$, $j = 1, \ldots, m$, or $\mu_{j_1} = \mu_{j_2} = m_X$ for at least two distributions μ_{j_1}, μ_{j_2} and the other μ_j are arbitrary.*

PROOF. a) It is sufficient to prove the theorem for the group $X = \mathbb{R}^n + D$. Let $\nu_j = \mu_j * \overline{\mu_j}$. By Proposition 10.3 the characteristic functions of the

---

(*) It should be noted that any group X satisfying $X^{(p)} = \{0\}$, where p is a prime number, is topologically isomorphic to a group of the form $\mathbb{Z}(p)^{\mathfrak{M}} + \mathbb{Z}(p)^{\mathfrak{N}*}$, where $\mathfrak{M}$ and $\mathfrak{N}$ are some cardinalities and $\mathbb{Z}(p)^{\mathfrak{N}*}$ is the subgroup of $\mathbb{Z}(p)^{\mathfrak{N}}$ that consist of elements (x_l) with only finitely many nonzero entries x_l. This group is endowed with discrete topology. (See [HR1, §25].)

distributions μ_j satisfy equation 10.3(i). Hence, by 2.10(d),(e) the characteristic functions of the distributions ν_j satisfy the same equation. Now observe that $\hat{\nu}_j(y) \geq 0$, $y \in Y$, and consider the restriction of the functions $\hat{\nu}_j(y)$ to the subgroup $H = D^* \subset Y$. By 2.10(h) these restrictions are the characteristic functions of some distributions $\delta_j \in \mathcal{M}^1(D)$. Since the functions $\delta_j(h)$ clearly satisfy equation 10.3(i), by Proposition 10.3 and Lemma 10.4 we have that $\delta_j = E_0$, $j = 1, \ldots, m$. This implies $\hat{\nu}_j(y) \equiv 1$ for $y \in H$, and then by virtue of Proposition 2.13 $\sigma(\nu_j) \subset A(X, H) = \mathbb{R}^n$. Using Proposition 2.5 we can replace the distributions μ_j by their shifts μ_s' so that $\sigma(\mu_j') \subset \mathbb{R}^n$. Since the characteristic functions of μ_j' clearly also satisfy equation 10.3(i), one can easily obtain from Proposition 10.3 and Theorem 10.1 that $\mu_j' \in \Gamma(\mathbb{R}^n)$. Hence $\mu_j \in \Gamma(X)$, $j = 1, \ldots, m$.

b) $X^{(2)} = \{0\}$ implies that the sets admissible for X consists of odd integers. Observe also that $Y^{(2)} = \{0\}$. By virtue of Proposition 10.3 the characteristic functions of the distributions μ_j satisfy equation 10.3(i), which takes the form

$$\prod_{j=1}^{m} \hat{\mu}_j(u + v) = \prod_{j=1}^{m} \hat{\mu}_j(u) \prod_{j=1}^{m} \hat{\mu}_j(v), \qquad u, v \in Y. \tag{1}$$

Setting $u = v = y$ in (1) we obtain $|\hat{\mu}_j(y)| \equiv 1$ on Y. Hence, by 2.10(h) we have $\hat{\mu}_j(y) = (x_j, y)$, i.e., $\mu_j = E_{x_j}$, $j = 1, \ldots, m$.

c) Since the sets admissible for the group X consist of integers indivisible by 3, as can easily be seen one may assume that $L_1 = \xi_1 + \cdots + \xi_m$ and $L_2 = \xi_1 + \cdots + \xi_l - \xi_{l+1} - \cdots - \xi_m$. By Proposition 10.3 the characteristic functions of the distributions μ_j satisfy equation 10.3(i), which takes the form

$$\prod_{j=1}^{l} \hat{\mu}_j(u + v) \prod_{j=l+1}^{m} \hat{\mu}_j(u - v)$$

$$= \prod_{j=1}^{m} \hat{\mu}_j(u) \cdot \prod_{j=1}^{l} \hat{\mu}_j(v) \cdot \prod_{j=l+1}^{m} \hat{\mu}_j(-v), \qquad u, v \in Y.$$

Let $\xi = \xi_1 + \cdots + \xi_l$, $\eta = \xi_{l+1} + \cdots + \xi_m$. The random variables ξ and η have the distributions

$$\mu = \mathop{*}_{j=1}^{l} \mu_j, \qquad \nu = \mathop{*}_{j=l+1}^{m} \mu_j. \tag{2}$$

By 2.10(d) the characteristic functions of the distributions μ and η clearly satisfy equation 9.17(i). It follows from Proposition 9.17 and Theorems 9.19 and 5.5 that $\mu, \nu \in I(X)$. Hence if $\mu \in D(X)$, $j = 1, \ldots, m$, then 9.17(i) implies that $\nu \in D(X)$; in this case $\mu_j \in D(X)$, $j = 1, \ldots, m$. If $\mu = m_X$, then 9.17(i) implies that $\nu = m_X$ as well. Observe now that if $\gamma_1 * \gamma_2 = m_X$ on a group X, then at least one of the distributions γ_1, γ_2 is equal to m_X.

For definiteness let $\mu_1 = \mu_{l+1} = m_X$. Then it follows from (2) that the other distributions μ_j are arbitrary. $\square$

Theorem 10.5 is in a certain sense sharp (see Theorem 10.9). To prove this we shall need some lemmas.

10.6. LEMMA. *Let* K *be a compact Corwin group,* $K \not\approx \mathbb{Z}(3)$. *Then there exist* K*-valued independent random variables* ξ_j, $j = 1, \ldots, 4$, *with distributions* μ_j, *such that the linear forms* $L_1 = \xi_1 + \xi_2 + \xi_3 + \xi_4$ *and* $L_2 = \xi_1 + \xi_2 + \xi_3 + \xi_4$ *are independent and* $\mu_j \notin I(K) * \Gamma(K)$, $j = 1, \ldots, 4$.

PROOF. From Proposition 9.2 and 9.5 it follows that the characteristic function $\hat{\mu}_K(y)$ satisfies equation 9.2(i). Let $y_1, y_2 \in Y = K^*$ be such that the elements $y_1, y_2, -y_1$, and $-y_2$ are pairwise distinct. Consider the functions

$$\rho_j(x) = 1 + \frac{1}{2}[(x, y_j) + \overline{(x, y_j)}], \qquad j = 1, 2,$$

on the group K. Clearly, $\rho_j(x) \geq 0$. Denote by μ_j the distribution on the group K having the density ρ_j with respect to m_K. One can easily see that $\mu_1 * \mu_2 = m_K$ and that the characteristic functions $\hat{\mu}_1(y)$ and $\hat{\mu}_2(y)$ satisfy the equation

$$\hat{\mu}_1(u + v)\hat{\mu}_2(u + v)\hat{\mu}_1(u - v)\hat{\mu}_2(u - v)$$
$$= \hat{\mu}_1^2(u)\hat{\mu}_2^2(u)|\hat{\mu}_1(v)|^2|\hat{\mu}_2(v)|^2. \tag{1}$$

Let ξ_j, $j = 1, \ldots, 4$, be K-valued independent variables with distributions μ_j such that $\mu_1 = \mu_3$, $\mu_2 = \mu_4$. It follows from (1), by 10.3, that the linear forms $L_1 = \xi_1 + \xi_2 + \xi_3 + \xi_4$ and $L_2 = \xi_1 + \xi_2 - \xi_3 - \xi_4$ are independent. That $\mu_j \notin I(K) * \Gamma(K)$ is obviously. $\square$

10.7. LEMMA. *Let* $X = \Delta_p$. *Then there exist* X*-valued independent random variables* ξ_1 *and* ξ_2 *with distributions* μ_1 *and* μ_2, *such that the linear forms* $L_1 = p\xi_1 - \xi_2$ *and* $L_2 = \xi_1 + p\xi_2$ *are independent and* $\mu_1, \mu_2 \notin I(X)$.

PROOF. Since $\Delta_p^* \approx \mathbb{Z}(p^\infty)$, then, without loss of generality, we may assume $Y = \mathbb{Z}(p^\infty)$. After the natural embedding of the group $\mathbb{Z}(p)$ into $\mathbb{Z}(p^\infty)$ consider the functions $f_j(y)$ on the group $\mathbb{Z}(p^\infty)$:

$$f_j(y) = \begin{cases} \hat{\nu}_j(y), & y \in \mathbb{Z}(p), \\ 0, & y \notin \mathbb{Z}(p), \ j = 1, 2, \end{cases}$$

where $\hat{\nu}_j(y)$ are arbitrary characteristic functions on the group $\mathbb{Z}(p)$. Theorem 2.8 and Remark 2.9 imply that the functions f_j are positive definite, and, since the group $\mathbb{Z}(p^\infty)$ is discrete, the functions f_j are continuous. Then, by Theorem 2.8, there exist distributions $\mu_j \in \mathscr{M}^1(x)$ such that $\hat{\mu}_j(y) = f_j(y)$. Let ξ_j be X-valued independent random variables with distributions μ_j, $j = 1, 2$. By Proposition 10.3, to verify the independence of the linear forms L_1 and L_2, it suffices to prove that the characteristic functions $\hat{\mu}_j(y)$

satisfy equation 10.3(i). This equation has the form

$$f_1(pu+v)\cdot f_2(-u+pv) = f_1(pu)f_2(-u)f_1(v)f_2(pv), \qquad u,v \in \mathbb{Z}(p^\infty). \quad (1)$$

If $u,v \in \mathbb{Z}(p)$, then equality (1), obviously, is true. If either $u \in \mathbb{Z}(p)$, $v \notin \mathbb{Z}(p)$ or $u \notin \mathbb{Z}(p)$, $v \in \mathbb{Z}(p)$, then both sides of (1) vanish. Let $u,v \notin \mathbb{Z}(p)$. Then the right-hand side of equality (1) vanishes. If $pu+v$, $-u+pv \in \mathbb{Z}(p)$, then $(p^2+1)u \in \mathbb{Z}(p)$ and we obtain $u \in \mathbb{Z}(p)$, contrary to the assumption. This means that either $pu+v \notin \mathbb{Z}(p)$ or $-u+pv \notin \mathbb{Z}(p)$ and the left-hand side of (1) also vanishes. So we have proved that the characteristic functions $\hat{\mu}_j(y)$ satisfy equation 10.3(i). □

10.8. LEMMA. *Let G be a closed subgroup of the group X, $\mu \in \mathscr{M}^1(G)$. If $\mu \notin I(G)*\Gamma(G)$, then $\mu \notin I(X)*\Gamma(X)$.*

This assertion follows directly from Proposition 2.5.

Now we shall prove that Theorem 10.5 is sharp in the following sense.

10.9. THEOREM. *Let the group X be topologically nonisomorphic to the groups enumerated in Theorem 10.5. Then there exist X-valued independent random variables ξ_j, $j = 1,\ldots,m$, $m \geq 2$, with distributions μ_j and admissible for X sets of integers $\{a_j\}_{j=1}^m$ and $\{b_j\}_{j=1}^m$ such that the linear forms $L_1 = a_1\xi_1 + \cdots + a_m\xi_m$ and $L_2 = b_1\xi_1 + \cdots + b_m\xi_m$ are independent and $\mu_j \notin I(X)*\Gamma(X)$, $j = 1,\ldots,m$.*

PROOF. There are four possibilities.

1. The group X contains a subgroup $G \approx \mathbb{Z}(2)$. Let ξ_1 and ξ_2 be G-valued independent random variables whose distributions μ_1,μ_2 are not in $I(G)$. By Proposition 5.5, $\Gamma(G) = D(G)$, and so $I(G)*\Gamma(G) = I(G)$. According to Lemma 10.8, $\mu_1,\mu_2 \notin I(X)*\Gamma(X)$. Clearly, the linear forms $L_1 = 2\xi_1 - \xi_2$ and $L_2 = \xi_1 + 2\xi_2$ are independent and the sets $\{2,-1\}$ and $\{1,2\}$ are admissible for the group X, because, by the condition, $X^{(2)} \neq \{0\}$.

2. The group X contains a subgroup $G \approx \mathbb{Z}(3)$. Suppose $X(3) \neq \{0\}$. Let ξ_1 and ξ_2 be G-valued independent random variables whose distributions μ_1,μ_2 are not in $I(G)$. Considering the linear forms $L_1 = 3\xi_1 - \xi_2$ and $L_2 = \xi_1 + 3\xi_2$ in the same way as in the case 1, we convince ourselves that ξ_1 and ξ_2 are the desired random variables. If $X^{(3)} = \{0\}$, then, since $X \not\approx \mathbb{Z}(3)$, the group X contains a subgroup $K \approx (\mathbb{Z}(3))^2$. One can easily see that the random variables ξ_j and the linear forms L_1 and L_2 constructed in Lemma 10.6 are the needed ones.

3. The group X contains a subgroup $K \approx \mathbb{Z}(p)$ for any prime $p > 3$. In this case the random variables ξ_j and the linear forms L_1 and L_2 constructed in Lemma 10.6 are the needed ones.

4. X is a torsion-free group. By Theorem 1.14, $X \approx \mathbb{R}^n + G$, where $n \geq 0$ and the group G contains a compact open subgroup K. Since the group X

is topologically nonisomorphic to a group of the form 10.5(i), then $K \neq \{0\}$. Since K is a compact torsion-free group, by virtue of Theorem 1.16, K is topologically isomorphic to a group of the form 1.16(i). According to Remark 1.17, the group $\Sigma_{\mathbf{a}}$, $\mathbf{a} = (2, 3, 4, \ldots)$, contains a subgroup $G_1 \approx \Delta_p$ for any prime p. Since the group Δ_p is totally disconnected, by Proposition 5.5 we have $\Gamma(G_1) = D(G_1)$ and, hence, $I(G_1) * \Gamma(G_1) = I(G_1)$. The existence of the needed random variables ξ_1, ξ_2 as well as the linear forms L_1 and L_2, follows from Lemmas 10.7 and 10.8. $\square$

Let ξ_j, $j = 1, \ldots, m$, $m \geq 2$, be independent random variables with values in a group X with distributions μ_j, and let $\{a_j\}_{j=1}^m$, $\{b_j\}_{j=1}^m$ be admissible sets of integers for X. When proving Theorems 10.5 and 10.9 we saw that, in contrast to the case $X = \mathbb{R}$, the independence of the linear forms $L_1 = a_1\xi_1 + \cdots + a_m\xi_m$ and $L_2 = b_1\xi_1 + \cdots + b_m\xi_m$ does not imply (see the proof of the Skitovich-Darmois Theorem 10.1) that the characteristic function $\hat{\mu}_j(y)$ does not vanish. (From the proof of Theorems 10.5 and 10.9 it follows that a group X, for which the independence of the linear forms L_1 and L_2 implies the nonvanishing of the characteristic function $\hat{\mu}_j(y)$, is a group either of the form 10.5(i) or such that $X^{(2)} = \{0\}$.)

Now we pass to the main problem of this section, that is, the problem of complete description of those groups X for which the independence of linear forms L_1 and L_2 under additional restriction $\hat{\mu}_j(y) \neq 0$ for all $y \in Y$ $j = 1, \ldots, m$, implies $\mu_j \in \Gamma(X)$, $j = 1, \ldots, m$.

10.10. LEMMA. *Let K be a compact group, $a, b \in \mathbb{Z}$, $K^{(a)} = K$, and let $f(x)$ be a continuous nonnegative function on K. Then the inequality*

(i)
$$\int_{K^2} f(au + bv)\, dm_{K^2}(u, v) \leq \int_{K^2} f(au)\, dm_{K^2}(u, v)$$
$$+ \int_{K^2} f(bv)\, dm_{K^2}(u, v)$$

takes place. Equality takes place if and only if $f(x) \equiv 0$ for $x \in K^{(b)}$.

PROOF. Since $K^{(a)} = K$, one can easily see that

$$\int_{K^2} f(au + bv)\, dm_{K^2}(u, v) = \int_K f(x)\, dm_K(x), \qquad (1)$$

$$\int_{K^2} f(au)\, dm_{K^2}(u, v) = \int_K f(x)\, dm_K(x). \qquad (2)$$

Since $f(x) \geq 0$ on K, inequality (i) follows from (1) and (2), and, moreover, it becomes equality if and only if $f(bx) \equiv 0$ on K, that is, $f(x) \equiv 0$ for $x \in K^{(b)}$. $\square$

10.11. LEMMA. *Let $X = \mathbb{Z}(p^\infty)$, and let ξ_j, $j = 1, \ldots, m$, $m \geq 2$, be X-valued independent random variables with distributions μ_j such that*

$\hat{\mu}_j(y) > 0$ *for all* $y \in Y$, $j = 1, \ldots, m$, *and*

$$\text{(i)} \quad \prod_{j=1}^{m} \hat{\mu}_j(y) = 1 \quad \text{only for } y = 0.$$

Let $\{a_j\}_{j=1}^{m}$, $\{b_j\}_{j=1}^{m}$ *be sets of nonzero integers. Then the linear forms* $L_1 = a_1\xi_1 + \cdots + a_m\xi_m$ *and* $L_2 = b_1\xi_1 + \cdots + b_m\xi_m$ *are dependent.*

PROOF. Observe that $Y \approx \Delta_p$. Replacing, if needed, the random variables ξ_j by $\xi_j' = d_j\xi_j$ we may assume the numbers a_j and b_j to be relatively prime for every $j = 1, \ldots, m$. This does not violate condition (i). Indeed, let $\hat{\mu}_j(d_j y) = 1$, $j = 1, \ldots, m$, for some $y \in Y$. Then $\hat{\mu}_j(d_1 \cdots d_m y) = 1$, $j = 1, \ldots, m$. Then $d_1 \cdots d_m y = 0$ by the condition of the lemma. Since Y is a torsion-free group, we obtain $y = 0$.

Let $f_j(y) = -\ln \hat{\mu}_j(y)$. If the linear forms L_1 and L_2 are independent, then, by Proposition 10.3, condition 10.3(i) is satisfied. Hence

$$\sum_{j=1}^{m} f_j(a_j u + b_j v) = \sum_{j=1}^{m} f_j(a_j u) + \sum_{j=1}^{m} f_j(b_j v), \qquad u, v \in Y.$$

After integrating this equality over the group Y^2 with respect to the measure dm_{Y^2}, changing the order of integration and summing we obtain

$$\sum_{j=1}^{m} \int_{Y^2} f_j(a_j u + b_j v) \, dm_{Y^2}(u, v)$$
$$= \sum_{j=1}^{m} \int_{Y^2} f_j(a_j u) \, dm_{Y^2}(u, v) + \sum_{j=1}^{m} \int_{Y^2} f_j(b_j v) \, dm_{Y^2}(u, v). \tag{1}$$

Since a_j and b_j are relatively prime, at least one of them is not divisible by p, and therefore either $Y^{(a_j)} = Y$ or $Y^{(b_j)} = Y$, $j = 1, \ldots, m$. By Lemma 10.10, (1) implies that, when $f(y) = f_j(y)$, $j = 1, \ldots, m$, inequality 10.10(i) becomes equality. Hence

$$\prod_{j=1}^{m} \hat{\mu}_j(y) = 1 \quad \text{for } y \in Y^{(c)}, y \neq 0$$

and

$$c = \prod_{j=1}^{m} a_j b_j.$$

We have obtained a contradiction. □

10.12. LEMMA. *Let* $X = \mathbb{Z}(p)$, *where* p *is a prime number; let* ξ_j, $j = 1, \ldots, m$, $m \geq 2$, *be* X-*valued independent random variables with distributions* μ_j *such that* $\hat{\mu}_j(y) > 0$ *for all* $y \in Y$, $j = 1, \ldots, m$, *and*

$$\prod_{j=1}^{m} \hat{\mu}_j(y) = 1 \quad \text{only for } y = 0;$$

and let $\{a_j\}_{j=1}^m$, $\{b_j\}_{j=1}^m$ *be sets of integers* a_j, b_j, *none of which is divisible by* p. *Then the linear forms* $L_1 = a_1\xi_1 + \cdots + a_m\xi_m$ *and* $L_2 = b_1\xi_1 + \cdots + b_m\xi_m$ *are dependent.*

The proof of the lemma is completely similar to that of Lemma 10.11 and so is omitted.

Now let us prove the main result of this section.

10.13. THEOREM. *Let* ξ_j, $j = 1, \ldots, m$, $m \geq 2$, *be independent random variables with values in a group* X *and with distribution* μ_j *such that*

$$\prod_{j=1}^m \hat{\mu}_j(y) \neq 0 \quad \text{for all } y \in Y,$$

and let $\{a_j\}_{j=1}^m$, $\{b_j\}_{j=1}^m$ *be admissible for* X *sets of integers. The independence of the linear forms* $L_1 = a_1\xi_1 + \cdots + a_m\xi_m$ *and* $L_2 = b_1\xi_1 + \cdots + b_m\xi_m$ *implies* $\mu_j \in \Gamma(X)$, $j = 1, \ldots, m$, *if and only if either* X *is a torsion-free group or* $X^{(p)} = \{0\}$, *where* p *is a prime number.*

PROOF. *Necessity.* Suppose that the group X contains an element x_0 of order p, where p is a prime number, and $X^{(p)} \neq \{0\}$. Let G be a subgroup of X generated by the element x_0, and let ξ_1, ξ_2 be G-valued independent variables with nondegenerate distributions μ_1, μ_2 such that $\hat{\mu}_1(h)\hat{\mu}_2(h) \neq 0$ for $h \in G^*$. It follows from Proposition 10.3 that the linear forms $L_1 = p\xi_1 - \xi_2$ and $L_2 = \xi_1 + p\xi_2$ are independent and the sets $\{p, -1\}$, $\{1, p\}$ are admissible for the group X. Clearly, $\mu_1, \mu_2 \notin \Gamma(X)$.

Sufficiency. We give the proof in several stages.

1. According to Proposition 10.3, we need to verify that if the characteristic functions of the distributions μ_j satisfy equation 10.3(i), then $\mu_j \in \Gamma(X)$, $j = 1, \ldots, m$. Let $\nu_j = \mu_j * \bar{\mu}_j$. Then, by 2.10(d),(e), the characteristic functions of the distributions ν_j also satisfy equation 10.3(i) and, in addition, $\hat{\nu}_j(y) > 0$ for all $y \in Y$. Since no subgroup of the group X is topologically isomorphic to $\mathbb{T}$, then by virtue of Theorem 5.22 it suffices to check that $\nu_j \in \Gamma(X)$, $j = 1, \ldots, m$. So we may assume $\hat{\mu}_j(y) > 0$ for all $y \in Y$, $j = 1, \ldots, m$, from the outset.

First suppose that X is a torsion-free group. It follows from Proposition 2.13 that we may also assume that the condition

$$\prod_{j=1}^m \hat{\mu}_j(y) = 1 \quad \text{only for } y = 0 \qquad (1)$$

is satisfied. We shall verify that in this case $X \approx \mathbb{R}^n + (\Sigma_{\mathbf{a}})^n$, where $n \geq 0$, $\mathbf{a} = (2, 3, 4, \ldots)$, and $\mathfrak{n}$ is a cardinal number.

2. By Theorem 1.14, we have $X \approx \mathbb{R}^n + G$, where $n \geq 0$ and the group G contains a compact open subgroup K. Let $Y = \mathbb{R}^n + H$, $H \approx G^*$. Since C_H is a connected compact group, considering the restriction of 10.3(i) to

C_H and applying Lemma 10.4 we conclude that $\hat{\mu}_j(y) \equiv 1$ on C_H and then
(1) implies $C_H = \{0\}$, i.e., the group H is totally disconnected.

Since K is a compact torsion-free group and, by virtue of Theorem 1.16,
K is topologically isomorphic to the group 1.16(i), $C_X \approx \mathbb{R}^n + (\Sigma_\mathbf{a})^n$, $\mathbf{a} = (2, 3, 4 \ldots)$. Since X is a torsion-free group, the subgroup C_X is a topological direct summand in X [HR1, §25], that is, $X = C_X + A$. The group
A already is totally disconnected, and then, by Corollary 1.10, the group
$B = A(Y, C_X) \approx A^*$ consists of compact elements. Suppose $B \neq \{0\}$. As
one can easily see, the group B is topologically isomorphic to a closed subgroup of H. Since H is totally disconnected, B is totally disconnected as
well. Let $h \in B$, and let M_h be a closed subgroup generated by h. It follows
from the properties of B that M_h is a totally disconnected compact group.
Hence [HR1, §25]

$$M_h = \mathsf{P}_{p \in \mathscr{P}} F_p, \tag{2}$$

where $\mathscr{P}$ is the set of all prime integers and for each group F_p we have either
$F_p = \{0\}$ or $F_p \approx \mathbb{Z}(p^{r_p})$ (r_p is a natural number) or $F_p \approx \Delta_p$. Consider the
restriction of equation 10.3(i) to the group F_p. By Lemma 10.11 the case
$F_p \approx \Delta_p$ is not possible for any p. Consider the set of all $p \in \mathscr{P}$ such that
the subgroup $F_p \approx \mathbb{Z}(p^{r_p})$ may be encountered in decomposition (2) for some
element $h \in B$. According to Lemma 10.12 this set is finite and each of its
elements divides at least one of the numbers from the set $\{a_j, b_j\}_{j=1}^m$. If, in
addition, all exponents r_p are bounded by some constant C, then $B^{(n)} = \{0\}$
for some n and, hence, $A^{(n)} = \{0\}$. This is impossible, because the subgroup
$A \subset X$ consists of elements of infinite order. Therefore, at least for one
$p \in \mathscr{P}$ the group B contains subgroups that are topologically isomorphic to
$\mathbb{Z}(p^{r_p})$ with arbitrary large r_p. Let such a p be fixed. Denote $a_j = p^{l_j} a_j'$,
$b_j = p^{m_j} b_j'$, and $n_j = \min\{l_j, m_j\}$. Choose $q > \max_{j=1, \ldots, m}\{l_j, m_j\}$ such
that the group B contains a subgroup F topologically isomorphic to $\mathbb{Z}(p^q)$.
Let $f_j(y) = -\ln \hat{\mu}_j(y)$. We note that by virtue of (1),

$$\sum_{j=1}^m f_j(y) = 0 \quad \text{only for } y = 0. \tag{3}$$

Let us integrate the logarithm of equality 10.3(i) and integrate over the group
F^2 with respect to the measure dm_{F^2}; changing the order summation and
integration we get

$$\sum_{j=1}^m \frac{1}{p^{2q}} \sum_{(u,v) \in F^2} f_j(a_j u + b_j v)$$
$$= \sum_{j=1}^m \frac{1}{p^{2q}} \sum_{(u,v) \in F^2} f_j(a_j u) + \sum_{j=1}^m \frac{1}{p^{2q}} \sum_{(u,v) \in F^2} f_j(b_j v). \tag{4}$$

Consider the homomorphism $\varphi_j : F^2 \to F$ defined by the formula $\varphi_j(u, v)$ $= a_j u + b_j v$. One can easily see that $\operatorname{Im} \varphi_j \approx \mathbb{Z}(p^{q-n_j})$. Since $F^2/\ker \varphi_j \approx$ $\operatorname{Im} \varphi_j$, the number of elements in the subgroup $\ker \varphi_j$ is p^{q+n_j}. Hence,

$$
\sum_{(u,v)\in F^2} f_j(a_j u + b_j v) = \sum_{y\in\operatorname{Im}\varphi_j} \sum_{(u,v)\in\varphi_j^{-1}(y)} f_j(a_j u + b_j v)
$$
$$
= p^{q+n_j} \sum_{y\in F^{(p^{n_j})}} f_j(y). \tag{5}
$$

Similarly

$$
\sum_{(u,v)\in F^2} f_j(a_j u) = p^{q+l_j} \sum_{y\in F^{(p^{l_j})}} f_j(y), \tag{6}
$$

$$
\sum_{(u,v)\in F^2} f_j(b_j v) = p^{q+m_j} \sum_{y\in F^{(p^{m_j})}} f_j(y). \tag{7}
$$

Observe also that, by the choice of q, we have $F^{(p^{l_j})} \neq \{0\}$, $F^{(p^{m_j})} \neq \{0\}$, $j = 1, \ldots, m$. Substituting (5)–(7) into (4) we obtain

$$
\sum_{j=1}^m \frac{1}{p^{q-n_j}} \sum_{y\in F^{(p^{n_j})}} f_j(y) = \sum_{j=1}^m \frac{1}{p^{q-l_j}} \sum_{y\in F^{(p^{l_j})}} f_j(y)
$$
$$
+ \sum_{j=1}^m \frac{1}{p^{q-m_j}} \sum_{y\in F(p^{m_j})} f_j(y). \tag{8}
$$

Since either $n_j = l_j$ or $n_j = m_j$, equality (8) contradicts to (3). Hence $B = \{0\}$ and consequently $A = \{0\}$ and $X \approx \mathbb{R}^n + (\Sigma_{\mathbf{a}})^n$, $\mathbf{a} = (2, 3, 4, \ldots)$. Since the group X satisfies the second axiom of countability, this implies that either $X \approx \mathbb{R}^n + (\Sigma_{\mathbf{a}})^m$ or $X \approx \mathbb{R}^n + (\Sigma_{\mathbf{a}})^\infty$.

3. The problem is thus reduced to proving the theorem for a group X of the type just indicated. In this case either $Y \approx \mathbb{R}^n + \mathbb{Q}^m$ or $Y \approx \mathbb{R}^n + \mathbb{Q}_0^\infty$, where $\mathbb{Q}_0^\infty$ is the additive discrete group of all finitary sequence of rational numbers. Therefore, for every natural k, X and Y are groups with single-valued division by k. Without loss of generality, we can assume that $a_j = 1$, $j = 1, \ldots, m$, and that all b_j are distinct rational numbers. Indeed, setting $\xi'_j = a_j \xi_j$, we may assume $L_1 = \xi_1 + \cdots + \xi_m$, $L_2 = c_1 \xi_1 + \cdots + c_m \xi_m$, $c_j \in \mathbb{Q}$. Suppose that among the coefficients c_j there are exactly q pairwise distinct. Renumbering them if necessary, we may assume that

$$
c_1 = \cdots = c_{k_1}, \qquad c_{k_1+1} = \cdots = c_{k_2}, \qquad \cdots \qquad , c_{k_{q-1}} = \cdots = c_{k_q},
$$

where $k_q = m$ and $c_1, \ldots, c_{k_q}$ are distinct rational numbers. Let

$$
\eta_1 = \sum_{j=1}^{k_1} \xi_j \qquad \cdots \qquad \eta_q = \sum_{j=k_{q-1}+1}^m \xi_j,
$$

and consider the two independent linear forms $\sum_{j=1}^{q} \eta_j$ and $\sum_{j=1}^{q} d_j \eta_j$, $d_j = c_{k_j}$, $j = 1, \ldots, q$. If we prove that the random variables η_j, $j = 1, \ldots, q$, are Gaussian, then the assertion of the theorem would follow from Theorem 5.22.

Fix $y_0 \in Y$ and denote $L_{y_0} = \{ y \in Y : y = r y_0, \ r \in \mathbb{Q} \}$, $\psi_j(r) = f_j(r y_0)$. Considering a restriction of equation 10.3(i) to the subgroup L_{y_0} we obtain that the functions $\psi_j(r)$ satisfy the equation

$$\sum_{j=1}^{m} \psi_j(u + b_j v) = A(u) + B(v), \qquad u, v \in \mathbb{Q}, \tag{9}$$

where $A(u) = \sum_{j=1}^{m} \psi_j(u)$, $B(v) = \sum_{j=1}^{m} \psi_j(b_j v)$.

4. The reasoning on this step coincides with the proof of the Skitovich-Darmois Theorem 10.1 by the method of finite differences (see [Ram, Chapter VIII, §8.2]). All the increments considered of the variables are to be assumed rational.

Let $h \in \mathbb{Q}$, and let $g(u)$ be a function on $\mathbb{Q}$. Denote by Δ_h the finite difference operator $\Delta_h g(u) = g(u + h) - g(u)$.

First consider the case $m = 2$. Equation (9) has the form

$$\psi_1(u + b_1 v) + \psi_2(u + b_2 v) = A(u) + B(v), \qquad u, v \in \mathbb{Q}. \tag{10}$$

Choose increments h, $k \in \mathbb{Q}$ of u and v respectively such that $h + b_2 k = 0$. Since $b_1 \neq b_2$, $l = h + b_1 k \neq 0$. Hence

$$\psi_1(u + b_1 v + l) + \psi_2(u + b_2 v) = A(u + h) + B(v + k). \tag{11}$$

Subtracting (10) from (11) we obtain

$$\psi_1(u + b_1 v + l) - \psi_1(u + b_1 v) = \Delta_h A(u) + \Delta_k B(v). \tag{12}$$

Setting $v = 0$ we have

$$\psi_1(u + l) - \psi_1(u) = \Delta_h A(u) + \Delta_k B(0). \tag{13}$$

Subtracting (13) from (12) and setting $v = 1/b_1$ we have

$$\psi_1(u + 2l) - 2\psi_1(u + l) + \psi_1(u) = \Delta_k B\left(\frac{1}{b_1}\right) - \Delta_k B(0),$$

i.e., $\Delta_l^2 \psi_1(u) = \varepsilon(l)$, and, hence $\Delta_l^3 \psi_1(u) = 0$. Observe now that $l = h(b_1 - b_1)/b_2$ is an arbitrary rational number, since h is arbitrary.

Let $m > 2$. Consider equation (9) and choose increments h_1, $k_1 \in \mathbb{Q}$ of u and v such that $h_1 + b_m k_1 = 0$. One can see that $l_{1j} = h_1 + b_j k_1 \neq 0$, $j = 1, \ldots, m - 1$. From equation (9) we obtain

$$\sum_{j=1}^{m-1} \Delta_{l_{1j}} \psi_j(u + b_j v) = A_1(u) + B_1(v), \qquad u, v \in \mathbb{Q}.$$

Further, for $q = 2, \ldots, m - 1$, let us choose the increments h_q, $k_q \in \mathbb{Q}$ of u and v respectively such that $h_q + b_{m-q+1}k_q = 0$. Then $l_{qj} = h_q + b_j k_q \neq 0$ for $j = 1, \ldots, m - q$. We have a relation similar to (12), namely,

$$\Delta_{l_{m-1,1}} \Delta_{l_{m-2,1}} \cdots \Delta_{l_{11}} \psi_1(u + b_1 v) = A_{m-1}(u) + B_{m-1}(v). \qquad (14)$$

The right-hand side of (14) depends on h_j, k_j, $j = 1, \ldots, m-1$. Put $v = 0$ in (14) and subtract the equality obtained from (14) itself. We have

$$\Delta_{l_{m-1,1}} \Delta_{l_{m-2,1}} \cdots \Delta_{l_{11}} [\psi_1(u + b_1 v) - \psi_1(u)] = B_m(v). \qquad (15)$$

We may set the numbers $l_{ql} = h_q(b_{m-q+1} - b_1)/b_{m-q+1}$, $q = 1, \ldots, m - 1$, together with $b_1 v$ equal to an arbitrary rational number l. Then we obtain the equation $\Delta_l^m \psi_1(u) = \varepsilon(l)$ that implies

$$\Delta_l^{m+1} \psi_1(u) = 0, \qquad l, u \in \mathbb{Q}. \qquad (16)$$

Naturally, such an equation is obtained for each function ψ_j, $j = 1, \ldots, m$.

5. Rewrite equation (16) in the form

$$\sum_{j=1}^{m+1} (-1)^{m-k+1} C_k^{m+1} \psi_j(u + kl) = 0, \qquad u, l \in \mathbb{Q}. \qquad (17)$$

Fix $l \in \mathbb{Q}$. Let $M(l)$ be a subgroup generated by l. Consider equation (17) on $M(l)$, and observe that if $\psi_j(u) = 0$ for any $u \in \mathbb{Z} \cap M(l)$, then $\psi_j(u) \equiv 0$ on $M(l)$. Indeed, every solution of equation (17) on the group $M(l)$ has the form $\psi_j(u) = c_0 + c_1 u + \cdots + c_m u^m$, where c_j are some constants [Ge, p. 327]. Since the set $\mathbb{Z} \cap M(l)$ is infinite, $c_0 = \cdots = c_m = 0$.

Observe now that any polynomial of the degree $\leq m$ satisfies equation (17). Choose a polynomial $\lambda_j(u) = d_0 + d_1 u + \cdots + d_m u^m$ such that $\psi_j(p) = \lambda_j(p)$ for $p = 0, 1, \ldots, m$. Let $\delta_j(p) = \psi_j(p) - \lambda_j(p)$. The function $\delta_j(p)$ also satisfies equation (17) for every $l \in \mathbb{Q}$. Besides, we have $\delta_j(p) = 0$, $p = 0, 1, \ldots, m$. Therefore, putting $l = 1$, $u = 0, \pm 1, \ldots$ in (17) we obtain successively that $\delta_j(u) \equiv 0$ on $\mathbb{Z}$. As noted above, this implies that $\delta_j(u) \equiv 0$ on $M(l)$. Since l is an arbitrary rational number, $\delta_j(u) \equiv 0$ on $\mathbb{Q}$. Thus on $\mathbb{Q}$ the function $\psi_j(u)$ is a restriction of some polynomial $\lambda_j(u)$. By Theorem 2.8, $g_j(u) = \exp\{-\psi_j(u)\}$ is a positive definite function on $\mathbb{Q}$. Therefore, its continuous extension to $\mathbb{R}$ (we preserve the same notation) $g_j(u)$, $u \in \mathbb{R}$, is a continuous positive definite function on $\mathbb{R}$ as well. According to Theorem 2.8, $g_j(u)$ is a characteristic function on $\mathbb{R}$. Obviously, the functions $g_j(u)$ satisfy equation 10.3(i). By Proposition 10.3 and Theorem 10.1, we obtain $g_j(u) = \exp\{-\sigma_j u^2\}$, $\sigma_j \geq 0$, $u \in \mathbb{R}$. In particular, for every natural n the equality $g_j(nu)(g_j(u))^{n^2}$ is valid. Putting $u = 1$ here and returning to the characteristic functions $\hat{\mu}_j(y)$ we get

$$\hat{\mu}_j(ny_0) = (\hat{\mu}_j(y_0))^{n^2}, \qquad n = 2, 3, \ldots. \qquad (18)$$

Since y_0 is an arbitrary element of Y, (18) implies that $\mu_j \in \Gamma_U(X)$, $j = 1, \ldots, m$. The group X obviously satisfies the condition 5.36(ii). Therefore, by Theorem 5.36, $\mu_j \in \Gamma(X)$, $j = 1, \ldots, m$.

6. Assume $X^{(p)} = \{0\}$, where p is a prime. Then also $Y^{(p)} = \{0\}$. Since the sets $\{a_j\}_{j=1}^m$, $\{b_j\}_{j=1}^m$ are admissible for the group X, none of a_j, b_j is a multiple of p. Consider the restriction of equation 10.3(i) to the subgroup $G \subset Y$ generated by an arbitrary element y_0, $y_0 \neq 0$. Clearly, $G \approx \mathbb{Z}(p)$. Lemma 10.12 implies that the subgroup

$$E = \left\{ y \in G \colon \prod_{j=1}^m \hat{\mu}_j(y) = 1 \right\} \neq \{0\}.$$

Therefore, $E = G$, i.e., $\hat{\mu}_j(y) \equiv 1$ on G; hence, $\hat{\mu}_j(y) \equiv 1$ on Y, that is, $\mu_j = E_0 \in D(X) = \Gamma(X)$, $j = 1, \ldots, m$. $\square$

10.14. REMARK. Comparing Theorems 5.22, 10.5, and 10.9 one can see that the class of the groups for which the group analog of the Cramér theorem is valid is much wider than that for which the group analog of the Skitovich-Darmois theorem is valid. The circumstance does not change if it is additionally required that the characteristic functions $\hat{\mu}_j(y)$ of the distributions considered do not vanish (Theorem 10.13).

As in the case $X = \mathbb{R}$, the Cramér theorem for arbitrary groups is implied by the following special case of the Skitovich-Darmois theorem.

(α) Let ξ_j, $j = 1, \ldots, 4$, be X-valued independent random variables with distributions μ_j, and

    (i) $\mu_1 = \mu_3$, $\mu_2 = \mu_4$;
    (ii) $\hat{\mu}_1(y)\hat{\mu}_2(y) \neq 0$ for every $y \in Y$.

If the linear forms $L_1 = \xi_1 + \xi_2 + \xi_3 + \xi_4$ and $L_2 = \xi_1 + \xi_2 - \xi_3 - \xi_4$ are independent, then $\mu_1, \mu_2 \in \Gamma(X)$.

Indeed, let ξ_1 and ξ_2 be X-valued independent random variables with distributions μ_1, μ_2. Let the sum $\xi_1 + \xi_2$ have a Gaussian distribution, i.e., $\mu_1 * \mu_2 \in \Gamma(X)$. Choose random variables ξ_3, ξ_4 such that $\mu_1 = \mu_3$, $\mu_2 = \mu_4$. The random variables ξ_j, $j = 1, \ldots, 4$, obtained are independent. One can see that the linear forms $L_1 = (\xi_1 + \xi_2) + (\xi_3 + \xi_4)$ and $L_2 = (\xi_1 + \xi_2) - (\xi_3 + \xi_4)$ are independent. Thus (α) implies that $\mu_1, \mu_2 \in \Gamma(X)$.

In its turn, the assertion (α) itself follows from the Cramér theorem. In order to prove this, observe that, by Proposition 10.3, the independence of linear forms L_1 and L_2 is equivalent to the fact that the characteristic functions $\hat{\mu}_j(y)$ satisfy equation 10.6(1). Let $\gamma = \mu_1 * \mu_2 * \overline{\mu}_3 * \overline{\mu}_4$. Then the characteristic function $\hat{\gamma}(y)$, as it follows from 10.6(1), satisfies the equation

$$\hat{\gamma}(u + v) \cdot \hat{\gamma}(u - v) = \hat{\gamma}^2(u) \cdot \hat{\gamma}^2(v), \qquad u, v \in Y.$$

Conditions (i) and (ii), by 2.10(d), (e), imply $\hat{\gamma}(y) > 0$ for every $y \in Y$. Let $\varphi(y) = -\ln \hat{\gamma}(y)$. One can see that the function $\varphi(y)$ is the same as in

Definition 5.1, hence $\gamma \in \Gamma(X)$. Then, by the Cramér theorem, $\mu_1, \mu_2 \in \Gamma(X)$.

So, the class of the groups for which the Cramér theorem is valid coincides with the class of the groups for which statement (α) is valid. When removing condition (ii) in (α), as Lemma 10.6 shows, this coincidence ceases to take place.

10.15. Consider now the problem: for what groups X there exist admissible sets of integers $\{a_j\}_{j=1}^m$ and $\{b_j\}_{j=1}^m$ ($a_i/b_i \neq a_j/b_j$ for at least one pair of indexes i, j) such that, if ξ_j, $j = 1, \ldots, m$, $m \geq 2$, are X-valued independent random variables with distributions μ_j satisfying

$$\prod_{j=1}^m \hat{\mu}_j(y) \neq 0, \qquad y \in Y, \tag{1}$$

and the linear forms $L_1 = a_1\xi_1 + \cdots + a_m\xi_m$ and $L_2 = b_1\xi_1 + \cdots + b_m\xi_m$ are independent, then $\mu_j \in \Gamma(X)$, $j = 1, 2, \ldots, m$? It follows from Remark 9.24 that such admissible sets do exist if X has no subgroup topologically isomorphic to $\mathbb{T}$. In this case the linear forms L_1 and L_2 are $L_1 = \xi_1 + \xi_2$, $L_2 = \xi_1 - \xi_2$.

10.16. PROPOSITION. *Let the group X contain a subgroup topologically isomorphic to $\mathbb{T}$, and let $\{a_j\}_{j=1}^m$, $\{b_j\}_{j=1}^m$ be arbitrary sets of integers such that $a_i/b_i \neq a_j/b_j$ for at least one pair of indices i, j. Then there exist X-valued independent random variables ξ_j such that their distributions μ_j satisfy 10.15(i), and the linear forms $L_1 = a_1\xi_1 + \cdots + a_s\xi_s$ and $L_2 = b_1\xi_1 + \cdots + b_s\xi_s$ are independent, but $\mu_j \notin \Gamma(X)$ for at least one $j \in \{1, \ldots, m\}$.*

PROOF. Without loss of generality we may assume $X = \mathbb{T}$, $Y = \mathbb{Z}$. Two cases are possible.

1. $|a_j| = |b_j| = 1$, $j = 1, \ldots, m$. In this case we may assume $L_1 = \xi_1 + \cdots + \xi_m$, $L_2 = \xi_1 + \cdots + \xi_p - \xi_{p+1} - \cdots - \xi_m$. Consider the following functions on the group $\mathbb{Z}$:

$$f(n) = \begin{cases} \exp\{-an^2/p\}, & n \in \mathbb{Z}^{(2)}, \\ \exp\{-(an^2 - 1)/p\}, & n \notin \mathbb{Z}^{(2)}; \end{cases}$$

$$g(n) = \begin{cases} \exp\{-an^2/(m-p)\}, & n \in \mathbb{Z}^{(2)}, \\ \exp\{-(an^2 - 1)/(m-p)\}, & n \notin \mathbb{Z}^{(2)}. \end{cases}$$

One can easily see that, for sufficiently large $a > 0$, $f(n)$ and $g(n)$ are the characteristic functions of some distributions $\mu, \nu \in \mathcal{M}^1(\mathbb{T})$. Let ξ_j be $\mathbb{T}$-valued independent random variables valued in $\mathbb{T}$ having the distributions $\mu_j = \mu$ for $j = 1, \ldots, p$ and $\mu_j = \nu$ for $j = p+1, \ldots, m$. Clearly, the characteristic functions $\hat{\mu}_j(y)$ satisfy the equation

$$\prod_{j=1}^p \hat{\mu}_j(u+v) \prod_{j=p+1}^m \hat{\mu}_j(u-v) = \prod_{j=1}^m \hat{\mu}_j(u) \prod_{j=1}^p \hat{\mu}_j(v) \prod_{j=p+1}^m \hat{\mu}_j(-v), \qquad u, v \in \mathbb{Z}.$$

In view of Proposition 10.3 this means that the linear forms L_1 and L_2 are independent. By construction $\mu_j \notin \Gamma(X)$, $j = 1, \ldots, m$.

2. There is at least one j_0 such that either $|a_{j_0}| > 1$ or $|b_{j_0}| > 1$. For definiteness, let $|a_1| > 1$.

Consider an arbitrary random variable ξ_1 with values in the subgroup $\mathbb{Z}(|a_1|) \subset \mathbb{T}$ and with a nondegenerate distribution μ_1. Clearly, $\mu_1 \notin \Gamma(\mathbb{T})$. One can easily see that the characteristic function $\hat{\mu}_1(y)$ satisfies the equation

$$\hat{\mu}_1(a_1 u + b_1 v) = \hat{\mu}_1(a_1 u) \cdot \hat{\mu}_1(b_1 v), \qquad u, v \in \mathbb{Z}.$$

Let ξ_j, $j = 2, \ldots, m$, be $\mathbb{T}$-valued independent random variables with degenerate distributions $\mu_j \in D(\mathbb{T})$. Then, clearly, $\hat{\mu}_j(y)$ satisfy equation 10.3(i). By Proposition 10.3 the linear forms L_1 and L_2 are independent, but by construction $\mu_1 \notin \Gamma(\mathbb{T})$. $\square$

10.17. REMARK. It should be noted that one cannot strengthen Proposition 10.16 by demanding that all $\mu_j \notin \Gamma(X)$. Indeed, let $X = \mathbb{T}$, $Y = \mathbb{Z}$, and let ξ_1 and ξ_2 be $\mathbb{T}$-valued independent random variables, whose distributions μ_1, μ_2 satisfy $\hat{\mu}_1(y)\hat{\mu}_2(y) \neq 0$ for any $y \in \mathbb{Z}$. Let us verify that the independence of the linear forms $L_1 = \xi_1 + \xi_2$ and $L_2 = 2\xi_1 + \xi_2$ yields $\mu_2 \in D(\mathbb{T}) \subset \Gamma(\mathbb{T})$.

By Proposition 10.3 the characteristic functions $\hat{\mu}_1(y)$ and $\hat{\mu}_2(y)$ satisfy the equation

$$\hat{\mu}_1(u + 2v)\hat{\mu}_2(u + v) = \hat{\mu}_1(u)\hat{\mu}_2(u)\hat{\mu}_1(2v)\hat{\mu}_2(v), \qquad u, v \in \mathbb{Z}. \qquad (1)$$

Replacing the distributions μ_j by $\nu_j = \mu_j * \overline{\mu}_j$, if necessary, we may assume from the outset that $\hat{\mu}_j(y) > 0$ for all $y \in \mathbb{Z}$, $j = 1, 2$. Denote $f_j(y) = -\ln \hat{\mu}_j(y)$, $j = 1, 2$. Equation (1) yields

$$f_1(u + 2v) + f_2(u + v) = A(u) + B(v), \qquad u, v \in \mathbb{Z}, \qquad (2)$$

where $A(u) = f_1(u) + f_2(u)$, $B(v) = f_1(2v) + f_2(v)$.

Choose arbitrary elements $h_1, k_1 \in \mathbb{Z}$ such that $h_1 + 2k_1 = 0$. Replace u and v by $u + h_1$ and $v + k_1$ in equality (2) and subtract the initial equality (2) from what is obtained. We get

$$f_2(u + v + l) - f_2(u + v) = \Delta_{h_1} A(u) + \Delta_{k_1} B(v), \qquad u, v \in \mathbb{Z}, \qquad (3)$$

where $l = h_1 + k_1$. Now putting $v = 0$ in (3) and subtracting the result from (3) we obtain

$$f_2(u + v + l) - f_2(u + v) - f_2(u + l) + f_2(u) = \Delta_{k_1} B(v) - \Delta_{k_1} B(0), \qquad (4)$$
$$u, v \in \mathbb{Z}.$$

If k_1 is an arbitrary element of the group $\mathbb{Z}$, then $h_1 = -2k_1$ and $l = -k_1$. Hence in equation (4) u, v, l are arbitrary elements of the group $\mathbb{Z}$. Setting $v = l$ in (4) we get

$$\Delta_l^2 f_2(u) = \varepsilon(l), \qquad u, l \in \mathbb{Z},$$

and, hence,

$$\Delta_l^3 f_2(u) = 0, \qquad u, l \in \mathbb{Z}. \tag{5}$$

The latter relation implies

$$f_2(u) = c_0 + c_1 u + c_2 u^2, \qquad u \in \mathbb{Z}.$$

Since $f_2(0) = 0$ and $f_2(u) = f_2(-u)$, it follows that

$$f_2(u) = c_2 u^2, \qquad c_2 \geq 0, \ u \in \mathbb{Z}.$$

Now let us return to equation (2). Let h_2, k_2 be arbitrary elements of $\mathbb{Z}$ such that $h_2 + k_2 = 0$. Replace u and v by $u + h_2$ and $v + k_2$ and subtract the initial equality (2) from what is obtained. We get

$$f_1(u + 2v + l') - f_1(u + 2v) = \Delta_{h_2} A(u) + \Delta_{k_2} B(v), \qquad u, v \in \mathbb{Z}, \tag{6}$$

where $l' = h_2 + 2k_2$. Now putting $v = 0$ in (6) and subtracting the result from (6), we see that

$$f_1(u + 2v + l') - f_1(u + 2v) - f_1(u + l') + f_1(u) = \Delta_{k_2} B(v) - \Delta_{k_2} B(0), \tag{7}$$
$$u, v \in \mathbb{Z}.$$

If k_2 is an arbitrary element of $\mathbb{Z}$, then $h_2 = -k_2$ and $l' = k_2$ is also an arbitrary element of $\mathbb{Z}$. Hence in equation (7) the elements $u, v, l' \in \mathbb{Z}$ are arbitrary. Set $v' = 2v$ and consider equation (7) on the subgroup $\mathbb{Z}^{(2)}$. We have

$$f_1(u + v' + l') - f_1(u + v') - f_1(u + l') + f_1(u) = \varepsilon'(l'),$$
$$u, v', l' \in \mathbb{Z}^{(2)}.$$

Setting $v = l'$ here we get

$$\Delta_{l'}^2 f_1(u) = \varepsilon'(l'), \qquad u, l' \in \mathbb{Z}^{(2)},$$
$$\Delta_{l'}^3 f_1(u) = 0, \qquad u, l' \in \mathbb{Z}^{(2)}.$$

This implies that

$$f_1(u) = d_2 u^2, \qquad d_2 \geq 0, \ u \in \mathbb{Z}^{(2)},$$

on $\mathbb{Z}^{(2)}$. Substituting the expression obtained for $f_1(u)$ and $f_2(u)$ into (1) and considering this equation on the subgroup $\mathbb{Z}^{(2)}$ we have

$$d_2(u + 2v)^2 + c_2(u + v)^2 = d_2 u^2 + c_2 u^2 + 4d_2 v^2 + c_2 v^2, \qquad u, v \in \mathbb{Z}^{(2)}.$$

Hence $c_2 = d_2 = 0$, i.e., $\mu \in D(\mathbb{T})$. We have also proved that $\hat{\mu}_1(y) \equiv 1$ for $y \in \mathbb{Z}^{(2)}$. By Proposition 2.13, $\sigma(\mu_1)\hat{\mu}_1 \subset \mathbb{Z}^{(2)} \subset \mathbb{T}$.

§11. Characterization of Gaussian distribution by identical distribution of a monomial and a linear form

Yu. V. Linnik proved a theorem that characterizes Gaussian distributions on the real line by the identical distribution of a monomial and a linear form. In this section we give a complete description of those groups X to which this theorem may be extended. Some related topics are considered as well. The original result of Linnik is formulated as follows.

11.1. THEOREM. *Let $\xi_j, \ldots, \xi_m$, $m \geq 2$, be independent identically distributed random variables. If for some system $a_1, \ldots, a_m$ of real numbers satisfying $a_1^2 + \cdots + a_m^2 = 1$, the random variables ξ_j, and $a_1\xi_1 + \cdots + a_m\xi_m$ are identically distributed, then ξ_j are Gaussian random variables* [KLR, *Chapter* II].

The case $m = 2$ was studied in 1923 by Pólya. It was the first result that characterized Gaussian distribution by identical distributions of linear statistics of a repeated sample. Theorem 11.1 will be called the generalized Pólya theorem.

11.2. PROPOSITION. *Let ξ_j, $j = 1, \ldots, m$, $m \geq 2$, be independent identically distributed random variables with values in a group X and with distribution μ. For the linear forms $a_0\xi_1$ and $a_1\xi_1 + \cdots + a_m\xi_m$, $a_j \in \mathbb{Z}$, to be identically distributed, it is necessary and sufficient that the characteristic function $\hat{\mu}(y)$ satisfy the equation*

$$\text{(i)} \quad \hat{\mu}(a_0 y) = \prod_{j=1}^{m} \hat{\mu}(a_j y)$$

for any $y \in Y$.

The proof of this proposition is similar to that of Proposition 9.2, and we omit it.

Let $A = \{a_j\}_{j=0}^{m}$, $m \geq 2$, be a set of integers. By $\Gamma_A(X)$ we mean the class of distributions $\mu \in \mathscr{M}^1(X)$ with the following property: if ξ_j, $j = 1, \ldots, m$, $m \geq 2$, are X-valued independent identically distributed random variables with distribution μ, then the linear forms $a_0\xi_1$ and $a_1\xi_1 + \cdots + a_m\xi_m$ are identically distributed. By Proposition 11.2 μ belongs to the class $\Gamma_A(X)$ if and only if its characteristic function $\hat{\mu}(y)$ satisfies equation (i). According to 2.10(d), Proposition 11.2 implies that $\Gamma_A(X)$ is a subsemigroup of $\mathscr{M}^1(X)$.

11.3. PROPOSITION. *Let K be a compact subgroup of the group X and $A = \{a_j\}_{j=0}^{m}$, $m \geq 2$, a set of integers, with $a_1, \ldots, a_m$ relatively prime. Then the following statements are equivalent:*

(i) $K^{(a_0)} = K$.

(ii) *If $a_0 y \in A(Y, K)$, then $y \in A(Y, K)$.*

(iii) $m_K \in \Gamma_A(X)$.

PROOF. The equivalence of (i) and (ii) is a consequence of Lemma 9.3 because $K^{(a_0)} = \overline{K^{(a_0)}}$. Observe now that since the integers $a_1, \ldots, a_m$ are relatively prime, there exist integers $b_1, \ldots, b_m$, such that $\sum_{j=1}^{m} b_j a_j = 1$. Hence

$$\sum_{j=1}^{m} b_j a_j y = y. \tag{1}$$

(ii) $\Rightarrow$ (iii). According to Proposition 11.2 it suffices to prove that the characteristic function $\hat{m}_K(y)$ satisfies equation 11.2(i). Let us use the representation 2.14(i). If $y \in A(Y, K)$, then $a_j y \in A(Y, K)$ and $\hat{m}_K(a_j y) = 1$, $j = 0, 1, \ldots, m$, and, hence, both sides of equation 11.2(i) become unity. Suppose that $y \notin A(Y, K)$. Then by (ii) we have $a_0 y \notin A(Y, K)$ and therefore $\hat{m}_K(a_0 y) = 0$. If for each $j = 1, \ldots, m$ the element $a_j y$ belongs to $A(Y, K)$, then (1) yields $y \in A(Y, K)$. This contradiction proves that at least for one $j = j_0$ we have $a_{j_0} y \notin A(Y, K)$. Then $\hat{m}_K(a_{j_0} y) = 0$ and the right-hand side of equation 11.2(i) also is zero.

(iii) $\Rightarrow$ (ii). By Proposition 11.2 the characteristic function $\hat{m}_K(y)$ satisfies equation 11.2(i). Let $a_0 y \in A(Y, K)$. According to 2.14(i) we have $\hat{m}_K(a_0 y) = 1$ and equation 11.2(i) implies $m_K(a_j y) = 1$, that is, $a_j y \in A(Y, K)$, $j = 1, \ldots, m$. Relation (1) immediately leads to $y \in A(Y, K)$. □

11.4. For a group X, denote by $\mathscr{A}(X)$ the collection of all admissible sets $A = \{a_j\}_{j=0}^{m}$, $m \geq 2$, of relatively prime integers, satisfying the condition

$$a_0^2 = a_1^2 + \cdots + a_m^2. \tag{1}$$

Let $A \in \mathscr{A}(X)$. It follows from (1) that $\Gamma^s(X) \subset \Gamma_A(X)$. Denote by $I_A(X)$ the set of idempotent distributions from $\Gamma_A(X)$. Since $\Gamma_A(X)$ is a semigroup, we have

$$I_A(X) * \Gamma^s(X) \subset \Gamma_A(X). \tag{2}$$

Our next problem is to describe completely those group X for which

$$I_A(X) * \Gamma^s(X) = \Gamma_A(X) \tag{3}$$

for any $A \in \mathscr{A}(X)$. The solution is given below in Theorem 11.9. Accounting for 2.14 and Proposition 11.3, one may observe that the fulfillment of equality (3) for a group X means exactly the following: any distribution $\mu \in \Gamma_A(X)$ is invariant with respect to a compact subgroup $K \subset X$ such that $K^{(a_0)} = K$ and under the natural homomorphism $X \to X/K$ the distribution μ induces a Gaussian distribution on the factor group X/K.

Before passing to Theorem 11.9 let us prove several lemmas.

11.5. LEMMA. *Let X be a discrete torsion-free group and $A = \{a_j\}_{j=0}^{m}$, $m \geq 2$, an arbitrary set of integers satisfying 11.4(1). Then $\Gamma_A(X) \subset D(X)$.*

The proof of the lemma is completely similar to that of Lemma 10.4. The only difference is that one has to use the generalized Pólya Theorem 11.1 instead of the Skytovich-Darmois Theorem 10.1.

11.6. LEMMA. *Let* p *be a prime number, let the group* X *contain a subgroup* G, $G \approx Z(p)$, *and let* $X^{(p)} \neq \{0\}$. *Then for some set* $A \in \mathscr{A}(X)$ *the inclusion in* 11.4(2) *is strict.*

PROOF. Since the group G is discrete, we have, by Proposition 5.5, $\Gamma(G) = D(G)$. Therefore the class $I(G) * \Gamma(G)$ contains only degenerated distributions and the distribution m_G. Let μ_0 be nondegenerate symmetric distribution on G and $\mu_0 \neq m_G$. Then $\mu_0 \notin I(G) * \Gamma(G)$ and by Lemma 10.8 we have $\mu_0 \notin I(X) * \Gamma(X)$. Two cases are possible.

1. $p > 2$. Let $a_0 = (p^2 + 1)/2$, $a_1 = p$, and $a_2 = (p^2 - 1)/2$. Since $(p^2 + 1)/2 = p(p+1)/2 - (p-1)/2$ and $(p^2 - 1)/2 = p(p-1)/2 + (p-1)/2$, Proposition 11.2 implies $\mu_0 \in \Gamma_A(X)$, where $A = \{a_0, a_1, a_2\}$. Obviously, $A \in \mathscr{A}(X)$.

2. $p = 2$. Set $a_0 = 3$, $a_1 = a_2 = 2$, and $a_3 = 1$. Proposition 11.2 implies $\mu_0 \in \Gamma_A(X)$, where $A = \{a_0, \ldots, a_3\}$. Obviously $A \in \mathscr{A}(X)$. □

11.7. LEMMA. *Let* K *be a compact subgroup of the group* X *such that* $K^{(a_0)} = K$, *let* $A = \{a_j\}_{j=0}^m$, $m \geq 2$, *be a set of integers with* $a_1, \ldots, a_m$ *relatively prime, and let* $\nu \in \Gamma_A(X/K)$. *Then the function*

$$f(y) = \begin{cases} \hat{\nu}(y), & y \in A(Y, K), \\ 0, & y \notin A(Y, K), \end{cases}$$

is the characteristic function of some distribution $\mu \in \Gamma_A(X)$.

PROOF. Observe first that by Theorem 1.6 we have $(X/K)^* \approx A(Y, K)$. Therefore one may treat $\hat{\nu}(y)$ as a function defined on $A(Y, K)$. It follows from Theorem 2.8 and Remark 2.9 that $f(y)$ is a positive definite function. The subgroup K is compact. Therefore, by Remark 1.7 the subgroup $A(Y, K)$ is open and, hence, $f(y)$ is continuous. By Theorem 2.8 there exists a distribution $\mu \in \mathscr{M}^1(X)$ such that $\hat{\mu}(y) = f(y)$. Let us prove that $\mu \in \Gamma_A(X)$. According to Proposition 11.2, it suffices to prove that the function $f(y)$ satisfies equation 11.2(i). This is actually the case if $y \in A(Y, K)$, since, in this case $f(y) = \hat{\nu}(y)$ and $\nu \in \Gamma_A(X/K)$. If $y \notin A(Y, K)$, then, by Proposition 11.3, $a_0 y \notin A(Y, K)$. Hence $f(a_0 y) = 0$. The verification of the fact that there exists at least one $j = j_0$ such that $a_{j_0} y \notin A(Y, K)$ and, hence, $f(a_{j_0} y) = 0$, coincides with the proof of the implication (ii) $\Rightarrow$ (iii) in Proposition 11.3. Therefore equality 11.2(i) is fulfilled for any $y \in Y$, i.e., $\mu \in \Gamma_A(X)$. □

11.8. LEMMA. *Let* $X = \Delta_p$. *Then there exists a set* $A \in \mathscr{A}(X)$ *such that the inclusion* 11.4(2) *is strict.*

PROOF. The group X is totally disconnected. Therefore by Proposition 5.5, $\Gamma(X) = D(X)$. Since $\Delta_p^* \approx \mathbb{Z}(p^\infty)$, one may assume $Y = \mathbb{Z}(p^\infty)$. Let us embed the group $\mathbb{Z}(p)$ into $\mathbb{Z}(p^\infty)$ and set $K = A(X, \mathbb{Z}(p))$. Obviously, the group K is compact, and hence by Theorem 1.6 and 1.5(i) we have

$X/K \approx \mathbb{Z}(p)$ and then $(X/K)^* \approx \mathbb{Z}(p)$. If $p > 2$, set $a_0 = (p^2 + 1)/2$, $a_1 = p$, and $a_2 = (p^2 - 1)/2$. If $p = 2$, set $a_0 = 5$, $a_1 = 3$, and $a_2 = 4$. Consider on the group $\mathbb{Z}(p^\infty)$ the function

$$f(y) = \begin{cases} \hat{\nu}(y), & y \in \mathbb{Z}(p), \\ 0, & y \notin \mathbb{Z}(p), \end{cases}$$

where ν is an arbitrary nongenerated symmetric distribution on $\mathbb{Z}(p)$, $\nu \neq m_{\mathbb{Z}(p)}$. As can easily be seen, the characteristic function $\hat{\nu}(y)$ satisfies the equation

$$\hat{\nu}(a_0 y) = \hat{\nu}(a_1 y)\hat{\nu}(a_2 y)$$

for any $y \in \mathbb{Z}(p)$. By Proposition 11.2 $\nu \in \Gamma_A(X/K)$, where $A = \{a_0, a_1, a_2\}$. Lemma 11.7 being applicable here, $f(y) = \hat{\mu}(y)$ for some $\mu \in \Gamma_A(X)$. Obviously, $A \in \mathscr{A}(X)$ and by construction $\mu \notin I(X) * \Gamma(X)$. $\square$

11.9. THEOREM. *For equality* 11.4(3) *to hold for the group* X *and every* $A \in \mathscr{A}(X)$, *it is necessary and sufficient that the group* X *satisfies at least one of the conditions*:

(i) $X \approx \mathbb{R}^n + D$, *where* $n \geq 0$ *and* D *is a discrete torsion-free group.*
(ii) $X^{(p)} = \{0\}$, *where* p *is a prime number.*

PROOF. Note first that if the group X satisfies condition (i), then $I(X) = D(X)$. In this case Proposition 11.2 directly implies that $I_A(X) = D(X)$, when $a_0 = a_1 + \cdots + a_m$, and that $I_A(X) = \{E_0\}$, when $a_0 \neq a_1 + \cdots + a_m$. Therefore, for such groups, equality 11.4(3) is equivalent to having for any $A \in \mathscr{A}(X)$ either $\Gamma^s(X) = \Gamma_A(X)$ or $\Gamma(X) = \Gamma_A(X)$.

If the group X satisfies condition (ii), then X is totally disconnected and by Proposition 5.5, we have $\Gamma^s(X) = \{E_0\}$. For such groups equality 11.4(3) is equivalent to $I_A(X) = \Gamma(X)$ for any $A \in \mathscr{A}(X)$.

Necessity. Assume that the group X contains an element x_0 of finite order p, where p is a prime number. Let $M(x_0)$ be the subgroup of X generated by x_0. Then $M(x_0) \approx \mathbb{Z}(p)$ and, since equality 11.4(3) is true for every $A \in \mathscr{A}(X)$, Lemma 11.6 yields $X^{(p)} = \{0\}$, i.e., X satisfies condition (ii).

Now let X be a torsion-free group. By Theorem 1.14, $X \approx \mathbb{R}^n + G$, where $n \geq 0$ and the group G contains a compact open subgroup K. Since K is a compact torsion-free group, then, according to Theorem 1.16, the group K is isomorphic to the group 1.16(i). If $K \neq \{0\}$, then by Remark 1.17 the group K and, hence, the group X contains a subgroup G_1 that is isomorphic to Δ_p for some prime number p. According to Lemma 11.8, there exists a set $A \in \mathscr{A}(G_1)$ such that the inclusion $I(G_1) * \Gamma(G_1) \subset \Gamma_A(G_1)$ is strict. Obviously, $A \in \mathscr{A}(X)$. From here, with account taken of Proposition 2.5, it follows that 11.4(2) also is strict, which contradicts the hypothesis. The contradiction obtained shows that $K = \{0\}$, i.e., the gorup X satisfies condition (i).

Sufficiency. Let the group X satisfy condition (i). We consider $X = \mathbb{R}^n + D$. Let $A \in \mathscr{A}(X)$ and $\mu \in \Gamma_A(X)$. Since $\Gamma_A(X)$ is a semigroup, $\nu = \mu * \overline{\mu} \in \Gamma_A(X)$. Observe now that by 2.10(d), and 2.10(e) we have $\hat{\nu}(y) \geq 0$ and consider the restriction of the characteristic function $\hat{\nu}(y)$ to the subgroup $H = A(Y, \mathbb{R}^n)$, $H \approx D^*$. By 2.10(h) this restriction is the characteristic function of some distribution $\delta \in \mathscr{M}^1(D)$. Since the function $\hat{\delta}(h)$, clearly, satisfies equation 11.2(i), Proposition 11.2 and Lemma 11.5 yield $\delta = E_0$. Hence $\delta(y) \equiv 1$ for $y \in H$ and Proposition 2.13 yield $\sigma(\nu) \subset A(X, H) = \mathbb{R}^n$. It easily follows from the generalized Pólya theorem, that $\nu \in \Gamma(\mathbb{R}^n) \subset \Gamma(X)$. Applying Theorem 5.22, we conclude that $\mu \in \Gamma(X)$ and equality 11.4(3), clearly, is fulfilled.

Let us now suppose that the group X satisfies condition (ii). Let $A = \{a_j\}_{j=0}^m \in \mathscr{A}(X)$ and $\mu \in \Gamma_A(X)$. Since $\Gamma_A(X)$ is a semigroup, $\nu = \mu * \overline{\mu} \in \Gamma_A(X)$. Set $E = \{y \in Y : \hat{\nu}(y) = 1\}$. By Proposition 2.13 $\sigma(\nu) \subset A(X, E) = X_1$, and also $X_1^{(p)} = \{0\}$. Set $Y_1 = X_1^*$. The distribution ν possesses the following property on the group X_1:

$$0 \leq \hat{\nu}(h) < 1, \qquad h \in Y_1, \ h \neq 0. \tag{1}$$

The characteristic function $\hat{\nu}(h)$ satisfies equation 11.2(i). This equation yields the equality

$$\hat{\nu}(a_0^n h) = \prod_{p_1 + \cdots + p_m = n} [\hat{\nu}(a_1^{p_1} \cdots a_m^{p_m} h)]^{C_{p_1, \ldots, p_m}}, \qquad h \in Y_1, \tag{2}$$

where $C_{p_1, \ldots, p_m} = (p_1 + \cdots + p_m)!/(p_1! \cdots p_m!)$. Since the set A is admissible, no a_j is divisible by p. But $Y_1^{(p)} = \{0\}$. Therefore, for any $j = 1, \ldots, m$ the continuous homomorphism $f_{a_j}(h): Y_1 \to Y_1$ defined by the formula $f_{a_j}(h) = a_j h$ is an isomorphism. In particular, Y_1 is a group with single-valued division on a_0. From here with account taken of (2) we obtain

$$\hat{\nu}(h) = \prod_{p_1 + \cdots + p_m = n} \left[\hat{\nu}\left(\frac{a_1^{p_1} \cdots a_m^{p_m} h}{a_0^n}\right)\right]^{C_{p_1, \ldots, p_m}}, \qquad h \in Y_1. \tag{3}$$

Let us fix $h \in Y_1$, $h \neq 0$, consider $M(h)$ (the subgroup of Y_1, generated by h). Obviously, $M(h) \approx \mathbb{Z}(p)$. Observe that, whatever nonnegative integer $p_1, \ldots, p_m$, we have $(a_1^{p_1} \cdots a_m^{p_m} h)/a_0^n \in M(h)$, $(a_1^{p_1} \cdots a_m^{p_m} h)/a_0^n \neq 0$. Therefore (1) and (3) yield $\hat{\nu}(h) = 0$ for $h \neq 0$. Hence the subgroup X_1 is compact and $\nu = m_{X_1}$. Since 2.10(d) and 2.10(e) imply $\hat{\nu}(y) = |\hat{\mu}(y)|^2$, we, clearly, have $\mu \in I_A(X)$, i.e., equality 11.4(3) is fulfilled. $\square$

Let $A_m = \{a_j\}_{j=0}^{m^2}$, where $a_0 = m$, $a_1 = \cdots = a_{m^2} = 1$, $m \geq 2$. Let us now turn to the problem of complete description of those groups X for which

$$I_{A_m}(X) * \Gamma^s(X) = \Gamma_{A_m}(X) \tag{4}$$

for fixed m. The solution is given in Theorem 11.16. Later this theorem will be used to obtain a complete description of those groups X that admit characterization of Gaussian distributions by identical distribution of a monomial and a linear form (Theorem 11.23).

Before proving Theorem 11.16 we shall obtain some lemmas.

11.10. LEMMA. *Let* $\mu \in \Gamma_{A_m}$ *and* $\hat{\mu}(y) = 1$ *only for* $y = 0$. *If a subgroup* $Y_1 \subset Y$ *is such that* $|\hat{\mu}(y)| \equiv 1$ *on* Y_1, *then* $Y_1 \approx \mathbb{Z}(k)$, *where* k *is a divisor of* $m^2 - m$.

PROOF. Since $|\hat{\mu}(y)| \equiv 1$ on Y_1, by 2.10(i) we have $\hat{\mu}(y) = ([x], y)$, $y \in Y_1$, for some $[x] \in Y_1^*$. By Proposition 11.2 the characteristic function $\hat{\mu}(y)$ satisfies equation 11.2(i), which takes the form

$$\hat{\mu}(my) = (\hat{\mu}(y))^{m^2}, \qquad y \in Y. \tag{1}$$

Setting $\hat{\mu}(y) = ([x], y)$ in equation (1) and considering the restriction of this equation to the subgroup Y_1 we obtain $([x], my) = ([x], y)^{m^2}$, i.e., $((m^2 - m)[x], y) \equiv 1$ for all $y \in Y_1$. Hence $(m^2 - m)[x] = 0$. Consider the continuous homomorphism $p: Y_1 \to \mathbb{Z}(m^2 - m) \subset \mathbb{T}$ defined by the formula $p(y) = ([x], y)$. By hypothesis $\ker p = \{0\}$. Consequently p is a monomorphism and the group Y_1 is isomorphic to the group $\mathbb{Z}(m^2 - m)$. $\square$

11.11. LEMMA. *Let* $\mu \in \Gamma_{A_m}(X)$. *Then there exists an element* $x \in X$ *such that* $mx = 0$ *and* $\sigma(\mu * E_x) \subset X_1$, *where* X_1 *is a subgroup of* X, *having the structure* $X_1 \approx \mathbb{R}^n + K$, $n \geq 0$, *the group* K *being compact, and* $K^{(m)} = K$.

PROOF. Let $E = \{y \in Y: \hat{\mu}(y) = 1\}$. By Proposition 2.13 we have $\sigma(\mu) \subset A(X, E)$ and so the distribution μ may be treated as a distribution on $A(X, E)$. Therefore, we may assume from the outset that $\hat{\mu}(y) = 1$ for $y = 0$ only. Applying Theorem 1.14 we may also assume that $X = \mathbb{R}^n + G$, where $n \geq 0$ and the group G contains an open compact subgroup K, and $Y = \mathbb{R}^n + H$, $H \approx G^*$. By Proposition 11.2 the characteristic function $\hat{\mu}(y)$ satisfies equation 11.10(1). Consider the restriction of this equation to the subgroup C_H. Since C_H is a connected compact group, and by Lemmas 11.5 and 11.10 we obtain $C_H = \{0\}$, i.e., the group H is totally disconnected. Verify first that H is discrete. To do this consider a compact open subgroup $L \subset H$. Let V be a neighborhood of zero in L such that $|\hat{\mu}(y)| > 0$ for any $y \in V$. By Theorem 1.19 there exists a compact subgroup B of the group L, such that $B \subset V$ and

$$L/B \approx \mathbb{T}^q + F, \tag{1}$$

where $q \geq 0$ and F is a finite group. Consider the restriction of equation 11.10(1) to B. Let $|\hat{\mu}(y_0)| < 1$ at some point $y_0 \in B$. Since the group B

is compact, the sequence $\{m^l y_0\}$ has a limit point $y' \subset B$. From 11.10(1) we obtain

$$\hat{\mu}(m^l y) = (\hat{\mu}(y))^{m^{2l}}, \qquad y \in Y. \tag{2}$$

Setting $y = y_0$ in (2) and passing to the limit as $l \to \infty$ we get $\hat{\mu}(y') = 0$, which is impossible. Therefore $|\hat{\mu}(y)| \equiv 1$ on B. Consequently, by Lemma 11.10 we have $B \approx \mathbb{Z}(k)$, where k is a divisor of $m^2 - m$. Since the group H is totally disconnected, the group L is totally disconnected as well and (1) yields $q = 0$. Taking into account the possible structure of the group B, we obtain that L is finite and, hence, discrete. Therefore, the group H is discrete as well, and by Theorem 1.7 the group G is compact.

Let us consider the subgroup $H_1 = \{y \in Y : my = 0\}$ and verify that $Y^{(m)} \cap H_1 = \{0\}$. Let $\eta \in Y$, $m\eta = \zeta \in H_1$. We have

$$m^2 l\eta = lm\zeta = 0 \tag{3}$$

for any natural l. Therefore $1 = \hat{\mu}(m^2 l\eta) = (\hat{\mu}(l\eta))^{m^4}$. Hence $|\hat{\mu}(y)| \equiv 1$ for all y belonging to the subgroup $M(\eta)$, generated by η. Let d be the order of η. It follows from (3) that d is a divisor of m^2. On the other hand, by Lemma 11.10 d is a division of $m^2 - m$. Consequently, d is a divisor of m and $\zeta = m\eta = 0$. Thus the subgroup H_1 satisfies the condition: if $my \in H_1$, then $y \in H_1$. Consider the group $X_1 = A(X, H_1)$. Obviously $X_1 \approx \mathbb{R}^n + K$, where K is a compact group. Since by 1.5(i) $H_1 = A(Y, X_1)$, Lemma 9.3 implies $\overline{X_1^{(m)}} = X_1$. From here we conclude that $K^{(m)} = K$.

Consider the restriction of the characteristic function $\hat{\mu}(y)$ to H_1. Since equation 11.10(1) yields $|\hat{\mu}(y)| \equiv 1$ on H_1, we get from 2.10(i) that $\hat{\mu}(y) = ([x], y)$, $y \in H_1$, for some $[x] \in H_1^*$. Define a character $\psi(y)$ on the direct sum $H_1 + H^{(m)}$ by the formula $\psi(y_1, y_2) = (\overline{[x], y_1})$, $y_1 \in H_1$, $y_2 \in H^{(m)}$. According to Theorem 1.6, one may extend this character to a character ψ on the whole group Y, i.e., $\psi(y) = (x, y)$, $y \in Y$, for some $x \in X$. By construction $x \in A(X, H^{(m)})$ and, hence, $mx = 0$. Set $\hat{\mu}_1 = \mu * E_x$. Then $\hat{\mu}_1(y) = 1$ for $y \in H_1$. Therefore, by Proposition 2.13 we have $\sigma(\mu_1) \subset A(X, H_1) = X_1$. $\square$

11.12. LEMMA. *Let the group X satisfy the equation*

$$\text{(i)} \quad Y^{(m)} = Y.$$

If $\mu \in \Gamma_{A_m}(X)$ is such that $\hat{\mu}(y) \neq 0$ for all $y \in Y$, then $\mu \in \Gamma(X)$.

PROOF. It follows from the hypothesis and Proposition 1.22 that the group X does not contain a subgroup topologically isomorphic to $\mathbb{T}$. Since $\Gamma_{A_m}(X)$ is a semigroup, according to Theorem 5.22 it suffices to verify that $\nu = \mu * \overline{\mu} \in \Gamma(X)$. By 2.10(d) and 2.10(e) we have $\hat{\nu}(y) = |\hat{\mu}(y)|^2 > 0$. Therefore we may assume from the outset that $\hat{\mu}(y) > 0$ for all $y \in Y$. Set $E = \{y \in Y : \hat{\mu}(y) = 1\}$. By Proposition 2.13 $\sigma(\mu) \subset A(X, E) = X_1$. Let us treat μ as a distribution on X_1. Since by Theorems 1.6 and 1.5 we have $X_1^* \approx Y/E$, condition (i) implies $(X_1^*)^{(m)} = X_1^*$, i.e., condition (i) remains valid when passing from X to X_1. It allows one to consider from the outset that $\hat{\mu}(y) = 1$ only for $y = 0$.

In accordance with Proposition 11.2, the characteristic function $\hat{\mu}(y)$ satisfies equation 11.10(1) and, hence, equation 11.11(2). Consider the restriction of the latter equation to Y_0, namely, the subgroup of all compact elements of Y. Let $y_0 \in Y_0$. The sequence $\{m^l y_0\}$ has a limit point $y' \in Y_0$. If $\hat{\mu}(y_0) < 1$, then setting $y = y_0$ in 11.11(2) and passing to the limit as $l \to \infty$ we obtain $\hat{\mu}(y') = 0$. Hence $\hat{\mu}(y) \equiv 1$ on Y_0, that is, $Y_0 = \{0\}$. In view of Corollary 1.10 the group X is connected, and by 1.20 $X^{(k)} = X$ for any natural k. In particular,

$$X^{(m)} = X. \tag{1}$$

Conditions (i), (1), and 1.20 yield that both X and Y are groups with single-valued division on m. It should also be mentioned that by Corollary 1.10 Y is a torsion-free group.

Fix $y_0 \in Y$, and consider the subgroup $L(y_0) = \{py_0/m^l\}_{p,l=-\infty}^{\infty}$. This subgroup may be naturally embedded in $\mathbb{R}$: $(py_0/m^l) \to p/m^l$. Denote the image of $L(y_0)$ by H. Let $f(r) = \hat{\mu}(ry_0)$, $r \in H$. By Theorem 2.8 $f(r)$ is a positive definite function on H. Let us check that this function is uniformly continuous in the topology induced on H by the group $\mathbb{R}$. By the inequality 2.10(c), it suffices to verify that $f(r)$ is continuous at the point $r = 0$. Set $f(1) = \hat{\mu}(y_0) = a$. Equation 11.11(2) yields $f(1/m^l) = f(1)^{1/m^{2l}} = a^{1/m^{2l}}$. From here

$$1 - f\left(\frac{1}{m^l}\right) \le \frac{1-a}{a} \frac{1}{m^{2l}}. \tag{2}$$

One can easily check that the inequality

$$1 - \text{Re}(x, py) \le p^2(1 - \text{Re}(x, y)) \tag{3}$$

is fulfilled for any $x \in X$, $y \in Y$, and any natural p. Therefore, for an arbitrary distribution $\gamma \in \mathscr{M}^1(X)$, we obtain

$$1 - \text{Re}\,\hat{\gamma}(py) \le p^2(1 - \text{Re}(\hat{\gamma}(y))), \qquad y \in Y.$$

Taking into account that $\hat{\mu}(y) > 0$ for all $y \in Y$ we have

$$1 - \hat{\mu}(py) \le p^2(1 - \hat{\mu}(y)), \qquad y \in Y. \tag{4}$$

Setting $y = (1/m^l)y_0$ in (4) and using inequality (2) we conclude that

$$1 - f\left(\frac{p}{m^l}\right) \le p^2\left(1 - f\left(\frac{1}{m^l}\right)\right) \le \frac{1-a}{a}\left(\frac{p}{m^l}\right)^2. \tag{5}$$

This implies that the function $f(r)$ is uniformly continuous on H with respect to the topology induced on H by the group $\mathbb{R}$. Hence the function

$f(r)$ admits an extension to a continuous positive definite function $f(s)$ on $\mathbb{R}$. By Theorem 2.8 there exists a distribution $\delta \in \mathscr{M}^1(\mathbb{R})$ such that $f(s) = \hat{\delta}(s)$. Observe now that the equality $f(mr) = (f(r))^{m^2}$, $r \in H$, yields $f(ms) = (f(s))^{m^2}$ for any $s \in \mathbb{R}$ and, hence, $\hat{\delta}(ms) = (\hat{\delta}(s))^{m^2}$. By Proposition 11.2 we get $\delta \in \Gamma_{A_m}(\mathbb{R})$ and, consequently, by the generalized Pólya Theorem 11.1 we get $\delta \in \Gamma^s(\mathbb{R})$. Hence $f(s) = \hat{\delta}(s) = \exp\{-\sigma s^2\}$, $\sigma \geq 0$, and, in particular, $f(ns) = (f(s))^{n^2}$. After setting $s = 1$ here and returning to the characteristic function $\hat{\mu}(y)$ we obtain

$$\hat{\mu}(ny_0) = (\hat{\mu}(y_0))^{n^2}. \tag{6}$$

Since y_0 is an arbitrary element of Y, equality (6) implies that $\mu \in \Gamma_U(X)$. Since $Y^{(m)} = Y$, we also have $(Y/Y_1)^{(m)} = Y/Y_1$ for any factor-group Y/Y_1. Therefore, all elements of the factor group Y/Y_1 are infinitely divisible. Hence, the group X satisfies condition 5.36(ii), and using Theorem 5.36 we obtain $\mu \in \Gamma(X)$. □

11.13. LEMMA. *Let a group X satisfy the condition*

$$\text{(i)} \quad X^{(m)} = X, \qquad Y^{(m)} = Y.$$

If $\mu \in \Gamma_{A_m}(X)$ and $\hat{\mu}(y_0) = 0$ at some point $y_0 \in Y$, then μ has a nondegenerate idempotent divisor.

PROOF. It follows from condition (i) and 1.20 that X and Y are groups with single-valued division by m. Let an arbitrary natural l be fixed. Consider the function $\hat{\mu}(y/m^l)$ on the group Y. By Theorem 2.8 this is the characteristic function of some distribution $\mu_l \in \mathscr{M}^1(X)$. By Proposition 11.2 the characteristic function $\hat{\mu}(y)$ satisfies equation 11.10(1) and, hence, equation 11.11(2). The latter equation yields $\hat{\mu}(y) = (\hat{\mu}(y/m^l))^{m^{2l}}$. According to 2.10(b), this yields $\mu = \mu_l^{*m^{2l}}$.

By Theorem 2.2 any sequence of divisors of a given distribution is shift compact, i.e., it contains a subsequence that converges after appropriate shifts. Since $\{\mu_l\}$ are divisors of μ, let γ be some limit of shifts of the μ_l. Clearly, any power of γ also is a divisor of μ; consequently, the sequence $\{\gamma^{*n}\}$ is shift compact as well. Denote by λ some limit of shifts of the γ^{*n}. Obviously, λ is an idempotent divisor of μ. If $\hat{\mu}(y_0) = 0$ at some point $y_0 \in Y$, then $\hat{\mu}_l(y_0) = 0$, $l = 1, 2, \ldots$, and therefore $\hat{\gamma}(y_0) = 0$ and, hence, also $\lambda(y_0) = 0$, that is, λ is a nondegenerate distribution. □

11.14. LEMMA. *Let the group X satisfy condition 11.12(i). If $\mu \in \Gamma_{A_m}(X)$, then the set $E = \{y \in Y : \hat{\mu}(y) \neq 0\}$ is an open subgroup of Y.*

PROOF. Let $H = M(E)$ be the open subgroup of Y generated by E, and let $f(y)$ be the restriction of the characteristic function $\hat{\mu}(y)$ to H.

According to 2.10(h) we have $f(y) = \hat{\delta}(y)$ for some $\delta \in \mathscr{M}^1(X/A(X, H))$. Proposition 11.2 implies $\delta \in \Gamma_{A_m}(X/A(X, H))$. Since $Y^{(m)} = Y$, we conclude $H^{(m)} = H$. Indeed, any $h \in H$ has the form $h = k_1 y_1 + \cdots + k_l y_l$, $k_j \in \mathbb{Z}$, $y_j \in E$. By the hypothesis $y_j = mz_j$, $j = 1, \ldots, l$, and it follows from equation 11.10(1) that $z_j \in E$. Therefore, $h = m(k_1 z_1 + \cdots + k_l z_l) \in H^{(m)}$. By construction, all idempotent divisors of the distribution δ are degenerate. Indeed, if a distribution $\nu \in \mathscr{M}^1(X)$ has a nondegenerate idempotent divisor, then the group Y cannot be generated by the set $\{y \in Y : \hat{\nu}(y) \neq 0\}$. Finally, let us note that $X^{(m)} = X$ implies $(X/A(X, H))^{(m)} = X/A(X, H)$. Applying Lemma 11.13 to the group $X/A(X, H)$, we obtain $\hat{\delta}(y) \neq 0$ for $y \in H$. But for $y \in H$ we have $\hat{\mu}(y) = \hat{\delta}(y)$. Hence $E = H$. □

11.15. LEMMA. *For an arbitrary group* X *the following statements are equivalent*:

(i) *The equality* $(K^*)^{(m)} = K^*$ *is fulfilled for any compact subgroup* $K \subset X$ *such that* $K^{(m)} = K$.

(ii) $\{x \in C_X : mx = 0\} = \{0\}$.

PROOF. (i) $\Rightarrow$ (ii). By Theorem 1.15 we have $C_X \approx \mathbb{R}^n + G$, where $n \geq 0$ and G is a connected compact group. According to §1.20, $G^{(p)} = G$ for any natural p. Then (i) yields $(G^*)^{(m)} = G^*$ and, hence, $(C_X^*)^{(m)} = C_X^*$. By Lemma 9.3 the latter equality is equivalent to relation (ii).

(ii) $\Rightarrow$ (i). Let K be a compact subgroup in X such that $K^{(m)} = K$. Let us set $H = K^*$ and verify that $H^{(m)} = H$. Consider the subgroup $H' \subset H$ consisting of all the elements that are infinitely divisible by m. First let us check that the factorgroup $L = H/H'$ contains neither nonzero elements infinitely divisible by m nor elements of finite order. Let $[h] \in L$ be an element infinitely divisible by m, i.e., let the equation

$$m^l[t] = [h] \tag{1}$$

have a solution in L for any natural l. Equality (1) is equivalent to $m^l t - h \in H'$, i.e., $m^l t - h = m^l y$ for some $y \in H$. Hence $h = m^l(t - y)$, i.e., $h \in H'$, or $[h] = 0$. Therefore the subgroup L does not contain nonzero elements that are infinitely divisible by m. Now let $[h] \in L$ be an element of finite order p. Without loss of generality, one may assume that p is a prime number. If m is not a multiple of p, then the element h is infinitely divisible by m and, as is proved above, $[h] = 0$. Therefore we can assume that p is a divisor of m. We have $p[h] = 0$, i.e., $ph \in H'$. Consequently, for any natural l there exists an element $z \in H$ such that $ph = m^{l+1} z$. Then $p(h - m^{l+1} z/p) = 0$. Since $K^{(m)} = K$, we have by Lemma 9.3 that $\{h \in H : mh = 0\} = \{0\}$, and since p is a divisor of m, we conclude that $h = m^l mz/p$, i.e., $h \in H'$ and $[h] = 0$.

So L is a discrete torsion-free group. According to Corollary 1.10, the group L^* is connected. By Theorem 1.6 we have $L^* \approx A(K, H')$. Therefore $A(K, H') \subset C_X$ and (ii) implies that $\{x \in A(K, H'): mx = 0\} = \{0\}$. From here by Lemma 9.3 we obtain $L^{(m)} = L$, i.e., the group L consists of elements infinitely divisible by m. Hence $L = \{0\}$. Thus we have obtained $H = H'$ and finally $H^{(m)} = H$. $\quad\square$

11.16.   THEOREM. *For equality 11.9(4) to be fulfilled on a group X, it is necessary and sufficient that the group X satisfy condition 11.15(i).*

PROOF. *Necessity.* Let there exist a compact subgroup $K \subset X$ such that $K^{(m)} = K$ and $(K^*)^{(m)} \neq K^*$. Let us denote $H = K^*$ and consider the subgroup $H' \subset H$ consisting of all elements infinitely divisible by m. It was established of the (ii) $\Rightarrow$ (i) part of the proof of Lemma 11.15 that the factor group $L = H/H'$ is a discrete torsion-free group and that none of its nonzero elements is infinitely divisible by m. By the hypothesis $L \neq \{0\}$.

For any element $z \in L$, $z \neq 0$, there exists a nonnegative integer $p(z)$ such that the equation $m^{p(z)} t = z$ has a solution in L but $m^{p(z)+1} t = z$ does not. Set $B_z = \{m^p z\}_{p=-p(z)}^{\infty}$ for every $z \in L$, $z \neq 0$. The group L may be represented as a union of disjoint sets

$$L = \{0\} \cup \bigcup_{k=1}^{\infty} B_{z_k}.$$

Define on B_{z_k} the function $\psi(z) = m^{2p} a_k$, $z = m^p z_k$, $p = -p(z_k)$, $-p(z_k)$ $+1, \dots$, where the numbers a_k are such that $\psi(z) = \psi(-z)$ and

$$\sum_{k=1}^{\infty} \sum_{p=-p(z_k)}^{\infty} \exp\{-m^{2p} a_k\} < 1.$$

Consider on the group L^* the continuous nonnegative function

$$\rho(g) = 1 + \sum_{k=1}^{\infty} \sum_{p=-p(z_k)}^{\infty} \exp\{-m^{2p} a_k\} \overline{(g, m^p z^k)}, \qquad g \in L^*.$$

Let μ be a distribution on the group L^* having the density $\rho(g)$ with respect to m_{L^*}. Since by construction

$$\hat{\mu}(z) = \begin{cases} \exp\{-\psi(z)\}, & z \neq 0, \\ 1, & z = 0, \end{cases}$$

the characteristic function $\hat{\mu}(z)$ satisfies equation 11.10(1) and by Proposition 11.2 $\mu \in \Gamma_{A_m}(L^*)$. Clearly, it is always possible to choose the numbers a_k in such a way that the function $\psi(z)$ does not satisfy equation 5.1(ii) and then $\mu \notin \Gamma(L^*)$.

Since by Theorem 1.6 $L^* \approx A(K, H') \subset K \subset X$, we can treat the distribution μ as a distribution on X. Then $\mu \in \Gamma_{A_m}(X)$ and $\mu \notin I(X) * \Gamma(X)$.

Sufficiency. Let $\mu \in \Gamma_{A_m}(X)$. According to Lemma 11.1, one can replace the distribution μ by its shift $\mu' = \mu * E_x$ with $mx = 0$, so that $\sigma(\mu') \subset B$, where $B \approx \mathbb{R}^n + K$, $n \geq 0$, and the group K is compact and $K^{(m)} = K$. Clearly, $\mu' \in \Gamma_{A_m}(B)$. Let $H = B^*$. It follows from 11.15(i) that $H^{(m)} = H$. Applying Lemma 11.14 to the group B we verify that the set $E = \{h \in H : \widehat{\mu'}(h) \neq 0\}$ is an open subgroup of H. By Remark 1.7 the group $G = A(B, E)$ is compact. By Proposition 11.2 the characteristic function $\widehat{\mu'}(h)$ satisfies equation 11.10(1), which implies that if $my \in E$, then $y \in E$. By Lemma 9.3 we obtain that $G^{(m)} = G$.

Let us consider the restriction $f(h)$ of the characteristic function $\widehat{\mu'}(h)$ to E. By 2.10(h) we have $f(h) = \hat{\gamma}(h)$ for some $\gamma \in \mathcal{M}^1(B/G)$. According to Proposition 11.2, $\gamma \in \Gamma_{A_m}(B/G)$. One can easily see that $E^{(m)} = E$. By Theorem 1.6 $(B/G)^* \approx E$. Therefore Lemma 11.12, applied to the distribution γ, yields $\gamma \in \Gamma(B/G)$. Hence

$$\widehat{\mu'}(h) = ([b], h)\exp\{-\varphi(h)\}, \qquad h \in E,$$

where $[b] \in B/G$ and the function $\varphi(h)$ on the set E is the same as in Definition 5.1. Now observe that the function $([b], h)$ satisfies equation 11.10(1) because the functions $\widehat{\mu'}(h)$ and $\exp\{-\varphi(h)\}$ satisfy this equation. Therefore $(m^2 - m)[b] = 0$. (Since $E^{(m)} = E$, by Lemma 9.3 we have $\{[b] \in B/G : m[b] = 0\} = \{0\}$. Hence $(m^2 - m)[b] = 0$ implies $(m - 1)[b] = 0$.)

According to Proposition 5.4, we may extend the function $\varphi(h)$, with preservation of its properties, from the subgroup E onto the whole group H. We preserve the notation $\varphi(h)$ for the extended function. Let $\mu_0 \in \Gamma^s(B)$, and let the characteristic function $\hat{\mu}_0(h)$ have the form $\hat{\mu}_0(h) = \exp\{-\varphi(h)\}$. Then we have $\mu' = m_G * E_b * \mu_0 = \lambda * \mu_0$ where $b \in B$, $\lambda = m_G * E_b \in I_{A_m}(B)$, and $\gamma_0 \in \Gamma^s(B)$. Consequently $\mu = \mu' * E_{-x} \in I_{A_m}(X) * \Gamma^s(X)$. $\square$

11.17. REMARK. It follows from the proof of Theorem 11.16 that condition 11.5(i) is necessary and sufficient for the following statement to be valid: if $\mu \in \Gamma_{A_m}(X)$ and $\hat{\mu}(y) \neq 0$ for all $y \in Y$, then $\mu \in \Gamma(X)$.

Now let us present another proof of Lemma 11.12. This proof is based on a principally different reasoning and is of independent interest. We shall restrict ourselves to the case $X^{(m)} = X$. As was shown in the proof of Lemma 11.12, this does not involve loss of generality. Lemma 11.12 is a consequence of the following assertions.

11.18. LEMMA. *Let the group X satisfy condition 11.3(i). If $\mu \in \Gamma_{A_m}(X)$ and $\hat{\mu}(y) > 0$ for all $y \in Y$, then μ is an infinitely divisible distribution.*

PROOF. Let us consider the following triangular sequence of distributions: $\mu_{lj} = \mu_l$, $j = 1, \ldots, m^{2l}$, $l = 1, 2, \ldots$, where μ_l are those constructed in the proof of Lemma 11.13. Since $\hat{\mu}_l(y) = (\hat{\mu}(y))^{1/m^{2l}}$, the distributions μ_{lj}

clearly satisfy condition 2.23(i), i.e., they are infinitesimal distributions. By Theorem 2.24 the distribution $\mu = \mu_{l1} * \cdots * \mu_{lm^{2l}}$ is infinitely divisible. □

11.19. LEMMA. *Let X be an arbitrary group, let γ be an infinitely divisible distribution on X such that $\hat{\gamma}(y) \neq 0$ for all $y \in Y$, and let $\nu = \gamma * \overline{\gamma}$. Then, for any natural m,*

$$(i) \quad \hat{\nu}(my) \geq (\hat{\nu}(y))^{m^2}, \qquad y \in Y.$$

The equality sign occurs if and only if $\gamma \in \Gamma(X)$.

PROOF. Let us use representation 2.22(i). This representation, properties of the function $g(x, y)$ and 2.10(e), 2.10(d) together imply that

$$\hat{\nu}(y) = \exp\left\{ \int_{X\backslash\{0\}} [\mathrm{Re}(x, y) - 1] \, d\Phi(X) - \varphi(y) \right\}. \tag{1}$$

Observe now that inequality 11.12(3) becomes equality if and only if $(x, y) = 1$. Hence we obtain

$$\int_{X\backslash\{0\}} [\mathrm{Re}(x, my) - 1] \, d\Phi(x) \geq m^2 \int_{X\backslash\{0\}} [\mathrm{Re}(x, y) - 1] \, d\Phi(x). \tag{2}$$

Since $\varphi(my) = m^2 \varphi(y)$, inequality (2) implies (i). If equality occurs in (i) for any $y \in Y$, then equality occurs in (2) for any $y \in Y$ as well. Therefore the measure Φ is concentrated on the set $\{x: \mathrm{Re}(x, my) - 1 = m^2(\mathrm{Re}(x, y) - 1)$ for any $y \in Y\}$, i.e., on the set $\{x \in X: (x, y) = 1$ for any $y \in Y\}$. Hence the measure Φ is degenerate in zero, i.e., $\nu \in \Gamma(X)$. Then $\gamma \in \Gamma(X)$ as well. The converse statement is evident. □

11.20. SECOND PROOF OF LEMMA 11.12. It follows from the hypothesis and Proposition 1.22 that no subgroup of X is topologically isomorphic to $\mathbb{T}$. Since $\Gamma_{A_m}(X)$ is a semigroup, according to Theorem 5.22 it suffices to check that $\gamma = \mu * \overline{\mu} \in \Gamma(X)$. The distribution γ satisfies the condition $\hat{\gamma}(y) = |\hat{\mu}(y)|^2 > 0$ for all $y \in Y$. By Lemma 11.18 γ is an infinitely divisible distribution. According to Proposition 11.2 the characteristic function $\hat{\gamma}(y)$ satisfies equation 11.10(1). Set $\nu = \gamma * \overline{\gamma}$. By 2.10(d) the characteristic function $\hat{\nu}(y)$ also satisfies equation 11.10(1) and by Lemma 11.19 we have $\gamma \in \Gamma(X)$. □

11.21. REMARK. The second proof of Lemma 11.12 does not use the generalized Pólya Theorem 11.1.

11.22. REMARK. By using Corollary 2.18, Lemma 11.19 and the line of reasoning in the proof of the sufficiency part of Theorem 11.16, one can easily prove that if an infinitely divisible distribution μ belongs to $\Gamma_{A_m}(X)$, then $\mu \in I_{A_m}(X) * \Gamma^s(X)$.

We say that, on a group X, a Gaussian distribution is characterized by identical distributions of a monomial and a linear form, if for some $A \in \mathscr{A}(X)$ equality 11.4(3) is valid. Theorem 11.6 allows one to obtain a complete description of such groups X.

11.23. THEOREM. *For equality 11.4(3) to be valid on a group X for some $A \in \mathscr{A}(X)$, it is necessary and sufficient that the group C_X have no elements of order p, for some prime p.*

PROOF. *Necessity.* Let the group C_X contain an element of order p for every prime p, i.e., $\{x \in C_X : px = 0\} \neq \{0\}$. By Theorem 1.15 we may assume that $C_X = K$ is a connected compact group. Set $H = K^*$. By Lemma 9.3 $H^{(p)} \neq H$ for each prime p. By Corollary 1.10 H is a discrete torsion-free group. Let $A = \{a_j\}_{j=0}^m \in \mathscr{A}(X)$. Consider the equation

$$f(a_0 y) = \prod_{j=1}^{m} f(a_j y), \qquad y \in H. \tag{1}$$

Let $E = \{p_1, \ldots, p_k\}$ be the set of common prime divisors of the numbers $a_0, a_1, \ldots, a_m$. Since $H^{(p_i)} \neq H$ for all $i = 1, \ldots, k$, there exists an element $h_0 \in H$ such that $(L_{h_0})^{(p_i)} \neq L_{h_0}$, $i = 1, \ldots, k$, where L_h is the subgroup of H that consists of elements depending on h. Indeed, since $H^{(p_i)} \neq H$, we have $(L_{h_i})^{(p_i)} \neq L_{h_i}$ for some $h_i \in H$. Consider the subgroup $H_1 \subset H$ that consists of all elements depending on $h_1, \ldots, h_k$. Set $r(H_1) = l$. Let $z_1, \ldots, z_n$, $n = (l-1)k + 1$, be chosen in H_1 in such a way that any l elements $\{z_{j_1}, \ldots, z_{j_l}\}$ generate a subgroup $M\{z_{j_1}, \ldots, z_{j_l}\}$ of rank l. Now observe that if an element qz, where $q \in \mathbb{Z}$ and $z \in H$, is infinitely divisible by p_i, then z is infinitely divisible by p_i as well. Therefore the set $\{z_1, \ldots, z_n\}$ contains no more than $(l-1)$ elements infinitely divisible by p_i for every $i = 1, \ldots, k$. Hence there is at least one element $h_0 = z_j$, which is not infinitely divisible by any p_i, $i = 1, \ldots, k$, i.e., $(L_{h_0})^{(p_i)} \neq L_{h_0}$, $i = 1, \ldots, k$.

The group L_{h_0} is topologically isomorphic to some subgroup B of the group $\mathbb{Q}$. We may assume that B possesses the following property: if $m/n \in B$, then no p_i, $i = 1, \ldots, k$, is a divisor of n. Denote by π the isomorphism $\pi : L_{h_0} \to B$.

Let $B_E \subset B$ be the set consisting of all integers q of the form $q = p_1^{b_1} \cdots p_k^{b_k}$, where $p_i \in E$ and $b_i \geq 0$. If $q \in B_E$, then, clearly, $a_j q \in B_E$ for all $j = 0, 1, \ldots, m$. If $q \in B$ and $a_j q \in B_E$ for some j, then $q \in B_E$. Consider the function

$$f(y) = \begin{cases} 1 & \text{if } y = 0, \\ \exp\{-2q^2\} & \text{if } y \in \pi^{-1}(B_E), \\ 0 & \text{if } y \in H \backslash (\pi^{-1}(B_E) \cup \{0\}) \end{cases}$$

on H. By construction this function satisfies equation (1). Set

$$\rho(g) = \sum_{y \in H} f(y)\overline{(g, y)}, \qquad g \in K.$$

The function $\rho(g)$ is continuous and $\rho(g) > 0$, since $\sum_{y \in H, \ y \neq 0} f(y) < 1$.
Let $\mu \in \mathscr{M}^1(K)$ be the distribution having the density $\rho(g)$ with respect
to m_K. Its characteristic function has the form $\hat{\mu}(y) = f(y)$. According
to Proposition 11.2 we have $\mu \in \Gamma_A(K) \subset \Gamma_A(X)$. By construction $\mu \notin I(X) * \Gamma(X)$.

Sufficiency. If for some prime p the group C_X has no elements of order
p, then $\{x \in C_X : px = 0\} = \{0\}$. By Lemma 11.15 statement 11.15(i) is
valid and then by Theorem 11.16 equality 11.9(4) is satisfied for $m = p$. If
$X^{(p)} \neq 0$, then the set $A = \{a_j\}_{j=0}^{p^2}$, where $a_0 = p$, $a_1 = \cdots = a_{p^2} = 1$,
belongs to $\mathscr{A}(X)$. If $X^{(p)} = 0$, then by Theorem 11.9 equality 11.4(3) is
true for any $A \in \mathscr{A}(X)$. $\square$

11.24. REMARK. The group $\mathbb{T}$ is the simplest example of a connected
compact group that contains elements of order p for any prime p. Another
example may be obtained as follows. Let the subgroup $H \subset \mathbb{Q}$ consist of
all elements of the form m/n, where n is not a multiple of the square of
any prime number. Then, for any prime p we have $H^{(p)} \neq H$. By Theorem
1.15 and Lemma 9.3 the group $X = H^*$ is a connected compact group that
contains an element of order p for any prime p.

11.25. Let $\Gamma_\infty(X) = \bigcap_A \Gamma_A(X)$ where the intersection is taken over all
collections $A = \{a_j\}_{j=0}^m$ of integers satisfying condition 11.4(1). The class
$\Gamma_\infty(X)$ is closely connected with the class $\Gamma_U(X)$ of all Gaussian distributions in the sense of Urbanik (see §5.23). Clearly, $\Gamma^s(X) \subset \Gamma_\infty(X)$. Denote
by $I_\infty(X)$ the set of all idempotent distributions from $\Gamma_\infty(X)$. Since $\Gamma_\infty(X)$
is a semigroup, we have

$$I_\infty(X) * \Gamma^s(X) \subset \Gamma_\infty(X). \tag{1}$$

The next characterization problem we want to consider is the problem of
complete description of groups X such that

$$I_\infty(X) * \Gamma^s(X) = \Gamma_\infty(X). \tag{2}$$

The solution is given in Theorem 11.32. First, let us establish some properties
of the class $\Gamma_\infty(X)$.

11.26. LEMMA. *Let $X \approx \mathbb{T}$. Then $\Gamma_\infty(X) = \{m_X\} \cup \Gamma^s(X) \cup E_\zeta * \Gamma^s(X)$,
where ζ is an element of order two in the group X.*

PROOF. By Proposition 11.2 the characteristic function of a distribution
$\mu \in \Gamma_\infty(X)$ on any group X satisfies the system of equations

$$\hat{\mu}(my) = (\hat{\mu}(y))^{m^2}, \qquad m = 2, 3, 4, \dots, \ y \in Y. \tag{1}$$

Since $\mathbb{T}^* \approx \mathbb{Z}$, without loss of generality we may assume $Y = \mathbb{Z}$. Let
$\mu \in \Gamma_\infty(X)$. If $\hat{\mu}(1) = 0$, then $\hat{\mu}(-1) = 0$ and (1) implies $\hat{\mu}(m) = 0$ for all

$m \in \mathbb{Z}$, $m \neq 0$, i.e., $\mu = m_X$. If $\hat{\mu}(1) \neq 0$, let us denote $\hat{\mu}(1) = \exp\{a+ib\}$, where $a \leq 0$ and $b \in \mathbb{R}$. Then $\hat{\mu}(-1) = \exp\{a - ib\}$ and (1) yields

$$\hat{\mu}(m) = \begin{cases} \exp\{am^2 + ibm^2\}, & m \geq 0, \\ \exp\{am^2 - ibm^2\}, & m < 0. \end{cases}$$

By Proposition 11.2 the characteristic function $\hat{\mu}(m)$ satisfies also the equation $\hat{\mu}(2m) = (\hat{\mu}(m))^3 \hat{\mu}(-m)$. Therefore $\exp\{2ibm^2\} = 1$ for every $m \in \mathbb{Z}$. Hence $b = \pi k$ for some $k \in \mathbb{Z}$ and so $\mu \in \Gamma^s(X) \cup E_\zeta * \Gamma^s(X)$. The reverse inclusion is obvious. $\square$

11.27. PROPOSITION. *Let μ be a distribution on X. Then the following statements are equivalent:*

   (i) *$\mu \in \Gamma_\infty(X)$.*

   (ii) *$\mu = \overline{\mu}$ and for each character $y \in Y$ either $y(\mu) \in \Gamma(\mathbb{T})$ or $y(\mu) = m_\mathbb{T}$.*

   (iii) *The characteristic function $\hat{\mu}(y)$ is real valued and satisfies the system of equations 11.26(i).*

PROOF. (i) $\Rightarrow$ (ii). Let $\mu \in \mathscr{M}^1(X)$ and $y \in Y$. Then the characteristic function of the distribution $y(\mu)$ has the form

$$\widehat{y(\mu)}(m) = \hat{\mu}(my), \qquad m \in \mathbb{Z}. \tag{1}$$

Hence $y(\mu) \in \Gamma_\infty(\mathbb{T})$ when $\mu \in \Gamma_\infty(X)$ and (ii) follows from Lemma 11.26.

   (ii) $\Rightarrow$ (iii). Let $y \in Y$. As can easily be seen, the characteristic function of $y(\mu)$ satisfies the conditions

$$\widehat{y(\mu)}(m) = (\widehat{y(\mu)}(1))^{m^2}, \qquad m = 2, 3, 4, \ldots.$$

From this and (1) we obtain condition 11.26(1).

   (iii) $\Rightarrow$ (i). Evident. $\square$

11.28. COROLLARY. *The following equality is fulfilled:*

$$\{\mu \in \Gamma_\infty(X): \hat{\mu}(y) > 0 \text{ for all } y \in Y\}$$
$$= \{\mu \in \Gamma_U(X): \hat{\mu}(y) > 0 \text{ for all } y \in Y\}.$$

11.29. PROPOSITION. *Let K be a compact subgroup of X. Then the following statements are equivalent:*

   (i) *K is a connected group.*

   (ii) *$m_K \in \Gamma_\infty(X)$.*

PROOF. (i) $\Rightarrow$ (ii). Let K be connected. Then by Corollary 1.10 and §1.20 $K^{(m)} = K$, $m = 2, 3, 4, \ldots$. By Proposition 11.3 $m_K \in \Gamma_{A_m}(X)$, where $A_m = \{a_j\}_{j=1}^{m^2}$, $a_0 = m$, $a_1 = \cdots = a_{m^2} = 1$ for any $m = 2, 3, 4\ldots$. Hence, according to Proposition 11.2 the characteristic function $\hat{m}_K(y)$ satisfies system 11.26(1). Now Proposition 11.27 yields $m_K \in \Gamma_\infty(X)$.

(ii) = (i). Reverse the order in the previous argument. □

11.30. REMARK. Clearly, $E_x \in \Gamma_\infty(X) \Leftrightarrow 2x = 0$. Let $\lambda = m_K * E_{x_0} \in$ $\Gamma_\infty(X)$. Since $\Gamma_\infty(X)$ is a semigroup, $m_K = \lambda * \bar{\lambda} \in \Gamma_\infty(X)$ and by Proposition 11.29 K is connected. From 2.14 and 2.10 it follows that the restriction of the characteristic function $\hat{\lambda}(y)$ to $A(Y, K)$ may be represented in the form $\hat{\lambda}(y) = ([x_0], y)$, where $[x_0] \in X/K$ and $2[x_0] = 0$, i.e., $2x_0 = x' \in K$. Since the group K is connected, we have, in particular, $K^{(2)} = K$. Then $x' = 2x''$, $x'' \in K$. Set $x = x_0 - x''$. We have $2x = 0$ and $[x_0] = [x]$. Therefore, $\lambda = m_K * E_x$.

11.31. PROPOSITION. *Let $\mu \in \Gamma_\infty(X)$. Then there exists an element $x \in X$ with $2x = 0$ such that $\sigma(\mu * E_x) \subset C_X$.*

PROOF. By Proposition 11.2 the characteristic function $\hat{\mu}(y)$ satisfies system 11.26(1). Consider the subgroup Y_0 of all compact elements of the group Y. Let $y_0 \in Y_0$. Then there exists a sequence of natural numbers $n_j \to \infty$ such that $n_j y_0 \to 0$. If $|\hat{\mu}(y_0)| < 1$, then 11.26(1) implies

$$1 = \hat{\mu}(0) = \lim_{n_j \to \infty} \hat{\mu}(n_j y_0) = \lim_{n_j \to \infty} (\hat{\mu}(y_0))^{n_j^2} = 0.$$

Therefore, $|\hat{\mu}(y)| \equiv 1$ on Y_0. According to Theorems 1.6 and 1.9 we have $Y_0 \approx (X/C_X)^*$. This yields $\hat{\mu}(y) = ([x_0], y)$, $y \in Y_0$, for some $[x_0] \in X/C_X$ with $2[x_0] = 0$. Arguing now as in Remark 11.30, we obtain that there exists an element $x \in X$ with $2x = 0$ such that $\hat{\mu}(y) = (x, y)$ for $y \in Y_0$. Set $\mu' = \mu * E_x$. Then $\widehat{\mu'}(y) \equiv 1$ on Y_0 and by Proposition 2.13 we have $\sigma(\mu') \subset A(X, Y_0)$. But by Proposition 1.9 $A(X, Y_0) = C_X$. □

11.32. THEOREM. *The equality 11.25(2) holds on a group X if and only if X satisfies the following condition:*

(i) *Any two non-infinitely divisible elements of the group C_X^* are dependent.*

PROOF. *Necessity.* Suppose condition (i) is not fulfilled for the group X. Then $C_X \neq \{0\}$. Clearly, it suffices to construct a distribution $\mu \in \Gamma_\infty(C_X)$ such that $\mu \notin I(C_X) * \Gamma(C_X)$. Therefore without loss of generality we may assume from the outset that X itself is connected. By Theorem 1.15 $X \approx \mathbb{R}^n + K$, where $n \geq 0$ and K is a connected compact group. Then $Y \approx \mathbb{R}^n + H$, $H = K^*$, and, according to 1.10, H is a discrete torsion-free group. Two cases are possible.

1. The group X is not compact and in the group Y there exists an element that is not infinitely divisible.

Assume $X = \mathbb{R}^n + K$, where $n \geq 1$, and $Y = \mathbb{R}^n + H$. The elements of the group Y will be denoted by $y = (s, h)$, where $s \in \mathbb{R}^n$, $h \in H$. If an element $y_0 = (s_0, h_0)$ is not infinitely divisible, then neither is the element $\eta = (0, h_0)$. Denote by L_η the subgroup of Y consisting of all elements depending on η. Clearly $L_\eta \approx \mathbb{Z}$. Set $H_1 = \mathbb{R}^n + L_\eta$ and $G_1 = H_1^*$. Then

$G_1 \approx \mathbb{R}^n + \mathbb{T}$. By Lemma 5.35 there exists a distribution μ_0 on G_1 such that $\mu_0 \in \Gamma_U(G_1)$ and $\mu_0 \notin \Gamma(G_1)$ and $\mu_0(y) > 0$ for all $y \in H_1$. Consider the function

$$f(y) = \begin{cases} \hat{\mu}_0(y) & \text{if } y \in H_1, \\ 0 & \text{if } y \notin H_1 \end{cases}$$

on the group Y. It follows from Theorem 2.8 and Remark 2.9 that the function $f(y)$ is positive definite. Since the subgroup H_1 is open, the function $f(y)$ is continuous. By Theorem 2.8 there exists a distribution $\mu \in \mathcal{M}^1(X)$ such that $\hat{\mu}(y) = f(y)$. Let us check that $\mu \in \Gamma_\infty(X)$. By Proposition 11.27 it suffices to prove that the function $f(y)$ satisfies the system of equations 11.26(1). According to Corollary 11.28, $\mu_0 \in \Gamma_\infty(G_1)$. Therefore each of the equations 11.26(1) is satisfied for $y \in H_1$. If $y \notin H_1$, then by construction $my \notin H_1$ for any $m \in \mathbb{Z}$, so each of the equations 11.26(1) is fulfilled as well. Thus $\mu \in \Gamma_\infty(X)$. Since $\mu_0 \notin \Gamma(G_1)$ and $\hat{\mu}_0(y) \neq 0$ for $y \in H_1$, reasoning as in the proof of the necessity part of Theorem 9.10, one can easily see that $\mu \notin I(X) * \Gamma(X)$.

2. The group X is compact and in the group Y there exist two independent elements η and ζ that are not infinitely divisible. According to 1.10, Y is a discrete torsion-free group. Consider the subgroups L_η and L_ζ. We have $L_\eta \approx L_\zeta \approx \mathbb{Z}$ and $L_\eta \cap L_\zeta = \{0\}$. Without loss of generality we may assume that η and ζ are generators of the group L_η and L_ζ, respectively. Consider the function $g(y)$ on the group Y

$$g(y) = \begin{cases} \exp\{-2m^2\} & \text{if } y = m\eta \text{ or } y = m\zeta, \ m \in \mathbb{Z}, \\ 0 & \text{if } y \notin L_\eta \cup L_\zeta. \end{cases}$$

By construction this function satisfies the system of equations 11.26(1). Set

$$\rho(x) = \sum_{y \in Y} g(y)(\overline{x, y}), \qquad x \in X.$$

The function $\rho(x)$ is continuous and $\rho(x) > 0$, since $\sum_{x \in Y, y \neq 0} g(y) < 1$. The characteristic function of the distribution $\mu \in \mathcal{M}^1(X)$, having the density $\rho(x)$ with respect to m_X, has the form $\hat{\mu}(y) = g(y)$. According to Proposition 11.27, $\mu \in \Gamma_\infty(X)$, and by construction $\mu \notin I(X) * \Gamma(X)$.

Sufficiency. If $C_X = \{0\}$, then by virtue of Proposition 11.31 $\Gamma_\infty(X) = \{E_x : x \in X, \ 2x = 0\}$ and equality 11.25(2) holds on the group X. Let $C_X \neq \{0\}$. In view of Proposition 11.31 we may assume that X is connected. Let $\mu \in \Gamma_\infty(X)$. Denote $\nu = \mu * \bar{\mu}$ and $E = \{y \in Y : \hat{\mu}(y) \neq 0\}$. Since $\Gamma_\infty(X)$ is a semigroup, $\nu \in \Gamma_\infty(X)$. By Proposition 11.27 the characteristic function $\hat{\nu}(y)$ satisfies the system of equations 11.26(1). Two cases are possible.

1. All elements of Y are infinitely divisible.

We first check that E is an open subgroup of Y. Let $y_1, y_2 \in E$, $y_1 \neq 0$, $y_2 \neq 0$. There are the following possibilities.

a) $L_{y_1} = L_{y_2}$. Then $py_1 = qy_2$ for some $p, q \in \mathbb{Z}$. Hence $\hat{\mu}(y_1 + y_2) \neq 0 \Leftrightarrow \hat{\mu}(p(y_1 + y_2)) \neq 0 \Leftrightarrow \hat{\mu}((p + q)y_2) \neq 0 \Leftrightarrow \hat{\mu}(y_2) \neq 0$, provided that $p + q \neq 0$. If $p + q = 0$, then, clearly, $\hat{\mu}(y_1 + y_2) \neq 0$. In this reasoning we did not use the fact that the elements y_1, y_2 are infinitely divisible.

b) $L_{y_1} \neq L_{y_2}$. Then $L_{y_1} \cap L_{y_2} = \{0\}$. Consider the group $L = L_{y_1} + L_{y_2}$ endowed with the discrete topology. Define a continuous monomorphism $\psi: L \to \mathbb{R}^2$ in the usual way. Namely, for any $y \in L$, $y \neq 0$ there exist $n, n_1, n_2 \in \mathbb{Z}$, $n \neq 0$, such that $ny = n_1 y_1 + n_2 y_2$. Set $\psi(y) = (r_1, r_2)$, where $r_j = n_j/n$, $j = 1, 2,$, for $y \neq 0$ and $\psi(0) = 0$. Consider the function $f(r_1, r_2) = \hat{\nu}(\psi^{-1}(r_1, r_2))$ on the group $\psi(L)$. By Theorem 2.8 the function $f(r_1, r_2)$ is positive definite on $\psi(L)$. From inequality 2.13(1) (valid for the characteristic function $\hat{\nu}(y)$) we obtain

$$1 - f(r_1, r_2) \leq 2[(1 - f(r_1, 0)) + (1 - f(0, r_2))], \qquad (r_1, r_2) \in \psi(L). \quad (1)$$

Consider the groups $\psi(L_{y_j})$ with the discrete topology and set $K_j = (\psi(L_{y_j}))^*$, $j = 1, 2$. In view of Theorem 2.8 the functions $f(r_1, 0)$, $(r_1, 0) \in \psi(L_{y_1})$, and $f(0, r_2)$, $(0, r_2) \in \psi(L_{y_2})$, are characteristic functions of some distributions $\mu_j \in \mathcal{M}^1(K_j)$, and using Proposition 11.27 we conclude that $\mu_j \in \Gamma_\infty(K_j)$, $j = 1, 2$. Since $\hat{\nu}(y_j) > 0$, it follows from a) that $\hat{\mu}_j(h) > 0$ for any $h \in \psi(L_{y_j})$. By Corollary 11.28 we have $\mu_j \in \Gamma_U(K_j)$. Since y_1 and y_2 are infinitely divisible, all elements of the groups $\psi(L_{y_j})$ are infinitely divisible as well. Clearly, any factor-group of L_{y_j} also possesses this property. Applying Theorem 5.36 to the groups K_j and distributions μ_j, we see that $\mu_j \in \Gamma^s(K_j)$. Consequently, $f(r_1, 0) = \exp\{-a_1 r_1^2\}$ and $f(0, r_2) = \exp\{-a_2 r_2^2\}$, we here $a_1, a_2 \geq 0$, $(r_1, 0) \in \psi(L_{y_1})$, $(0, r_2) \in \psi(L_{y_2})$. Since the subgroups $H_1 = \{r \in \mathbb{R}: \psi(y) = (r, 0), y \in L_{y_1}\}$, $H_2 = \{r \in \mathbb{R}: \psi(y) = (0, r), y \in L_{y_2}\}$, clearly, are dense in $\mathbb{R}$, it follows from (1) that the function $f(r_1, r_2)$ is continuous at zero on $\psi(L)$ in the topology induced on $\psi(L)$ by $\mathbb{R}^2$. From inequality 2.10(c) applied to the characteristic function $\hat{\nu}(y)$ it follows that $f(r_1, r_2)$ is uniformly continuous on $\psi(L)$. Since the subgroup $\psi(L)$ is dense in $\mathbb{R}^2$, the function $f(r_1, r_2)$ can be extended to a continuous positive definite function $f(s_1, s_2)$ on $\mathbb{R}^2$, also satisfying the system of equations 1.26(1) because the function $\hat{\nu}(y)$ satisfies this system. As can easily be seen, the function $f(s_1, s_2)$ is a characteristic function of a Gaussian distribution on $\mathbb{R}^2$. Hence, in particular, $f(r_1, r_2) \neq 0$ for $(r_1, r_2) \in \psi(L)$. Therefore, $f(1, 1) = \hat{\nu}(y_1, y_2) = |\hat{\mu}(y_1, y_2)|^2 \neq 0$. We have thus proved that E is a subgroup of Y. Obviously, E is open.

Let $G = A(X, E)$. It follows from Theorem 1.7 that G is a compact group. The characteristic function $\hat{\mu}(y)$ satisfies the system 11.26(1).

Therefore, if $my \in E$ for some $m \in \mathbb{Z}$, $m \neq 0$, then $y \in E$ as well. By Lemma 9.3 it follows that $G^{(m)} = G$ for $m \in \mathbb{Z}$, $m = 1, 2, \ldots$. By Corollary 1.10 and §1.20 we conclude that G is a connected group. By Theorem 1.6 $(X/G)^* \approx E$.

Consider the restriction $g(y)$ of the characteristic function $\hat{\mu}(y)$ to E. In view of 2.10(h), $g(h) = \hat{\delta}(h)$, $h \in E$, for some $\delta \in \mathscr{M}^1(X/G)$. By Proposition 11.27 $\delta \in \Gamma_\infty(X/G)$ and then also $\gamma = \delta * \bar{\delta} \in \Gamma_\infty(X/G)$ and, in addition, $\hat{\gamma}(h) = |\hat{\delta}(h)|^2 > 0$ for all $h \in E$. By Corollary 11.28 $\gamma \in \Gamma_U(X/G)$. Since every element of Y is infinitely divisible, one can easily verify that the same property is enjoyed by the group E and any factor-group of it. Applying Theorem 5.36 to the group X/G, we conclude that $\gamma \in \Gamma(X/G)$. One can easily see that the factor-group X/G contains no subgroup topologically isomorphic to $\mathbb{T}$. By Theorem 5.22 $\delta \in \Gamma(X/G)$. Now the inclusion $\mu \in I_\infty(X) * \Gamma^s(X)$ may be proved in the same way as the sufficiency in Theorem 11.16.

2. The group Y contains a nonzero element that is not infinitely divisible.

In this case one can easily see that the group X is compact and then by Theorem 1.10 Y is a discrete torsion-free group.

As above, let us check that E is an open subgroup of Y. Let $y_1, y_2 \in E$, $y_1 \neq 0$, $y_2 \neq 0$. If the elements y_1 and y_2 are either dependent or infinitely divisible, then $y_1 + y_2 \in E$, as have been proved when studying the case 1. It remains to consider the case when y_1 is an infinitely divisible element and y_2 is not. Let $L = L_{y_1} + L_{y_2}$.

We shall prove that there exists an element $y_3 \in (L \backslash (L_{y_1} \cup L_{y_2})) \cap E$. Assume not. Then for any element $y \in L \backslash (L_{y_1} \cup L_{y_2})$ we have $y \notin E$. Let the continuous monomorphism $\psi : L \to \mathbb{R}^2$, the function $f(r_1, r_2)$ on $\psi(L)$, and the subgroups $H_1, H_2 \subset \mathbb{R}$ be defined in the same way as in case 1b). Here the subgroup H_1 is dense in $\mathbb{R}$ and $H_2 \approx \mathbb{Z}$. Without loss of generality we may assume that $H_2 = \mathbb{Z}$. Just as in case 1b), we conclude that $f(r_1, 0) = \exp\{-a_1 r_1^2\}$, $a_1 \geq 0$, $(r_1, 0) \in \psi(L_{y_1})$. It is also clear that $f(0, m) = \exp\{-a_2 m^2\}$, $a_2 \geq 0$, $(0, m) \in \psi(L_{y_2})$. The function $f(r_1, m)$ is continuous at zero on the group $\psi(L)$ with respect to the topology induced on $\psi(L)$ by $\mathbb{R} + \mathbb{Z} \subset \mathbb{R}^2$ and with inequality 2.10(c) taken into account we obtain that it is uniformly continuous on $\psi(L)$. Since the subgroup $\psi(L)$ is dense in $\mathbb{R} + \mathbb{Z}$, the function $f(r_1, m)$ may be extended to a continuous positive definite function $f(s, m)$ on the group $\mathbb{R} + \mathbb{Z}$. Since by assumption $(L \backslash (L_{y_1} \cup L_{y_2})) \cap E = \varnothing$, it follows that $f(r_1, m) = 0$ for any $(r_1, m) \in \psi(L)$ such that $r_1 m \neq 0$. But this contradicts the continuity of the function $f(s, m)$ at the point $(0, 1)$. Hence there exists an element

$$y_3 \in (L \backslash (L_{y_1} \cup L_{y_2})) \cap E.$$

Since $y_3 \in L \backslash (L_{y_1} \cup L_{y_2})$, it follows from the condition (i) that y_3 is infinitely divisible. Since $y_3 \in E$, then $L_{y_1} + L_{y_3} \subset E$; we can prove this just as in case 1b). Thus we obtain $ny_3 = n_1 y_1 + n_2 y_2$ for some nonzero $n, n_1, n_2 \in \mathbb{Z}$. Therefore, $(n_2 - n_1)y_1 + ny_3 = n_2(y_1 + y_2) \in L_{y_1} + L_{y_3} \subset E$, and it follows from 10.26(1) that $y_1 + y_2 \in E$.

Just as in the case 1 let $\delta \in \mathscr{M}^1(X/G)$, $G = A(X, E)$, and let the characteristic function $\hat{\delta}(h)$ coincide with the restriction of the characteristic function $\hat{\mu}(y)$ to E. Let us verify that $\delta \in \Gamma(X/G)$.

Suppose $X/G \not\approx \mathbb{T}$. Then, as can easily be seen any nonzero factor-group of the group E contains a nonzero infinitely divisible element and the factor-group X/G does not contain a subgroup topologically isomorphic to $\mathbb{T}$. The proof in this case is the same as in the case 1. If $X/G \approx \mathbb{T}$, then by Lemma 11.26 $\delta \in \Gamma(X/G)$. Now we may complete the proof just as in the case 1. $\square$

11.33. REMARK. For a group X to possess the following property: if $\mu \in \Gamma_\infty(X)$ and $\hat{\mu}(y) \neq 0$ for any $y \in Y$, then $\mu \in \Gamma(X)$, it is necessary and sufficient that the group X satisfy either 5.36(ii) and 5.36(iii). Indeed, if the group X satisfies neither of these conditions, then as follows from the proof of the necessity in Theorem 5.36, there exists a distribution $\mu \in \Gamma_U(X)$ such that $\mu \notin \Gamma(X)$ and $\hat{\mu}(y) > 0$ for every $y \in Y$. By Corollary 11.28 we obtain $\mu \in \Gamma_\infty(X)$.

If the group X satisfies either 5.36(ii) or 5.36(iii) and $\mu \in \Gamma_\infty(X)$ and $\hat{\mu}(y) \neq 0$ for every $y \in Y$, then by Proposition 11.27 $\mu \in \Gamma_U(X)$ and Theorem 5.36 yields $\mu \in \Gamma(X)$.

11.34. REMARK. Combining Corollary 2.18 and Proposition 5.31 one can easily obtain that if an infinitely divisible distribution μ belongs to $\Gamma_\infty(X)$, then $\mu \in I_\infty(X) * \Gamma^s(X)$.

APPENDIX 1

Group Analogs of the Marcinkiewicz Theorem
and the Lukacs Theorem

A. 1.1. Marcinkiewicz (see [LinO, Chapter II, §5]) proved the following theorem.

THEOREM. *Let $P(s)$ be a polynomial, $P(0) = 0$. If $f(s) = \exp\{P(s)\}$ is a characteristic function then $P(s) = -\sigma s^2 + i\beta s$, $\sigma \geq 0$, $\beta \in \mathbb{R}$, i.e., $f(s)$ is the characteristic function of a Gaussian distribution.*

This result was obtained by Marcinkiewicz as a consequence of his more general theorem that gave a necessary condition for an entire function of finite order to be characteristic.

We give a complete description of the class of those groups X for which a similar theorem takes place. After that we consider the group analog of the Lukacs theorem, which generalizes that of Marcinkiewicz.

A. 1.2. DEFINITION. A continuous function $\psi(y)$ on the group Y is called a polynomial if for some m,

$$\Delta_h^{m+1} \psi(y) = 0 \tag{i}$$

for all h, $y \in Y$ (here Δ_h is the finite difference operator: $\Delta_h \psi(y) = \psi(y + h) - \psi(y)$).

The degree of the polynomial $\psi(y)$ is the minimal m for which (i) is fulfilled.

EXAMPLES.

1. Let $l : Y \to \mathbb{C}$ be a nonzero continuous homomorphism. Then $l(y)$ is a polynomial of degree 1.

2. Let $\varphi(y)$ be a continuous function on Y satisfying equation 5.1(ii) ($\varphi(y) \not\equiv 0$). Then $\varphi(y)$ is a polynomial of degree 2.

Let us establish some properties of polynomials.

A. 1.3. LEMMA. *Let $\psi(y)$ be a polynomial on a group Y. Then*

$$\psi(y + \zeta) = \psi(y) \tag{i}$$

for all $y \in Y$, $\zeta \in Y_0$.

PROOF. Let $y \in Y$ and let $\zeta \in Y_0$ be fixed. Define the function $P(l)$ on the group $\mathbb{Z}$ by the relation $P(l) = \psi(y + l\zeta)$, $l \in \mathbb{Z}$. Suppose that the degree of the polynomial $\psi(y)$ is m. Then $\Delta_1^{m+1} P(l) = 0$, and therefore $P(l) = a_0 + a_1 + \cdots + a_m l^m$ (see [Ge, Chapter 5, §3]) where the coefficients a_j depend on y and ζ. Consider the compact subgroup M_ζ generated by the element ζ. The function $\psi(\eta)$ is continuous and bounded on the compact set $y + M_\zeta$. Therefore $P(l) = \text{const}$, i.e., $\psi(y + \zeta) = \psi(y)$. □

A. 1.4. COROLLARY. *Let Y be a compact group. If $\psi(y)$ is a polynomial on Y, then $\psi(y) = \text{const}$.*

A. 1.5. REMARK. Let $\psi(y)$ be a polynomial on the group Y. By Lemma A.1.3 the function $\psi(y)$ is invariant with respect to the subgroup Y_0. Therefore $\psi(y)$ defines the polynomial $\tilde{\psi}[y] = \psi(y)$ on the factor-group Y/Y_0.

The following statement is a generalization of Proposition 5.4.

A. 1.6. PROPOSITION. *Let $\mu \in \mathscr{M}^1(X)$ and let the characteristic function $\hat{\mu}(y)$ have the form*

$$\hat{\mu}(y) = \exp\{\psi(y)\}, \quad \psi(0) = 0, \tag{i}$$

where $\psi(y)$ is a polynomial. Then $\sigma(\mu) \subset C_X$.

PROOF. Let L be a compact subgroup of the group Y. Corollary A.1.4 implies that $\psi(y) \equiv 0$ on L and so $\psi(y) \equiv 0$ on Y_0. Hence $\hat{\mu}(y) \equiv 1$ on Y_0. By Theorem 1.9 we have $A(X, Y_0) = C_X$ and Proposition 2.13 implies $\sigma(\mu) \subset C_X$. □

Let us prove a group analog of the Marcinkiewicz theorem.

A.1.7. THEOREM. *Let a group X contain no subgroup topologically isomorphic to $\mathbb{T}$. Let $\mu \in \mathscr{M}^1(X)$, and let the characteristic function $\hat{\mu}(y)$ have form A.1.6(i). Then $\mu \in \Gamma(X)$.*

PROOF. According to Proposition 1.6 one may assume that X is a connected group. By Theorem 1.15 we may then assume that $X = \mathbb{R}^n + K$, where $n \geq 0$ and K is a connected compact group. Let us prove the theorem for the case $X = K$. (The general case is similar.) By Corollary 1.10 $Y = D$ is a discrete torsion-free group. We also assume that $\dim X = \infty$; $\dim X < \infty$ only simplifies the argument.

Set $\varphi(y) = -\text{Re } \psi(y)$, $l(y) = \text{Im } \psi(y)$. Then $\varphi(y)$ and $l(y)$ also are polynomials. Let us prove that the function $\varphi(y)$ satisfies equation 5.1(ii) and

$$l(y_1 + y_2) = l(y_1) + l(y_2) \tag{1}$$

for all $y_1, y_2 \in D$. This would show that $\mu \in \Gamma(X)$.

Fix arbitrary elements $a_1, a_2 \in D$. Let $f : D \to \mathbb{R}_0^\infty$ be the continuous monomorphism constructed in 5.9. Since no subgroup of the group X is topologically isomorphic to $\mathbb{T}$, by Lemma 9.7 there exists a subgroup $B \subset D$ of a finite rank q such that $a_1, a_2 \in B$ and $\overline{f(B)} \approx \mathbb{R}^l$. Denote $G = f(B)$.

Let $\{b_j\}_{j=1}^q$ be a maximal independent system of elements of $f(B)$ and $\{e_j\}_{j=1}^q$ the standard basis in $\mathbb{R}^q$. Let the continuous monomorphism

$g : f(B) \rightarrow \mathbb{R}^q$ be constructed in the same way as the monomorphism f in 5.6. We have $g(b_j) = e_j$, $j = 1, \ldots, q$, and g can be naturally extended to a continuous isomorphism of the groups G and $\mathbb{R}^q$. Let $h = g \circ f$, $A = h(B)$. Since $\overline{f(B)} = G$, then $\bar{A} = \mathbb{R}^q$.

Define the function $\zeta(a) = \psi(h^{-1}(a))$ on the group A. Obviously, it is a polynomial. Considering the group A in the discrete topology, by Theorem 2.8 $\exp\{\zeta(a)\}$ is a characteristic function. We have reduced the theorem to the following problem. Let A be a group such that $\mathbb{Z}^q \subset A \subset \mathbb{Q}^q \subset \mathbb{R}^q$, $\bar{A} = \mathbb{R}^q$, and $\zeta(a)$, $a \in A$, a polynomial such that $\exp\{\zeta(a)\}$ is a characteristic function. Then the functions $\zeta_1(a) = \text{Re } \zeta(a)$ and $\zeta_2(a) = \text{Im } \zeta(a)$ satisfy equations 5.1(ii) and (1), respectively.

Let $\eta(a)$ be a polynomial of degree m on the group A. Consider the restriction of the function $\eta(a)$ to the subgroup $\mathbb{Z}^q \subset A$. Since by hypothesis $\Delta_b^{m+1} \eta(a) = 0$ for all a, $b \in \mathbb{Z}^q$, then one can easily verify that the function $\eta(a)$ admits the representation

$$\eta(a) = \sum_{p=0}^{m} \sum_{\|k\|=p} C_k a^k, \qquad (2)$$

$\mathbb{Z}^q$, where $k = (k_1, \ldots, k_q)$, $a = (n_1, \ldots, n_q) \in \mathbb{Z}^q$, $a^k = n_1^{k_1} \cdots n_q^{k_q}$. Hence if F is a subgroup of $\mathbb{Q}^q$ such that $F \approx \mathbb{Z}^q$ and $\mathbb{Z}^q \subset F \subset A$ and $\eta(a) = 0$ for all $a \in \mathbb{Z}^q$, then $\eta(a) = 0$ for all $a \in F$.

Let the group A be represented as the union of an increasing sequence of subgroups A_j:

$$A = \bigcup_{j=1}^{\infty} A_j, \qquad A_1 = \mathbb{Z}^q, \qquad A_{j+1} \supset A_j, \qquad A_j \approx \mathbb{Z}^q, \qquad j = 1, 2, \ldots.$$

Denote by $\eta(a)$ the restriction of the polynomial $\zeta(a)$ to the subgroup $\mathbb{Z}^q$. It was noted above that the function $\eta(a)$ may be represented in the form (2) on $\mathbb{Z}^q$. This relation extends $\eta(a)$ to a function on the group $\mathbb{R}^q$ and, in particular, on the group A. This extension will be also denoted by $\eta(a)$. By construction, the polynomial $\delta(a) = \zeta(a) - \eta(a)$, $a \in A$, vanishes on $\mathbb{Z}^q$. So $\delta(a) = 0$ when $a \in A_j$, $j = 1, 2, \ldots$, and, hence, $\delta(a) \equiv 0$ on the group A. Therefore the function $\zeta(a)$ may be represented in the form

$$\zeta(a) = \sum_{p=0}^{m} \sum_{\|k\|=p} C_k a^k, \qquad (3)$$

on the group A, where $k = (k_1, \ldots, k_q)$, $a = (r_1, \ldots, r_q) \in A$, $a^k = r_1^{k_1} \cdots r_q^{k_q}$.

Formula (3) extends $\zeta(a)$ from A to $\mathbb{R}^q$: this extension is also denoted by ζ. Since $\exp\{\zeta(a)\}$ is a positive definite function on the group A and A is dense in $\mathbb{R}^q$, the extended function $\exp\{\zeta(s)\}$ on the group $\mathbb{R}^q$ is positive definite as well. By Theorem 2.8 it is a characteristic function. From the Marcinkiewicz theorem A.1.1 one can now easily conclude that $\exp\{\zeta(s)\}$ is a characteristic function of a Gaussian distribution. Hence $\zeta_1(a)$ and $\zeta_2(a)$ satisfy equations 5.1(ii) and (1), respectively. $\square$

A. 1.8. REMARK. It follows from Theorem A.1.7 that for every $t \geq 0$ the function $\exp\{t\psi(y)\}$ is the characteristic function of some distribution $\mu_t \in \Gamma(X)$. Moreover $\mu_0 = E_0$, $\mu_1 = \mu$, $\mu_{t+s} = \mu_t * \mu_s$ for all $t, s \geq 0$ and $\mu_t \Rightarrow E_0$ as $t \to 0$. Thus the distribution μ belongs to a continuous one-parameter semigroup of Gaussian distributions.

Theorem A.1.7 is sharp. Namely, the following theorem is true.

A. 1.9. THEOREM. *Let the group X contain a subgroup topologically iso-morphic to $\mathbb{T}$. Then for every $m > 2$ there exists a distribution $\mu \notin \Gamma(X)$ with characteristic function of the form A.1.6(i), where $\psi(y)$ is a polynomial of degree m.*

PROOF. Clearly, it suffices to prove the theorem for the group $X = \mathbb{T}$. Consider the following polynomials of degree m on the group $\mathbb{Z}$,

$$\psi(n) = -n^2 + in^m, \quad n \in \mathbb{Z}, \text{ if } m = 2l + 1;$$
$$\psi(n) = -n^m, \quad n \in \mathbb{Z}, \text{ if } m = 2l.$$

Set $f(n) = \exp\{\psi(n)\}$, $n \in \mathbb{Z}$. As $\sum_{n \neq 0} f(n) < 1$, then the function

$$\rho(t) = \sum_{n=-\infty}^{\infty} f(n) \exp\{-int\} > 0.$$

Therefore the characteristic function of the distribution $\mu \in \mathcal{M}^1(\mathbb{T})$ with the density ρ with respect to m_T has the form $\hat{\mu}(n) = f(n)$. Obviously, $\mu \in \Gamma(\mathbb{T})$ and the characteristic function $\hat{\mu}(n)$ has the desired form. $\square$

A. 1.10. REMARK. If the characteristic function of a distribution μ on the real axis is of the form A.1.6(i), where $\psi(y)$ is a polynomial of degree 2, then μ is Gaussian. This statement ceases to be true for general groups. Namely, on the group $X = \mathbb{T}^2$ there exists a distribution $\mu \notin \Gamma(X)$ with characteristic function of the form A.1.6(i) where $\psi(y)$ is a polynomial of degree 2.

Indeed, set

$$\psi(n, m) = -a(n^2 + m^2) + i\pi(n^2 + m^2 + nm), \quad (n, m) \in \mathbb{Z}^2,$$

where $a > 0$ is such that inequality 9.9(1) is fulfilled. Consider the function

$$\rho(t, s) = \sum_{(n, m) \in \mathbb{Z}^2} \exp\{\psi(n, m) - i(tn + sm)\}$$

on the group $\mathbb{T}^2$. By construction $\rho(t, s) > 0$. Therefore the distribution $\mu \in \mathcal{M}^1(X)$ with density ρ with respect to m_X has the characteristic function $\hat{\mu}(y) = \hat{\mu}(n, m) = \exp\{\psi(n, m)\}$. Since $\exp\{i\pi(n^2 + m^2 + nm)\}$ is not a character of the group $\mathbb{Z}^2$, $\mu \notin \Gamma(X)$.

The Marcinkiewicz theorem was generalized by Lukacs, who proved the following assertion ([Lu, Chapter 7]).

A. 1.11. THEOREM. *Let $P(s)$ be a polynomial, $P(0) = 0$. If*

$$f(s) = \exp\{\lambda_1(e^{is} - 1) + \lambda_2(e^{-is} - 1) + P(s)\}$$

is a characteristic function, then $\lambda_j \geq 0$, $j = 1, 2$, and $P(s) = -\sigma s^2 + i\beta s$, $\sigma \geq 0$, $\beta \in \mathbb{R}$.

Below (Theorem A.1.14) we completely describe the class of groups X for which a similar theorem is true. To do this, we need the following lemmas.

A. 1.12. LEMMA. *Let $X = \mathbb{R}^n$, and let $P(s)$ be a polynomial defined on the group Y, $P(0) = 0$. If*

$$f(s) = \exp\{\lambda_1(e^{i\langle t_0, s\rangle} - 1) + \lambda_2(e^{-i\langle t_0, s\rangle} - 1) + P(s)\}$$

is a characteristic function, then $\lambda_j \geq 0$, $j = 1, 2$, and $P(s) = -\langle As, s\rangle + i\langle \beta, s\rangle$, where A is a symmetric positive semidefinite matrix and $\beta \in \mathbb{R}^n$.

We omit the proof since it follows from Theorem A.1.11 in a standard way.

A. 1.13. LEMMA. *Let $x_0 \in X$ be an element of infinite order, and let*

$$f(y) = \exp\{\lambda_1[(x_0, y) - 1] + \lambda_2[(-x_0, y) - 1]\} \qquad (i)$$

be a characteristic function. Then $\lambda_j \geq 0$, $j = 1, 2$. If x_0 is an element of order n, $n > 2$, then there exists a characteristic function $f(y)$ of form (i) with $\lambda_1 > 0$, $\lambda_2 < 0$.

PROOF. Consider the subgroup $F = \{\alpha \in \mathbb{T} : \alpha = (x_0, y), y \in Y\} \subset \mathbb{T}$. Since x_0 is an element of infinite order, then by Lemma 6.4 the subgroup F is dense in $\mathbb{T}$. Define the function $\Phi(\alpha) = f(y)$, $\alpha = (x_0, y)$, on F. This definition is meaningful because if $(x_0, y_1) = (x_0, y_2)$, then $f(y_1) = f(y_2)$. The relation

$$\Phi(\alpha) = \exp\{\lambda_1(\alpha - 1) + \lambda_2(\bar{\alpha} - 1)\}, \qquad \alpha \in F, \qquad (1)$$

extends the function Φ from F to the whole group $\mathbb{T}$. This extension is also denoted by Φ. Since the function $\Phi(\alpha)$ is positive definite on the group F and F is dense in $\mathbb{T}$, the extended function $\Phi(\xi)$ is positive definite on the group $\mathbb{T}$ as well. Therefore by Theorem 2.8 $\Phi(\xi)$ is a characteristic function on $\mathbb{T}$. This immediately implies $\lambda_j \geq 0$, $j = 1, 2$.

Let x_0 be an element of order n. Without loss of generality one may assume that $X = \mathbb{Z}(n)$. Let $\{1, \eta, \ldots, \eta^{n-1}\}$ be the elements of the group $\mathbb{Z}(n)$, $\eta = \exp\{\frac{2\pi i}{n}\}$, $\lambda \in \mathbb{R}$. Consider the charge $e(\lambda E_\eta)$ on the group $\mathbb{Z}(n)$. Since $e(E_\eta)(\{\eta^k\}) \geq \delta > 0$, $k = 0, 1, \ldots, n - 1$ (see Proposition 6.6) and $e(\lambda E_\eta) \Rightarrow E_1$ as $\lambda \to 0$, then for sufficiently small $\lambda_2 < 0$ the convolution $\nu = e(E_\eta) * e(\lambda_2 E_{\eta^{n-1}}) \in \mathscr{M}^1(\mathbb{Z}(n))$ and $\hat{\nu}(y) = f(y)$, where $\lambda_1 = 1$ and $\lambda_2 < 0$. $\quad \square$

We now prove the group analog of the Lukacs theorem.

A. 1.14. THEOREM. *Let a group X has no subgroup topologically isomorphic to $\mathbb{T}$, let $\mu \in \mathscr{M}^1(X)$, and let the characteristic function $\hat{\mu}(y)$ have the form*

$$\hat{\mu}(y) = \exp\{\lambda_1[(x_0, y) - 1] + \lambda_2[(-x_0, y) - 1] + \psi(y)\}, \qquad \psi(0) = 1, \quad \text{(i)}$$

where $\psi(y)$ is a polynomial. Then $\exp\{\psi(y)\}$ is the characteristic function of a Gaussian distribution. If x_0 is an element of infinite order, then $\lambda_j \geq 0$, $j = 1, 2$.

PROOF. If $C_X = \{0\}$, then by Corollary 1.10 $Y_0 = Y$. Then by Lemma A.1.3 $\psi(y) \equiv 0$ on Y and the assertion of the theorem follows from Lemma A.1.13.

Let $C_X \neq \{0\}$. By Theorem 1.15 $C_X \approx \mathbb{R}^n + K$, where $n \geq 0$ and K is a connected compact group. We shall confine ourselves to the case $C_X = K$. The general case is quite similar. Then by Corollary 1.10 $D = K^*$ is a discrete torsion-free group. Assume also that $\dim K = q < \infty$. Then, according to Theorem 1.13, $r(D) = q$. The case $\dim K = \infty$ is similar (compare with the proof of Theorem A.1.7).

Let $f : D \to \mathbb{R}^q$ be the continuous monomorphism constructed in 5.6. According to Remark A.1.5, the function $\psi(y)$ defines the function $\tilde{\psi}([y]) = \psi(y)$ on the factor-group Y/Y_0. By Theorems 1.6 and 1.9 we have $Y/Y_0 \approx D$. We shall keep the notation $[y]$ for elements of the group D and assume that the function $\tilde{\psi}([y])$ is defined on D. Let $A = f(D)$ and $\zeta(a) = \zeta(f([y])) = \tilde{\psi}([y])$, $a = f([y]) \in A$. The function $\zeta(a)$ is a polynomial on the group A. As was established in the proof of Theorem A.1.7, the function $\zeta(a)$ admits the representation A.1.7(3). This representation extends $\zeta(a)$ from the group A onto $\mathbb{R}^q$.

Since no subgroup of C_X is topologically isomorphic to $\mathbb{T}$, then, due to Remark 5.18, the subgroup A is dense in $\mathbb{R}^q$. Set $p = \tilde{f}$, $p : \mathbb{R}^q \to C_X$. Then by 1.20 p is a continuous monomorphism. Denote by π the embedding $\pi : C_X \to X$, i.e., $\pi(x) = x$, $x \in C_X$.

The following three cases are possible:

(1) For all $n \in \mathbb{Z}$, $n \neq 0$, we have $nx_0 \notin p(\mathbb{R}^q)$. In this case x_0 is an element of infinite order. Let us extend the monomorphism p to the continuous monomorphism $p_1 : \mathbb{R}^q + \mathbb{Z} \to X$ by setting $p_1(t, k) = \pi(p(t)) + kx_0$, $(t, k) \in \mathbb{R}^q + \mathbb{Z}$, and consider the adjoint homomorphism $\tilde{p}_1 : Y \to \mathbb{R}^q + \mathbb{T}$. We have

$$((t, k), \tilde{p}_1(y)) = (p_1(t, k), y) = (\pi(p(t)) + kx_0, y)$$

$$= (\pi(p(t)), y)(kx_0, y) = (t, f([y]))(x_0, y)^k.$$

Therefore $\tilde{p}_1(y) = (f([y]), (x_0, y))$.

Observe now that if $\tilde{p}_1(y_1) = \tilde{p}_1(y_2)$, then $y_1 - y_2 \in Y_0$ and $(x_0, y_1) = (x_0, y_2)$. Therefore, by Lemma A.1.3 $\hat{\mu}(y_1) = \hat{\mu}(y_2)$. Define the function $\Phi(\tilde{p}_1(y)) = \hat{\mu}(y)$ on the group $\tilde{p}_1(Y)$. It follows from the above-said that this definition is meaningful. The relation

$$\Phi(a, \alpha) = \exp\{\lambda_1(\alpha - 1) + \lambda_2(\bar{\alpha} - 1) + \zeta(a)\}, \qquad (a, \alpha) = \tilde{p}_1(y), \qquad (1)$$

extends the function $\Phi(a, \alpha)$ from the subgroup $\tilde{p}_1(Y)$ to the whole group $\mathbb{R}^q + \mathbb{T}$. This extension also is denoted by Φ. The mapping $\tilde{p}_1$ is a monomorphism. Therefore, according to 1.20 the subgroup $\tilde{p}_1(Y)$ is dense in $\mathbb{R}^q + \mathbb{T}$. The function $\Phi(a, \alpha)$ is positive definite on the group $\tilde{p}_1(Y)$. Hence the extended function $\Phi(s, \xi)$, $(s, \xi) \in \mathbb{R}^q + \mathbb{T}$, is positive definite on the group $\mathbb{R}^q + \mathbb{T}$ and, due to Theorem 2.8, $\Phi(s, \xi)$ is a characteristic function. By 2.10(h), $\Phi(s, 1) = \exp\{\zeta(s)\}$ is a characteristic function on $\mathbb{R}^q$. Theorem 2.8 implies now that $\exp\{\psi(y)\}$ is a characteristic function and then, according to Theorem 1.7, $\exp\{\psi(y)\}$ is the characteristic function of a Gaussian distribution. Since $\Phi(0, \xi) = \exp\{\lambda_1(\xi - 1) + \lambda_2(\bar{\xi} - 1)\}$ is a characteristic function on $\mathbb{T}$, then $\lambda_j \geq 0$, $j = 1, 2$. We have proved the theorem for this case.

(2) Let, for some natural $n \geq 2$, the relations $x_0, \ldots, (n-1)x_0 \notin p(\mathbb{R}^q)$ and $nx_0 \in p(\mathbb{R}^q)$ hold. Suppose $nx_0 = p(t_0)$. Set $x_1 = x_0 - \pi(p(\frac{t_0}{n}))$. Then x_1 is an element of order n. Let us extend the monomorphism p to a continuous monomorphism $p_1 : \mathbb{R}^q + \mathbb{Z}(n) \to X$ by setting $p_1(t, \eta^k) = \pi(p(t)) + kx_0$, $(t, \eta^k) \in \mathbb{R}^q + \mathbb{Z}(n)$. The elements of the group $(\mathbb{R}^q + \mathbb{Z}(n))^* \approx \mathbb{R}^q + \mathbb{Z}(n)$ will be denoted by (s, δ^l), where $\delta = \exp\{\frac{2\pi i}{n}\}$, $l = 0, 1, \ldots, n-1$. One can easily see that the formula $\tilde{p}_1(y) = (f([y]), (x_1, y))$ defines the adjoint homomorphism $\tilde{p}_1 : Y \to \mathbb{R}^q + \mathbb{Z}(n)$. Note that $p_1(\frac{t_0}{n}, \eta) = x_0$. Let $\tilde{p}_1(y_1) = \tilde{p}_1(y_2)$. Then $y_1 - y_2 \in Y_0$ and $(x_1, y_1) = (x_1, y_2)$. Since by Theorem 1.9 we have $Y_0 = A(Y, C_X)$ and $p(\mathbb{R}^q) \subset C_X$, then $(\pi(p(t)), y_1) = (\pi(p(t)), y_2)$ for all $t \in \mathbb{R}^q$. Therefore

$$(x_0, y_1) = \left(x_1 + \pi\left(p\left(\frac{t_0}{n}\right)\right), y_1\right) = (x_1, y_1)\left(\pi\left(p\left(\frac{t_0}{n}\right)\right), y_1\right)$$

$$= (x_1, y_2)\left(\pi\left(p\left(\frac{t_0}{n}\right)\right), y_2\right) = \left(x_1 + \pi\left(p\left(\frac{t_0}{n}\right)\right), y_2\right) = (x_0, y_2).$$

Therefore, by Lemma A.1.3 $\hat{\mu}(y_1) = \hat{\mu}(y_2)$. Define the function $\Phi(\tilde{p}_1(y)) = \hat{\mu}(y)$ on the group $\tilde{p}_1(Y)$. It follows from the arguments given above that this definition is meaningful. The relation

$$\Phi(a, \delta^l) = \exp\left\{\lambda_1\left(\exp\left\{i\left(\frac{2\pi l}{n} + \left\langle\frac{t_0}{n}, a\right\rangle\right)\right\} - 1\right)\right.$$

$$\left. + \lambda_2\left(\exp\left\{-i\left(\frac{2\pi l}{n} + \left\langle\frac{t_0}{n}, a\right\rangle\right)\right\} - 1\right) + \zeta(a)\right\},$$

$$(a, \delta^l) = \tilde{p}_1(y) \qquad (2)$$

extends the function Φ from the subgroup $\tilde{p}_1(Y)$ to the group $\mathbb{R}^q + \mathbb{Z}(n)$. We shall keep the same notation Φ for the extended function. The mapping $\tilde{p}_1$ is a monomorphism. Therefore, according to §1.20, the subgroup $\tilde{p}_1(Y)$ is dense in $\mathbb{R}^q + \mathbb{Z}(n)$. The function $\Phi(a, \delta^l)$ is positive definite on $\tilde{p}_1(Y)$. Hence the extended function $\Phi(s, \delta^l)$, $(s, \delta^l) \in \mathbb{R}^q + \mathbb{Z}(n)$, is positive definite on the group $\mathbb{R}^q + \mathbb{Z}(n)$ and, by Theorem 2.8, Φ is a characteristic function. According to 2.10(h),

$$\Phi(s, 1) = \exp\left\{ \lambda_1 \left(\exp\left\{ i \left\langle \frac{t_0}{n}, s \right\rangle \right\} - 1 \right) \right.$$
$$\left. + \lambda_2 \left(\exp\left\{ -i \left\langle \frac{t_0}{n}, s \right\rangle \right\} - 1 \right) + \zeta(s) \right\}$$

is a characteristic function on $\mathbb{R}^q$. According to Lemma A.1.11 this implies that, in particular, $\exp\{\zeta(s)\}$ is a characteristic function on $\mathbb{R}^q$. Then by Theorem 2.8 $\exp\{\psi(y)\}$ is a characteristic function and by virtue of Theorem A.1.7 $\exp\{\psi(y)\}$ is the characteristic function of a Gaussian distribution.

Let x_0 be an element of infinite order. Then $t_0 \neq 0$. Applying Lemma A.1.12 to the characteristic function $\Phi(s, 1)$, we obtain that $\lambda_j \geq 0$, $j = 1, 2$. We have proved the theorem for the case 2 as well.

It should be noted that when x_0 is an element of finite order $n > 2$, it follows from Lemma A.1.13 that, in general, λ_j may be negative. But if $n = 2$, then $x_0 = -x_0$ and the initial characteristic function has the form

$$\hat{\mu}(y) = \exp\{\lambda[(x_0, y) - 1] + \psi(y)\}, \qquad \lambda = \lambda_1 + \lambda_2.$$

In this case the fact that

$$\Phi(0, (-1)^l) = \exp\{\lambda(\exp\{i\pi l\} - 1)\}$$

is a characteristic function on the group $\mathbb{Z}(2)$ implies $\lambda \geq 0$.

(3) $x_0 \in p(\mathbb{R}^q)$. In this case x_0 is an element of infinite order. Extend the monomorphism p to the continuous monomorphism $p_1 : \mathbb{R}^q \to X$ setting $p_1(t) = \pi(p(t))$. The adjoint monomorphism $\tilde{p}_1 : Y \to \mathbb{R}^q$ has the form $\tilde{p}_1(y) = f([y])$. As in the cases 1 and 2 one can verify that if $\tilde{p}_1(y_1) = \tilde{p}_1(y_2)$, then $\hat{\mu}(y_1) = \hat{\mu}(y_2)$. This permits the definition of the function $\Phi(\tilde{p}_1(y)) = \hat{\mu}(y)$ on the group $\tilde{p}(Y)$. The rest of the argument is the same as in the cases (1) and (2). □

APPENDIX 2

On Decomposition Stability of Distributions

A.2.1. Let d be some metric on $\mathscr{M}^1(X)$, and $F(\mu)$ the set of all divisors of a distribution μ. The distribution μ is said to possess decomposition stability with respect to the metric d if for any sequence of distributions $\{\mu_n\}$, $\mu_n \in \mathscr{M}^1(X)$, such that $d(\mu_n, \mu) \to 0$ as $n \to \infty$, and for every sequence $\{\nu_n\}$, $\nu_n \in F(\mu_n)$, $n = 1, 2, 3, \ldots$, there exists a sequence $\{\gamma_n\}$, $\gamma_n \in F(\mu)$, such that $d(\nu_n, \gamma_n) \to 0$ as $n \to \infty$.

To measure the decomposition stability of distributions with respect to the metric d one may use the following quantity, introduced by Zolotarev [Z2]:

$$\beta_d(\mu, \varepsilon) = \sup_{\nu \in B_\varepsilon(\mu)} \sup_{\alpha \in F(\nu)} \inf_{\beta \in F(\mu)} d(\alpha, \beta),$$

where $B_\varepsilon(\mu) = \{\nu \in \mathscr{M}^1(X) : d(\mu, \nu) \le \varepsilon\}$ is the ball of the radius ε centered at the point μ.

One can easily see that the distribution μ possesses decomposition stability with respect to the metric d if and only if $\beta_d(\mu, \varepsilon) \to 0$ as $\varepsilon \to 0$.

Suppose that X is a complete separable metric abelian group. Let us verify that any distribution on X is stable with respect to any metric realizing the weak topology.

Denote by π the Lévy-Prokhorov metric in the space $\mathscr{M}^1(X)$:

$$\pi(\mu, \nu) = \inf_{A \in \mathscr{B}(X)} \{\varepsilon : \mu(A) \le \nu(A^\varepsilon) + \varepsilon, \ \nu(A) \le \mu(A^\varepsilon) + \varepsilon\},$$

where $A^\varepsilon = \{x : \rho(x, y) < \varepsilon, \ y \in A\}$, ρ is the metric in the group X.

A.2.2. THEOREM. *Let X be a complete separable metric abelian group. Then for every $\mu \in \mathscr{M}^1(X)$,*

$$\beta_\pi(\mu, \varepsilon) \to 0 \quad as \ \varepsilon \to 0.$$

PROOF. To prove the theorem it suffices to establish the following. Let $\{\mu_n\}$ be a sequence of distributions, $\mu_n \in \mathscr{M}^1(X)$, $\mu_n = \mu_{n1} * \mu_{n2}$, $n = 1, 2, 3, \ldots$, and let $\pi(\mu_n, \mu) \to 0$, $n \to \infty$. Then:

(a) the sequences $\{\mu_{n1}\}$ and $\{\mu_{n2}\}$ are shift-compact,

(b) the limit distributions corresponding to these sequences are divisors of the distribution μ.

Statement (a) immediately follows from Theorem 2.2. The proof of statement (b) is similar to that in the one-dimensional case. We shall assume that all distributions μ_{n1} and μ_{n2} are already shifted in such a way that the sequences $\{\mu_{n1}\}$ and $\{\mu_{n2}\}$ are compact. Let

$$\pi(\mu_{n_k, i}, \nu_i) \to 0 \quad \text{as } k \to \infty, \; i = 1, 2.$$

We show that $\nu_1, \nu_2 \in F(\mu)$. Indeed, accounting for the weak regularity of the metric π, we obtain

$$\pi(\nu_1 * \nu_2, \mu) \leq \pi(\nu_1 * \nu_2, \mu_{n_k}) + \pi(\mu_{n_k}, \mu)$$
$$\leq \pi(\nu_1, \mu_{n_k, 1}) + \pi(\nu_2, \mu_{n_k, 2}) + \pi(\mu_{n_k}, \mu) \to 0$$

as $k \to \infty$, i.e. $\nu_1 * \nu_2 = \mu$. $\square$

Let d_1 and d_2 be metrics in $\mathcal{M}^1(X)$. In some cases to measure the decomposition stability of a distribution μ it is convenient to use another quantity, also introduced by Zolotarev [Z2]:

$$\beta_{d_1, d_2}(\mu, \varepsilon) = \sup_{\nu \in B_\varepsilon(\mu)} \sup_{\alpha \in F(\nu)} \inf_{\beta \in F(\mu)} d_2(\alpha, \beta),$$

where $B_\varepsilon(\mu) = \{\nu \in \mathcal{M}^1(X) : d_1(\mu, \nu) \leq \varepsilon\}$.

Let X be a locally compact abelian group satisfying the second axiom of countability. Choose the metric

$$\sigma(\mu, \nu) = \sup_{A \in \mathcal{B}(X)} \{|\mu(A) - \nu(A)|\}$$

as the metric d_1 and the metric

$$\chi_0(\mu, \nu) = \sup_{y \in Y}\{|\hat{\mu}(y) - \hat{\nu}(y)|\}$$

as the metric d_2.

Our aim is to give an estimate for $\beta_{\sigma, \chi_0}(e(\Phi), \varepsilon)$, where the measure $\Phi \in \mathcal{M}^1(X)$ satisfies the conditions of Theorem 6.13. The following theorem plays the main role in this respect.

A.2.3. THEOREM. *Let* $\Phi \in \mathcal{M}_+(X)$ *and let the measures* Φ^{*n} *and* Φ^{*m} *be pairwise singular for all natural* $n, m, n \neq m$. *Let distributions* $\mu_j, j = 1, 2,$ *be such that*

$$\sigma(\mu_1 * \mu_2, e(\Phi)) \leq \varepsilon, \quad 0 < \varepsilon < e^{-2}. \tag{1}$$

Then there exist elements $x_1, x_2 \in X$, $x_1 + x_2 = 0$, *depending on the distributions* μ_j *only and such that* $\mu_j(\{x_j\}) > c_0$, $j = 1, 2$ *and*

$$\chi_0(\mu_j, e(\Phi_j) * E_{x_j}) \leq c_1 \frac{\ln \ln(\frac{1}{\varepsilon})}{\ln(\frac{1}{\varepsilon})}, \qquad j = 1, 2,$$

where Φ_j *is the restriction of the measure* $(\mu_j(\{x_j\}))^{-1}(\mu_j * E_{-x_j})$ *to the set* A_1 *on which the measure* Φ *is concentrated and the constants* $c_0 > 0$, $c_1 > 0$ *depend on the distribution* $e(\Phi)$ *only.*

PROOF. Without loss of generality we assume that $\varepsilon > 0$ ($\varepsilon \leq \varepsilon(\Phi)$) is sufficiently small. In what follows all positive constants that depend on the distribution $e(\Phi)$ only are denoted by c. Estimate (1) obviously yields

$$(\mu_1 * \mu_2)(\{0\}) \geq \tfrac{1}{2}\exp\{-\Phi(X)\} = \tilde{c}_0.$$

Since the number of points $x \in X$ for which

$$\max(\mu_1(\{x\}), \mu_2(\{-x\})) \geq \tfrac{1}{4}\tilde{c}_0$$

is finite and depends on the measure Φ, one can easily verify that there exists an element $x_0 \in X$ such that $\mu_1(\{x_0\}) \geq c$, $\mu_2(\{-x_0\}) \geq c$. Without loss of generality we assume that $x_0 = 0$; otherwise we go over to the shifts $\mu_1 * E_{-x_0}$, $\mu_2 * E_{x_0}$. Using estimate (1), we obtain the relation

$$c\mu_j(A) \leq e(\Phi)(A) + \varepsilon, \qquad j = 1, 2, \tag{2}$$

for any $A \in B(X)$ (compare with Lemma 4.8). Let the measures Φ^{*n} be concentrated on the sets A_n, $n = 1, 2, 3, \ldots$. By hypothesis these sets may be chosen to be pairwise disjoint. Denote the restriction of the distribution μ_j to A_n by ν_{nj}, $n = 1, 2, 3, \ldots$; $j = 1, 2$.

One can easily see that inequality (2) and the fact that Φ^{*n} are pairwise singular yield that the measures ν_{nj} possess the property

$$c\nu_{nj}(X) \leq \exp\{-\Phi(X)\}\frac{(\Phi(X))^n}{n!} + \varepsilon, \qquad n = 1, 2, 3, \ldots. \tag{3}$$

Consider the measures

$$\mu_j^* = \sum_{n=0}^{T}\nu_{nj}, \qquad \nu_{0j} = \mu_j(\{0\})E_0, \qquad T = \left[\frac{1}{2}\frac{\ln(\frac{1}{\varepsilon})}{\ln\ln(\frac{1}{\varepsilon})}\right], \qquad j = 1, 2.$$

Set

$$B_T = \bigcup_{n=0}^{T}A_n.$$

By inequality (2) we obtain

$$\sigma(\mu_j, \mu_j^*) \leq \mu_j(B_T) \leq c(e(\Phi)(B_T) + \varepsilon)$$

$$\leq c\left(\sum_{n=T+1}^{\infty}\exp\{-\Phi(X)\}\frac{(\Phi(X))^n}{n!} + \varepsilon\right) \leq \varepsilon^{\frac{1}{4}}, \qquad j = 1, 2. \tag{4}$$

Consider the functions

$$\varphi_j(z, y) = \sum_{n=0}^{T}z^n\hat{\nu}_{nj}(y), \qquad z \in \mathbb{C}, \ y \in Y, \ j = 1, 2. \tag{5}$$

Inequalities (3) yield the estimate

$$|\varphi_j(z, y)| \leq \exp(c\,|\,z\,|)\,, \qquad y \in Y\,, \; j = 1, 2 \tag{6}$$

in the disk $|z| \leq \frac{T}{2}$. So we have

$$\varphi_1(z, y)\varphi_2(z, y) = \sum_{n=0}^{2T} z^n b_n(y)\,, \qquad z \in \mathbb{C}\,, \; y \in Y.$$

The quantities $b_n(y)$, $n = 0, 1, \ldots, 2T$ satisfy the inequalities

$$q_n(y) = \left| b_n(y) - \exp\{-\Phi(X)\} \frac{(\hat{\Phi}(y))^n}{n!} \right| \leq \sum_{m=0}^{n} \hat{\nu}_{m1}(y)\hat{\nu}_{n-m,2}(y)$$

$$- \exp\{-\Phi\}(X)\frac{(\hat{\Phi}(y))^n}{n!} + \sum_{\{0 \leq m < n-T\} \cup \{T < m \leq n\}} |\,\hat{\nu}_{m1}(y)\,|\,|\,\hat{\nu}_{n-m,2}(y)\,|\,.$$

From this, with the aid of the inequalities (1)–(3) and taking into consideration the mutual singularity of the measures Φ^{*n}, we obtain

$$q_n(y) \leq c\varepsilon + c \sum_{\{0 \leq m < n-T\} \cup \{T < m \leq n\}} \left(\varepsilon + \frac{(\Phi(X))^m}{m!} \right) \left(\varepsilon + \frac{(\Phi(X))^{n-m}}{(n-m)!} \right)$$

$$\leq c(\Phi(X) + 1)^n \left(T\varepsilon + \frac{\gamma(n)}{n!} \right)\,,$$

where $\gamma(n) = 0$ if $n \leq T$ and $\gamma(n) = 2^n$ if $T < n \leq 2T$.

From these inequalities for $\alpha = (e^5(F(X)) + 1)^{-1}$, $y \in Y$, we obtain

$$\sum_{n=0}^{T} (\alpha T)^n q_n(y) \leq cT(\alpha(\Phi(X) + 1)T)^{2T}\varepsilon + \sum_{n=T}^{2T} \frac{(2\alpha T(\Phi(X) + 1))^n}{n!} \tag{7}$$

$$\leq \exp\{-3T\}\,.$$

We will also need the following inequality (it may be checked directly)

$$\sum_{n=2T+1}^{\infty} (\alpha T)^n \frac{(\Phi(X))^n}{n!} \leq \exp\{-3T\}\,. \tag{8}$$

From inequalities (7) and (8) we obtain the estimate

$$|\,\varphi_1(z, y)\varphi_2(z, y) - \exp\{z\hat{\Phi}(y) - \hat{\Phi}(0)\}\,|$$

$$\leq \sum_{n=0}^{2T} |\,z\,|^n q_n(y) + \sum_{n=2T+1}^{\infty} \exp\{-\Phi(X)\}\frac{(\Phi(X))^n}{n!}|z|^n \leq 2\exp\{-3T\}$$

for all $|z| \leq \alpha T$, $y \in Y$. For these values of z and y we also have the obvious lower estimate

$$|\exp\{z\hat{\Phi}(y) - \hat{\Phi}(0)\}| \geq \exp\{-2\hat{\Phi}(0)\alpha T\} \geq \exp\{-T\}\,.$$

Comparing the two last estimates, we obtain

$$|\,\varphi_1(z, y)\varphi_2(z, y) - \exp\{z\hat{\Phi}(y) - \hat{\Phi}(0)\}\,| \leq \tfrac{1}{2}|\exp\{z\hat{\Phi}(y) - \hat{\Phi}(0)\}|$$

for all z, y under consideration. Therefore

$$\tfrac{1}{2}|\exp\{z\hat{\Phi}(y) - \hat{\Phi}(0)\}| \le |\varphi_1(z, y)\varphi_2(z, y)| \le \tfrac{3}{2}|\exp\{z\hat{\Phi}(y) - \hat{\Phi}(0)\}| \quad (9)$$

in the disk $|z| < \alpha T$.

For any fixed character $y \in Y$ the functions $\varphi_j(z, y)$, $j = 1, 2$, are analytic with respect to z in the whole complex z-plane and in view of (9) they do not vanish in the disk $|z| < \alpha T$. Therefore for $|z| < \alpha T$ the functions $\varphi_j(z, y)$ admit the representation

$$\varphi_j(z, y) = \exp\{f_j(z, y)\}, \quad f_d(0, y) < 0, \quad j = 1, 2,$$

where the function $f_j(z, y)$ is analytic in the disk $|z| < \alpha T$. We expand $f_j(z, y)$ into a Taylor series

$$f_j(z, y) = \sum_{n=0}^{\infty} a_{nj}(y)z^n, \quad j = 1, 2$$

in this disk. By estimate (6) we have

$$\operatorname{Re} f_j(z, y) \le c(|z| + 1), \quad |z| < \alpha T, \ y \in Y, \ j = 1, 2.$$

Let us apply the following analytical result (see [LinO, Chapter VIII,§2]):
If a function

$$f(z) = \sum_{n=1}^{\infty} b_n z^n$$

is analytic in the disk $|z| \le R$ and $\operatorname{Re} f(z) \le A$, $|z| \le R$, then

$$|b_n| \le \frac{2A}{R^n}, \quad n = 1, 2, 3, \ldots.$$

Setting $f(z) = f_j(z, y) - a_{0j}(y)$, $R = \alpha\frac{T}{2}$, $A = cT$ we obtain

$$|a_{nj}(y)| \le \left(\frac{c}{T}\right)^{n-1}, \quad y \in Y, \ n = 2, 3, \ldots.$$

These estimates imply the following representation for the functions $\varphi_j(z, y)$:

$$\varphi_j(z, y) = \exp\{a_{0j}(y) + a_{1j}(y)z\}(1 + H_j(z, y)), \quad j = 1, 2, \quad (10)$$

for $|z| \le cT$, $y \in Y$. Here the functions $H_j(z, y)$ $(H_j(0, y) = H_j'(0, y) = 0)$ are analytic in the disk $|z| \le cT$ for every fixed $y \in Y$ and admit the estimate

$$|H_j(1, y)| \le \frac{c}{T}, \quad y \in Y.$$

Comparing relations (5) and (10) as functions of z we conclude that for all $y \in Y$,

$$\exp\{a_{0j}(y)\} = \mu_j(\{0\}), \quad \exp\{a_{0j}(y)\} \cdot a_{1j}(y) = \hat{\nu}_{1j}(y). \quad (11)$$

By estimate (4) we have

$$|\varphi_j(1, 0) - 1| \le c\varepsilon^{\frac{1}{4}};$$

therefore relations (10) for $z = 1$, $y = 0$, and (11) yield

$$|\hat{\nu}_{1j}(0) + \mu_j(\{0\})\ln\mu_j(\{0\})| \le \frac{c}{T}. \tag{12}$$

Now relations (4) and (10)-(12) lead to the inequalities

$$\left|\hat{\mu}_j(y) - \exp\left\{\frac{\hat{\nu}_{1j}(y) - \hat{\nu}_{1j}(0)}{\mu_j(\{0\})}\right\}\right| \le \frac{c}{T}, \qquad y \in Y, \ j = 1, 2,$$

that prove the theorem. $\square$

A.2.4. COROLLARY. *Let $\Phi \in \mathcal{M}_+(X)$ and let the measure Φ satisfy the conditions of Theorem A.2.3. Then the following estimate is valid:*

$$\beta_{\sigma,\chi_0}(e(\Phi), \varepsilon) \le c\frac{\ln\ln(\frac{1}{\varepsilon})}{\ln(\frac{1}{\varepsilon})}, \tag{13}$$

where the constant c depends on the distribution $e(\Phi)$ only.

PROOF. We use the notation of Theorem A.2.3. As in the proof of Theorem A.2.3, positive constants depending on the distribution $e(\Phi)$ only will be denoted by a single letter c, regardless of their magnitude. To prove inequality (13) it suffices to verify that

$$\Phi_j(A) \le \Phi(A) + c\varepsilon, \qquad j = 1, 2, \tag{14}$$

for all $A \in \mathfrak{B}(X)$. Arguing as in the proof of Theorem A.2.3, we may assume $\mu_j(\{0\}) > c$, $j = 1, 2$, without loss of generality. From the definition of the measures ν_{nj}, $n = 1, 2, 3, \ldots$, $j = 1, 2$, the pairwise singularity of the measures Φ^{*n} for different n and inequalities (1) and (2) we easily obtain

$$\mu_2(\{0\})\nu_{11}(A) + \mu_1(\{0\})\nu_{12}(A) \le \exp\{-\Phi(X)\}\Phi(A) + c\varepsilon$$

for all $A \in \mathfrak{B}$ and also

$$|\mu_1(\{0\})\mu_2(\{0\}) - \exp\{-\Phi(X)\}| \le c\varepsilon.$$

These two inequalities imply (14) and hence also (13). $\square$

A.2.5. REMARK. Estimate (13) is sharp in the following sense. Let $X = \mathbb{R}$, and let Φ be a degenerated distribution concentrated at the point $x = 1$. In [Ch3] sequences of distributions $\mu_{nj} \in \mathcal{M}^1(\mathbb{R})$, $n = 1, 2, 3, \ldots$, $j = 1, 2$, are constructed such that

$$\sigma(\mu_{n1} * \mu_{n2}, e(\Phi)) \le \exp\{-c_1 n\ln n\}, \tag{15}$$

where the constant c_1 depends on the measure Φ only and is independent of n. The characteristic functions of the distributions μ_{n1} have the form

$$\hat{\mu}_{n1}(s) = \exp\left\{\frac{1}{2}\lambda(e^{is} - 1) + \frac{\delta(\lambda)}{n}(e^{2is} - 1)\right\}, \qquad s \in \mathbb{R},$$

where $\lambda = \Phi(\mathbb{R})$ and $\delta(\lambda) > 0$ is a sufficiently small constant that depends on λ only. Theorem 6.5 implies that the characteristic function of an arbitrary divisor ν of the distribution $e(\Phi)$ has the form

$$\hat{\nu}(s) = \exp\{\lambda_0(e^{is} - 1) + i\beta_0 s\}, \qquad 0 \leq \lambda_0 \leq \lambda, \ -\infty < \beta < \infty.$$

Clearly the characteristic functions $\hat{\mu}_{n1}(s)$ and $\hat{\nu}(s)$ satisfy the inequalities

$$\inf_{\nu \in F(e(\Phi))} \sup_{s \in \mathbb{R}} |\hat{\mu}_{n1}(s) - \hat{\nu}(s)| \geq \frac{c_2}{n}, \qquad n = 1, 2, 3, \ldots, \tag{16}$$

where the constant $c_2 > 0$ depends on the measure Φ only and is independent of n. Comparing the estimates (15) and (16) we obtain the lower estimate

$$\beta_{\sigma, \chi_0}(e(\Phi), \varepsilon) \geq c_3 \frac{\ln \ln(\frac{1}{\varepsilon})}{\ln(\frac{1}{\varepsilon})},$$

where $c_3 > 0$ is a constant depending on the measure Φ only and is independent of n. This proves that the estimate (13) is sharp in the class of all locally compact abelian groups.

APPENDIX 3

Structure of Infinitely Divisible Poisson Distributions

A.3.1. Let X be a nondiscrete group and $\mu \in \mathcal{M}^1(X)$ be an arbitrary distribution. It will be convenient for us to write the decomposition 5.13(i) of μ in the form

$$\mu = \alpha_1 \mu^{(1)} + \alpha_2 \mu^{(2)} + \alpha_3 \mu^{(3)},\tag{1}$$

where $\alpha_1 \mu^{(1)} = \mu_{ac}$, $\alpha_2 \mu^{(2)} = \mu_s$, $\alpha_3 \mu^{(3)}$, $\mu^{(j)} \in \mathcal{M}^1(X)$. Therefore $\alpha_j \geq 0$, $\alpha_1 + \alpha_2 + \alpha_3 = 1$. We introduce the following notation.

$\mathcal{L} = \mathcal{L}(X)$ — the set of all Lévy measures on the group X (see 2.21);

$\mathcal{L}_1 = \mathcal{L}_1(X)$ — the set of all absolutely continuous Lévy measures on the group X;

$\mathcal{D} = \mathcal{D}(X)$ — the set of all discrete Lévy measures on the group X.

The family of singular Lévy measures on the group X will be subjected to further, more detailed, classification. To this end denote $U_m = \{x : \rho(x, 0) \geq \frac{1}{m}\}$, where ρ is the distance on the group X. Let $\Phi \in \mathcal{L}$. Set $\Phi_m(A) = \Phi(A \cap U_m) < \infty$. We have $\Phi_m(A) < \infty$ for every $A \in \mathcal{B}(X)$, and the measures Φ_m, $m = 1, 2, \ldots$, weakly converge to Φ on every set $A \in \mathcal{B}(X)$ with boundary of Φ-measure zero.

A measure $\Phi \in \mathcal{L}$ belongs to the class $\mathcal{L}_{\frac{1}{n}} = \mathcal{L}_{\frac{1}{n}}(X)$, $n \geq 2$, if the $(n-1)$-fold convolution of each measure Φ_m is singular, but for some m the n-fold convolution of Φ_m contains a nonzero absolutely continuous component in its representation of type 5.13(i).

We denote by $\mathcal{L}_0 = \mathcal{L}_0(X)$ the class of all singular measures $\Phi \in \mathcal{L}$ such that any convolution power of Φ_m is also singular.

It should be noted that in the case $\Phi \in \mathcal{M}_+(X)$ we may replace Φ_m by Φ in definition of the classes $\mathcal{L}_{\frac{1}{n}}$ and $\mathcal{L}_0$.

We have

$$\mathcal{L} = \mathcal{L}_1 \cup \mathcal{L}_{\frac{1}{2}} \cup \cdots \cup \mathcal{L}_0 \cup \mathcal{D}.$$

Consider an infinitely divisible Poisson distribution μ, i.e., distribution with characteristic function of the form

$$\hat{\mu}(y) = \exp\left\{\int_{X \setminus \{0\}} [(x, y) - 1 - i g(x, y)] d\Phi(x)\right\}$$

189

(see 2.21). In this case we use the notation $\mu = \Pi(\Phi)$. Our purpose is to describe the possible structure (i.e., decomposition (1)) of the distribution μ subject to the properties of the Lévy measure Φ.

We say that a product $\prod_{j=1}^{\infty} g_j$ of complex numbers g_j, $j = 1, 2, 3, \ldots,$ strictly converges if for some natural n the product $\prod_{j=n}^{\infty} g_j$ converges to a nonzero value.

A.3.2. LEMMA ([Par]). *Let ξ_j be independent random variables with values in a group X and with distributions μ_j. The series $\sum_{j=1}^{\infty} \xi_j$ converges with probability 1 if and only if the product $\prod_{j=1}^{\infty} \hat{\mu}_j(y)$ strictly converges for every character $y \in Y$.*

A.3.3. LEMMA. *Let a distribution $\mu \in \mathcal{M}^1(X)$ be a convolution of infinitely many discrete distributions*

$$\mu = \underset{j=1}{\overset{\infty}{*}} \mu_j.$$

Then μ may only be either absolutely continuous or singular or discrete.

PROOF. We split the proof into several steps.

(a) Suppose that some characteristic function $f(y)$ is a product of infinitely many characteristic functions $f_j(y)$, $j = 1, 2, \ldots$, i.e.,

$$f(y) = \prod_{j=1}^{\infty} f_j(y).$$

Denote by Y_1 the set of characters $y \in Y$ on which the product $\prod_{j=1}^{\infty} f_j(y)$ strictly converges. We shall prove that Y_1 is an open subgroup of Y. Let $y_1, y_2 \in Y_1$. Then there exist numbers n_1 and n_2 such that

$$\prod_{j=n_i}^{\infty} f_j(y_i) \neq 0, \qquad i = 1, 2.$$

Let $n \geq \max\{n_1, n_2\}$. Set $\hat{\mu}(y) = |f_j(y)|^2$ in inequality 2.13(1). We have

$$1 - |f_j(y_1 + y_2)|^2 \leq 2[(1 - |f_j(y_1)|^2) + (1 - |f_j(y_2)|^2)].$$

This implies that

$$\sum_{j=n}^{\infty} (1 - |f_j(y_1 + y_2)|^2) \leq 2 \sum_{j=n}^{\infty} [(1 - |f_j(y_1)|^2) + (1 - |f_j(y_2)|^2)].$$

So $y_1 + y_2 \in Y_1$. Since the products $\prod_{j=n}^{\infty} f_j(y)$, $n = 1, 2, 3, \ldots$, are continuous, Y_1 is an open set.

(b) Let $\mu_j \in \mathcal{M}^1(X)$, $j = 1, 2, 3, \ldots$, be an arbitrary sequence of distributions and suppose that the infinite convolution

$$\mu = \underset{j=1}{\overset{\infty}{*}} \mu_j$$

exists. Suppose also that ξ_j, $j = 1, 2, 3, \ldots$, are X-valued independent random variables with distributions μ_j. Let us prove that the sequence of the partial sums $\zeta_n = \xi_1 + \cdots + \xi_n$, $n = 1, 2, 3, \ldots$, is compact with the probability 1.

Denote the characteristic functions of the distributions μ and μ_j by $f(y)$ and $f_j(y)$, respectively. According to part (a) the set of characters

$$Y_1 = \left\{ y : \prod_{j=1}^{\infty} f_j(y) \text{ strictly converges} \right\}$$

is an open subgroup of Y. Therefore the factor-group Y/Y_1 is discrete. Set $X_1 = A(X, Y_1)$. By Theorem 1.6, $X_1 \approx (Y/Y_1)^*$. Hence by Theorem 1.7 X_1 is a compact group. According to Theorem 1.6 $Y_1 \approx (X/X_1)^*$. If $\hat{x} = x + X_1 \in X/X_1$ and $y \in Y_1$, then $(\hat{x}, y) = (x, y)$. The relations $\hat{\xi}_n = \xi_n + X_1$, $n = 1, 2, \ldots$, define a sequence of independent random variables with values in the factor-group X/X_1. Observe that $\mathbf{E}(\hat{\xi}_n, y) = \mathbf{E}(\xi_n, y)$, $y \in Y_1$. Hence the product $\prod_{j=1}^{\infty} \mathbf{E}(\hat{\xi}_n, y)$, $y \in Y_1$ of characteristic functions strictly converges. By Lemma A.3.2, the sequence of the random variables $\hat{\zeta}_n = \hat{\xi}_1 + \cdots + \hat{\xi}_n$ converges with probability 1 to some random variable $\hat{\xi} \in X/X_1$.

Let U be some neighborhood of zero in the group X with compact closure. We have

$$\mathbf{P}\left(\bigcup_{k=1}^{\infty} \bigcap_{n=k}^{\infty} \{\hat{\zeta}_n - \hat{\xi} \in \hat{U}\} \right) = 1, \qquad \hat{U} = \{x + X_1, x \in U\}.$$

The latter means that for almost every elementary event $\omega \in \Omega$ there exists a number $N = N(\omega)$ and elements $b_n = b_n(\omega) \in a(\omega) + X_1$, $n \geq N$, such that $\zeta_n(\omega) + b_n(\omega) \in U$ for $n \geq N$. Since X_1 is a compact group and U has compact closure, the sequence $\{\zeta_n\}$ is compact with the probability 1.

(c) Suppose now that the set of values of each random variable ξ_n is at most countable. According to part (b) there exists a sequence $\zeta'_n = \xi_1 + \cdots + \xi_{m_n}$, $m_n < m_{n+1}$, converging with probability 1 to a random variable ζ with distribution μ. Let M be the algebraic group generated by the elements ξ_n, $n = 1, 2, 3, \ldots$. Clearly, the group M is at most countable. Let $E \in \mathscr{B}(X)$ be any fixed set. One can easily verify that the sets $A_k = \{\sum_{j=k}^{\infty} (\zeta'_{j+1} - \zeta'_j) \text{ converges and } \sum_{j=k}^{\infty} (\zeta'_{j+1} - \zeta'_j) \in E + M\}$, $k = 1, 2, 3, \ldots$, coincide. According to the zero-one criterion, the measure of the set $A_1 = A_2 = \ldots$ is either zero or one. Therefore if $\mu(E) > 0$, then $\mu(E + M) = 1$. This implies the assertion of the lemma. $\square$

A.3.4. THEOREM. *Let $\Phi \in \mathscr{L}(X)$, $V = \Phi(X)$. Then the structure of the distribution $\mu = \Pi(\Phi)$ (see in A.3.1(1)) depends on the properties of the measure Φ in the following way.*

	$\Phi \in \mathcal{L}_1$	$\Phi \in \mathcal{L}_{1/n}$, $n \geq 2$	$\Phi \in \mathcal{L}_0$	$\Phi \in \mathcal{D}$
$V < \infty$	$\alpha_1 = 1 - \alpha_3$ $\alpha_2 = 0$	$\alpha_1 > 0$ $\alpha_2 \geq \exp\{-V\} \sum_{k=1}^{n-1} \frac{V^k}{k!}$	$\alpha_1 = 0$ $\alpha_2 = 1 - \alpha_3$	$\alpha_3 = 1$
	$\alpha_3 = \exp\{-V\}$			
$V = \infty$	$\alpha_1 = 1$	$\alpha_3 = 0$		$(1 - \alpha_1)$ $\cdot (1 - \alpha_2) = 0$

PROOF. 1. The case $V < \infty$. Considering the function $g(x, y)$, one can ensure that there exists an element $z_0 \in X$ such that

$$(z_0, y) = \exp\left\{-i \int_{X \setminus \{0\}} g(x, y) d\Phi(x)\right\}.$$

Hence

$$\mu = \Pi(\Phi) = \exp\{-V\} E_{z_0} * \left[E_0 + \Phi + \frac{\Phi^{*2}}{2!} + \cdots + \frac{\Phi^{*n}}{n!} + \cdots\right].$$

Since a convolution of an absolutely continuous distribution with an arbitrary one is absolutely continuous, we obviously have:

if $\Phi \in \mathcal{L}_1$, then

$$\alpha_1 = 1 - \exp\{-V\}, \qquad \alpha_2 = 0, \qquad \alpha_3 = \exp\{-V\};$$

if $\Phi \in \mathcal{L}_{\frac{1}{n}}$, $n \geq 2$, then

$$\alpha_1 > 0, \qquad \alpha_2 \geq \exp\{-V\} \sum_{k=1}^{n-1} \frac{V^k}{k!}, \qquad \alpha_3 = \exp\{-V\};$$

if $\Phi \in \mathcal{L}_0$, then

$$\alpha_1 = 0, \qquad \alpha_2 = 1 - \exp\{-V\}, \qquad \alpha_3 = \exp\{-V\};$$

if $\Phi \in \mathcal{D}$, then

$$\alpha_1 = \alpha_2 = 0, \qquad \alpha_3 = 1.$$

2. The case $V = \infty$;

(a) $\Phi \in \mathcal{L}_1$. Let U be a neighborhood of zero in the group X such that the boundary of U has zero Φ-measure. Let us represent $\mu = \Pi(\Phi)$ as the convolution

$$\Pi(\Phi) = \nu_1 * \nu_2,$$

where the distribution ν_j are defined by their characteristic functions

$$\hat{\nu}_1(y) = \exp\left\{\int_{U \setminus \{0\}} [(x, y) - 1 - ig(x, y)] d\Phi(x)\right\},$$

$$\hat{\nu}_2(y) = (z_0, y) \exp\left\{\int_{X \setminus U} [(x, y) - 1] d\Phi(x)\right\},$$

$$(z_0, y) = \exp\left\{-i\int_{X\setminus U} g(x, y)d\Phi(x)\right\}.$$

Let us prove that if $\mu \in \mathscr{L}_1$, then $\mu = \Pi(\Phi)$ is absolutely continuous. Indeed, denote $\tilde{V} = \Phi(X\setminus U)$, and define the distribution $F_0 \in \mathscr{M}_1(X)$ by $F_0(A) = \tilde{V}^{-1}\Phi(A \cap (X\setminus U))$, $A \in \mathscr{B}(X)$, and verify that $\alpha_2 = \alpha_3 = 0$ in expansion A.3.1(1) of the distribution $\mu = \prod(\Phi)$. We have

$$\nu_2 = \exp\{-\tilde{V}\}E_{z_0} * \left[E_0 + \tilde{V}F_0 + \frac{\tilde{V}^2 F_0^{*2}}{2!} + \cdots + \frac{\tilde{V}^n F_0^{*n}}{n!} + \cdots\right]. \qquad (1)$$

Consider expansion A.3.1(1) for ν_2:

$$\nu_2 = \beta_1 \nu_2^{(1)} + \beta_3 \nu_2^{(3)}.$$

It follows from (1) that $\beta = \exp\{-\tilde{V}\}$. From the "smoothness preservation" principle it follows (after measure convolution is performed) that $\alpha_1 \geq \beta_1$. Since $\tilde{V} \to \infty$ as U shrinks to zero, we obtain $\alpha_1 = 1$.

(b) $\Phi \in \mathscr{L}_0$. The only assertion here is that $\alpha_3 = 0$. To prove this, consider expansion A.3.1(1) for ν_2,

$$\nu_2 = \beta_1 \nu_2^{(1)} + \beta_2 \nu_2^{(2)} + \beta \nu_2^{(3)}$$

where $\beta_j \geq 0$, $\beta_1 + \beta_2 + \beta_3 = 1$. It follows from (1) that $\beta_3 = \exp\{-\tilde{V}\}$. From this, by the "smoothness preserving" principle, under convolution of measures, it follows that $\alpha_3 \leq \exp\{-\tilde{V}\}$ and hence $\alpha_3 = 0$.

(c) $\Phi \in \mathscr{D}$. Let the measure Φ be concentrated on the set $S = \{a_1, a_2, a_3, \ldots\}$. Let us choose a zero neighborhood U whose boundary is free of points of S. Define the measures Φ_1 and Φ_2 by the condition: $\Phi_2(\{a\}) = \Phi(\{a\})$ if $\Phi(\{a\}) \leq 1$, and $\Phi_2(\{a\}) = 1$ if $\Phi(\{a\}) > 1$, $a \in S$, and $\Phi_1 = \Phi - \Phi_2$. In its turn, the measure Φ_2 may be represented as a sum of the two measures $\Phi_2 = \Phi_3 + \Phi_4$ where $\Phi_3(A) = \Phi_2(A\cap U)$, $\Phi_4(A) = \Phi_2(A\cap(X\setminus U))$, $A \in \mathscr{B}(X)$. Obviously,

$$\Pi(\Phi) = \Pi(\Phi_1) * \Pi(\Phi_3) * \Pi(\Phi_4).$$

Denote $v = \Phi_2(X\setminus U)$ and define a distribution $G \in \mathscr{M}^1(X)$ by setting $G(A) = v^{-1}\Phi_4(A)$, $A \in \mathscr{B}(X)$. Let us show that

$$v^n G^{*n}(\{a\}) \leq v^{n-1}, \qquad n = 2, 3, 4, \ldots$$

for every point $a \in X$. Indeed, in the case $n = 2$ we have

$$v^2 G^{*2}(\{a\}) = \int_X v G(a - u)d((vG)(x)) \leq v,$$

since, by the construction of G, $vG(a-u) \leq 1$. The rest is by induction. Applying representation (1) to the distribution $\Pi(\Phi_4)$, we obtain the resulting inequality

$$\Pi(\Phi_4)(\{a\}) \leq e^{-v}\left[1 + \sum_{n=1}^{\infty} \frac{v^{n-1}}{n!}\right] \leq e^{-v} + \frac{1}{v}.$$

Since the measure Φ_2 is unbounded in a neighborhood of zero, one may contract the chosen neighborhood U to make $e^{-v} + \frac{1}{v}$ as small as desired. Hence $\alpha_3 = 0$. On the other hand, denoting $G_k(A) = \Phi(A \cap \{a_k\})$, $A \in \mathscr{B}(X)$, $k = 1, 2, 3, \ldots$, we have

$$\Pi(\Phi) = \Pi(G_1) * \Pi(G_2) * \cdots.$$

Each distribution $\Pi(G_k)$ is discrete. According to Lemma A.3.4 the distribution $\Pi(\Phi)$ is either absolutely continuous or singular. $\square$

A.3.5. COROLLARY. (a) *For the distribution $\Pi(\Phi)$ to be continuous it is necessary and sufficient that $V = \infty$.*

(b) *For the distribution $\Pi(\Phi)$ to be discrete it is necessary and sufficient that the Lévy measure of Φ be discrete and $V < \infty$.*

(c) *If the Lévy measure of Φ is absolutely continuous and $V = \infty$, then $\Pi(\Phi)$ is absolutely continuous.*

On Distributions with Mutually Singular Powers

Let X be a nondiscrete locally compact abelian group, satisfying the second axiom of countability. In this Appendix we give a proof of the Lin-Saeki theorem (Theorem 6.15) on the existence of a distribution with mutually singular powers. We used this result in the study of the problem of whether a generalized Poisson distribution belongs to the class I_0 (Theorem 6.13 and Lemma 7.9) and in the proof that the class I_0 is dense in the class of all infinitely divisible distributions on a nondiscrete group (Proposition 8.3). Some related subjects are considered as well.

A.4.1. Let $\mathcal{M}(X)$ denote the convolution algebra of all finite regular Borel complex-valued measures on the group X. We shall use a description of the maximal ideal space $\Delta(\mathcal{M}(X))$ of $\mathcal{M}(X)$ in terms of generalized characters [Sr]. Define a generalized character of $\mathcal{M}(X)$ to be an element $f = (f_\mu)_{\mu \in \mathcal{M}(X)}$ of the product space $\prod_{\mu \in \mathcal{M}(X)} L^\infty(X, \mu)$ satisfying the following conditions:

(i) if ν is absolutely continuous with respect to μ, then $f_\nu(x) = f_\mu(x)$ (ν-a.e.),

(ii) $f_\mu(u + v) = f_\mu(u) f_\mu(v)$ ($\mu \times \mu$-a.e.),

(iii) $\sup_{\mu \in \mathcal{M}(X)} \| f_\mu \|_\infty = 1$.

Any complex homomorphism of $\mathcal{M}(X)$ has the form $\mu \to \int_X f_\mu(x) d\mu(x)$, where (f_μ) is a generalized character of $\mathcal{M}(X)$, and conversely, any generalized character of $\mathcal{M}(X)$ generates a complex homomorphism defined by this relation. The Gelfand topology on the maximal ideal space $\Delta(\mathcal{M}(X))$ corresponds to the topology induced on the set of generalized characters by the product of the $\sigma(L^\infty(X, \mu), L^1(X, \mu))$ topologies ($\mu \in \mathcal{M}(X)$).

A local base $\{U_n\}_{n=1}^\infty$ at zero in X is called admissible if (1) each U_n is a compact neighborhood of zero and (2) $U_{n+1} \subset U_n$ for all n. A sequence $\{(a_n, b_n, c_n)\}_{n=1}^\infty$ of triples of nonnegative real numbers is called admissible if $a_n + b_n + c_n = 1$ for all n.

In what follows, we fix an arbitrary admissible local base $\{U_n\}_{n=1}^\infty$ at zero in X and an arbitrary admissible sequence $\{(a_n, b_n, c_n)\}_{n=1}^\infty$. Let $U = \prod_{n=1}^\infty U_n$, and let L denote the set of all limit points of $\{(a_n, b_n, c_n)\}_{n=1}^\infty$ in

$[0, 1]^3$. For each $x = (x_1, x_2, \ldots) \in U$, we write

$$\nu_n = a_n E_0 + b_n E_{x_n} + c_n E_{-x_n}.$$

For each $\tilde{x} \in U$ the convolution

$$\nu_{\tilde{x}} = \nu(\tilde{x}) = \mathop{\mathbf{*}}_{n=1}^{\infty} \nu_n$$

converges in $\mathscr{M}(X)$, as will be shown in Lemma A.4.2. We define $\overline{Y}_{\tilde{x}}$ to be the weak* closure of Y in $L^{\infty}(X, \nu_{\tilde{x}})$. The set of all constant functions in $\overline{Y}_{\tilde{x}}$ is denoted by $\overline{S}_{\tilde{x}}$.

The group X is called an I-group if every neighborhood of zero contains an element of infinite order. Let $\mathscr{D}$ be the closed unit disk $\mathscr{D} = \{z \in \mathbb{C} : |z| \leq 1\}$.

We shall say that some property holds for quasi-all $\tilde{x} \in U$ if this property holds everywhere on U except, maybe, a set of the first category.

To prove the main result we need the following assertions.

A.4.2. LEMMA. *For any given $\tilde{x} \in U$, the convolution product $\mathop{\mathbf{*}}_{n=1}^{\infty} \nu_n$ converges to some $\nu_{\tilde{x}} \in \mathscr{M}(X)$. Moreover, the mapping $(\tilde{x}, y) \to \hat{\nu}_{\tilde{x}}(y)$ is a continuous function on $U \times Y$.*

PROOF. Let $x \in U$, and $y \in Y$. Given natural numbers $r, p, r > p$, we have

$$\left| \left(\mathop{\mathbf{*}}_{n=1}^{r} \nu_n \right)^{\wedge}(y) - \left(\mathop{\mathbf{*}}_{n=1}^{p} \nu_n \right)^{\wedge}(y) \right|$$

$$= \left| \prod_{n=1}^{r} \hat{\nu}_n(y) - \prod_{n=1}^{p} \hat{\nu}_n(y) \right| \leq \left| \prod_{n=p+1}^{r} \hat{\nu}_n(y) - 1 \right|$$

$$= \left| \int_X ((x, y) - 1) d\left(\mathop{\mathbf{*}}_{n=p+1}^{r} \nu_n \right)(x) \right| \leq \sup_{x \in U_p - U_p} |(x, y) - 1|$$

because (2) $U_{n+1} \subset U_n$ for all $n \geq 1$. Since $\{U_n\}_{n=1}^{\infty}$ is a local base at zero in X, it follows that the sequence $(\mathop{\mathbf{*}}_{n=1}^{r} \hat{\nu}_n)(y)$ converges uniformly with respect to $(\tilde{x}, y) \in U \times A$ for any compact subset A of Y. Therefore the product $\mathop{\mathbf{*}}_{n=1}^{\infty} \nu_n$ converges to some $\nu_{\tilde{x}} \in \mathscr{M}(X)$ for any $\tilde{x} \in U$ (notice that all the measures under consideration are concentrated on the compact set $2U_1 - 2U_1$). The second assertion of our lemma is obvious by the above arguments. $\square$

Let us choose and fix an arbitrary countable dense subset $\{\Psi_k\}_{k=1}^{\infty}$ of Y.

A.4.3. LEMMA. *Let $\alpha \in \mathscr{D}$ and $\nu \in \mathscr{M}(X)$. Suppose that for each $N \geq 1$ there is an $y_N \in Y$ such that $|\alpha\hat{\nu}(\Psi_k) - \hat{\nu}(y_N + \Psi_k)| < \frac{1}{N}$ for all $1 \leq k \leq N$. Then the constant α belongs to the weak* closure of Y in $L^{\infty}(X, \nu)$.*

PROOF. Let $\{y_N\}_{N=1}^{\infty}$ be as above. Then we have

$$\lim_{N \to \infty} \int_X (x, y_N)(x, \Psi_k) d\nu(x) = \int_X \alpha(x, \Psi_k) d\nu(x). \tag{1}$$

We get

$$\lim_{N\to\infty} \int_X (x, y_N)(x, y) d\nu(x) = \int_X \alpha(x, y_N) d\nu(x)$$

for all $y \in Y$ by (1), since the set $\{\Psi_k\}_{k=1}^\infty$ is dense in Y. Hence we have

$$\lim_{N\to\infty} \int_X (x, y_N) l(x) d\nu(x) = \int_X \alpha(x) d\nu(x)$$

for any $l(x) \in L^1(X, \nu)$, since the linear subspace of functions on X generated by $\{(x, y): y \in Y\}$ is dense in $L^1(X, \nu)$ (cf. [HeRa2, §31]). In other words, the sequence $\{(x, y_N)\}_{N=1}^\infty$ converges to α in the weak* topology of $L^\infty(X, \nu)$. $\square$

A.4.4. LEMMA. *Let* $(a, b, c) \in L$, $\quad |z| = 1$ *and* $\alpha = a + bz + c\bar{z}$ *be given. Then the set*

$$E(\alpha, N) = \bigcap_{y\in Y} \bigcup_{k=1}^N \left\{ \tilde{x} \in U : |\alpha\hat{\nu}_{\tilde{x}}(\Psi_k) - \hat{\nu}_{\tilde{x}}(y + \Psi_k)| \geq \frac{1}{N} \right\}, \qquad N = 1, 2, \ldots$$

is closed in U *and* $E(\alpha, N)$ *has no interior point.*

PROOF. By Lemma A.4.2, $\hat{\nu}_{\tilde{x}}(y)$ is a continuous function of $\tilde{x} \in U$ for each $y \in Y$. Therefore $E(\alpha, N)$ is closed in U.

Now suppose that X is an I-group. To force a contradiction, assume that $E(\alpha, N)$ has nonempty interior. Then there exist finitely many nonempty sets $V_n \subset U_n$, $1 \leq n \leq M - 1$, such that

$$V_1 \times V_2 \times \cdots \times V_{M-1} \times \prod_{n=M}^\infty U_n \subset E(\alpha, N).$$

We may assume that M satisfies the following conditions

$$\max\{|a - a_M|, |b - b_M|, |c - c_M|\} < \frac{1}{8N}. \tag{1}$$

$$\sup_{x\in U_M} |(x, \Psi_k) - 1| < \frac{1}{8N}, \qquad 1 \leq k \leq N. \tag{2}$$

Choose any points $x_n \in V_n$, $1 \leq n \leq M - 1$. Since X is an I-group, we can find $x_M \in U_M$ and $y \in Y$ such that

$$|(x_n, y) - 1| < \frac{1}{8MN}, \qquad 1 \leq n \leq M - 1, \tag{3}$$

$$|(x_M, y) - \bar{z}| < \frac{1}{8N}. \tag{4}$$

Setting $\tilde{x} = (x_1, x_2, \ldots, x_M, 0, 0, \ldots) \in E(\alpha, N)$ and $\nu_{\tilde{x}} = *_{n=1}^M \nu_n$, we have

$$|\alpha\hat{\nu}_M(\Psi_k) - \hat{\nu}_M(y + \Psi_k)| \le |\alpha\{\hat{\nu}_M(\Psi_k) - 1\}|$$
$$+ |a + bz + c\bar{z} - a_M - b_M(x_n, y) - c_M\overline{(x_M, y)}|$$
$$+ |\hat{\nu}_M(y) - \hat{\nu}_M(y + \Psi_k)|$$
$$< \frac{1}{8N} + \frac{5}{8N} + \frac{1}{8N} = \frac{7}{8N}, \qquad 1 \le k \le N$$

by (2), (1) and (4), since $\alpha = a + bz + c\bar{z}$ and $\hat{\nu}_M(y) = a_M + b_M(x_M, y) + c_M\overline{(x_M, y)}$. It follows from (3) that,

$$|\alpha\hat{\nu}_{\tilde{x}}(\Psi_k) - \hat{\nu}_{\tilde{x}}(y + \Psi_k)| \le \left|\alpha\prod_{n=1}^{M}\hat{\nu}(\Psi_n) - \prod_{n=1}^{M}\nu_n(y + \Psi_k)\right|$$

$$\le \sum_{n=1}^{M-1}|\hat{\nu}_n(\Psi_k) - \hat{\nu}_n(y + \Psi_k)| \le |\alpha\{\hat{\nu}_M(\Psi_k) - \hat{\nu}_M(y + \Psi_k)|$$

$$< \frac{M-1}{8MN} + \frac{7}{8N} < \frac{1}{N}$$

$1 \le k \le N$. Hence $x \notin E(\alpha, N)$, which contradicts our choice of x. $\square$

A.4.5. PROPOSITION. *If X is an I-group, then, for quasi-all $\tilde{x} \in U$, $S_{\tilde{x}}$ contains the multiplicative compact semigroup in $\mathscr{D}$ generated by the set $\{a + bz + c\bar{z} : (a, b, c) \in L$ and $|z| = 1\}$. If X is not an I-group, this conclusion fails for some admissible local base $\{U_n\}_{n=1}^{\infty}$ at zero in X and some admissible sequence $\{(a_n, b_n, c_n)\}_{n=1}^{\infty}$.*

PROOF. Suppose X is an I-group and take any countable dense subset A of the set

$$\{a + bz + c\bar{z} : (a, b, c) \in L \text{ and } |z| = 1\}. \tag{1}$$

If $\tilde{x} \in U$ does not belong to $\bigcup\{E(\alpha, N) : \alpha \in A$ and $N \ge 1\}$, then we have $A \subset S_{\tilde{x}}$ by Lemma A.4.3. On the other hand, it is easy to see that $S_{\tilde{x}}$ is a compact semigroup in $\mathscr{D}$ for every $\tilde{x} \in U$. Therefore, for each $\tilde{x}$ as above, $S_{\tilde{x}}$ contains the compact semigroup generated by set (1). Thus the first assertion of Proposition A.4.5 follows from Lemma A.4.4.

Now assume that X is not an I-group. Then X contains an open subgroup of the form $\mathbb{R}^n + K$, where $n \ge 0$ and K is a compact group. Since X is not an I-group, then $n = 0$ and the group K is periodic [HeRa1, §25]. So K is a compact open periodic subgroup of X. Let n_0 be a positive integer such that $n_0 x = 0$ for all $x \in K$. Set $a_n = 0$ and $b_n = c_n = \frac{1}{2}$ for all $n \ge 1$; then $L = \{(0, \frac{1}{2}, \frac{1}{2})\}$, and the multiplicative compact semigroup in $\mathscr{D}$ generated by the set $\{a + bz + c\bar{z} : (a, b, c) \in L$ and $|z| = 1\}$ is the segment $[-1, 1]$. If $y \in Y$ and $x \in K$, then we have either $(x, y) = 1$ or $|\text{Re}(x, y)| \le |\cos(\frac{2\pi}{n_0})|$, since $(x, y)^{n_0} = (n_0 x, y) = 1$. Let $\{U_n\}_{n=1}^{\infty}$ be any admissible local base at zero in K. Then, for every $\tilde{x} \in U$ and $y \in Y$, we have $\hat{\nu}_{\tilde{x}}(u) = \prod_{n=1}^{\infty}\text{Re}(x_n, y)$. Therefore either $|\hat{\nu}_{\tilde{x}}(y)| = 1$ (if $n_0 = 2$), or $|\hat{\nu}_{\tilde{x}}(y)| \le |\cos\frac{2\pi}{n_0}|$ (if $n_0 \ge 2$). Hence every $S_{\tilde{x}}$ is disjoint from the open interval $(-1, 1)$ if $n_0 = 2$ and from $(|\cos \frac{2\pi}{n_0}|, 1)$ if $n_0 \geq 3$. This proves Proposition A.4.5. $\square$

A.4.6. PROPOSITION. *Suppose that X is not an I-group, and define $q = q(X)$ as the largest natural number such that every neighborhood of zero in X contains an element of order q. Then, for quasi-all $\tilde{x} \in U$, $S_{\tilde{x}}$ contains the compact semigroup in $\mathscr{D}$ generated by all complex numbers of the form $a + bz + c\bar{z}$, where $(a, b, c) \in L$ and $z^q = 1$.*

The proof of Proposition A.4.6 is almost the same as the proof of the first assertion of Proposition A.4.5, and we omit the details. $\square$

A.4.7. THEOREM. *Let X be an arbitrary group. Suppose that the sequence $\{(a_n, b_n, c_n)\}_{n=1}^{\infty}$ has a limit point (a, b, c) such that $\max\{a, b, c\} < 1$. Then:*

(i) If every neighborhood of zero in X contains either an element of infinite order or an element of order ≥ 4, then quasi-all $\tilde{x} \in U$ have the property that, for any $t \in X$, $\nu_{\tilde{x}}^{*m} * E_t$ is singular with respect to $\nu_{\tilde{x}}^{*n}$ for $n \neq m$.

PROOF. Suppose that the hypothesis of Theorem A.4.7 holds. Then, for quasi-all $\tilde{x} \in U$, $S_{\tilde{x}}$ contains a complex number α with $0 < |\alpha| < 1$. By Propositions A.4.5 and A.4.6 there exists $f \in \Delta[\mathscr{M}(X)]$ such that $f_{\nu_{\tilde{x}}}(u) = \alpha(\nu_{\tilde{x}})$-a.e.

Suppose that there exist different positive integers m and n, an element $t \in X$ and a distribution $\mu \in \mathscr{M}^1(X)$ absolutely continuous with respect to both $\nu_{\tilde{x}}^{*m} * E_t$ and $\nu_{\tilde{x}}^{*n}$. We have

$$f_{\nu_{\tilde{x}}^{*m} * E_t}(t) = f_\mu(u) = f_{\nu_{\tilde{x}}^{*n}}(u) \qquad (\mu\text{-a.e.}) \tag{1}$$

by A.4.1(i). On the other hand

$$f_{\nu_{\tilde{x}}^{*m} * E_t}(u) = \alpha^m f(E_t) \qquad (\nu_{\tilde{x}}^{*m} * E_t\text{-a.e.}) \tag{2}$$

and

$$f_{\nu_{\tilde{x}}^{*n}}(u) = \alpha^n \qquad (\nu_{\tilde{x}}^{*n}\text{-a.e.}) \tag{3}$$

Since $|f(E_t)| = 1$ for all $t \in X$ and since $|\alpha^m| \neq |\alpha^n|$ unless $m = n$, relations (1)–(3) prove Theorem A.4.7. $\square$

A.4.8. Assume now that $X = \mathbb{R}$. We give some results about Bernoulli convolutions with mutually singular powers. Let A be the class of symmetric Bernoulli convolutions, i.e., distributions $\mu \in \mathscr{M}^1(\mathbb{R})$ of the form

$$\mu = \mathop{\ast}_{n=1}^{\infty} \tfrac{1}{2}[E_{-x_n} + E_{x_n}], \qquad x_n \in \mathbb{R}, \tag{1}$$

where $\sum_{n=1}^{\infty} x_n^2 < \infty$, and let B be the class of antisymmetric Bernoulli convolutions

$$\mu = \mathop{\ast}_{n=1}^{\infty} \tfrac{1}{2}[E_0 + E_{x_n}], \qquad x_n \in \mathbb{R}, \ x_n > 0, \tag{2}$$

where $\sum_{n=1}^{\infty} x_n < \infty$. Of course there is a close connection between these two classes. In fact, if $\sum_{n=1}^{\infty} x_n = x$, then (1) is the shift of $*_{n=1}^{\infty} \frac{1}{2}[E_0 + E_{2x_n}]$ by x. It is easy to verify that the distributions in A and B are continuous unless $x_n = 0$ for all sufficiently large n. We shall consider only continuous Bernoulli convolutions, so that A and B consist solely of continuous distributions.

A.4.9. THEOREM. *Let μ belong to A or to B. Then either $\mu^{*n} \in L^1(\mathbb{R})$ for some integer n or $\mu^{*m} * E_x$ is singular with respect to μ^{*n} for all $x \in \mathbb{R}$, unless $m = n$.*

In order to make this result workable, we require necessary conditions for $L^1(\mathbb{R})$ to contain a power of μ. One such necessary condition is that the support $\sigma(\mu)$ of μ contains a basis for $\mathbb{R}$. This follows from the equality

$$\sigma(\mu^{*n}) = (n)\sigma(\mu)$$

and from the fact that every set of positive Lebesgue measure contains basis for $\mathbb{R}$. Šreider [Sr] gave necessary conditions for certain perfect subsets of $\mathbb{R}$ to contain a basis. In our case when μ belongs either to A or to B, support $\sigma(\mu)$ does not contain a basis if $\lim_{n\to\infty} \frac{x_{n+1}}{x_n} = 0$. Using this fact, we obtain the following result.

A.4.10. COROLLARY. *Let μ be as in A.4.8(1) or A.4.8(2), and suppose that $\lim_{n\to\infty} \frac{x_{n+1}}{x_n} = 0$. Then, for any $x \in \mathbb{R}$, $\mu^{*m} * E_x$ is singular with respect to μ^{*n}, unless $n = m$.*

Another necessary condition for some power of μ to belong to $L^1(\mathbb{R})$ is that the Fourier-Stieltjes transform of μ vanishes at infinity. This gives us a result for distributions of the form (A.4.8.1), where $x_n = \alpha^{-n}$ and α is a Pisot number (i.e., α is an algebraic integer all of whose conjugates lie inside the unit circle) and $\alpha \neq 2$. Salem has shown that the Fourier-Stieltjes transforms of such distributions do not vanish at infinity [Sal, Chapter 4]. From these remarks we obtain the following result.

A.4.11. COROLLARY. *Let μ be the measure*

$$\mu = \mathop{*}_{n=1}^{\infty} [E_{-\alpha^{-n}} + E_{\alpha^{-n}}],$$

where α is a Pisot number and $\alpha \neq 1, 2$. Then, for any $x \in \mathbb{R}$, $\mu^{*m} * E_x$ is singular with respect to μ^{*n}, unless $n = m$.

As a special case of this result we see that the Lebesgue-Stieltjes distribution of the Cantor set

$$\mu_0 = \mathop{*}_{n=1}^{\infty} \frac{1}{2}[E_0 + E_{(\frac{2}{3})^n}]$$

must have the property that every shift of μ_0^{*m} is singular to μ_0^{*n} unless $m = n$. This follows since μ_0 is the shift by $\frac{1}{2}$ of

$$\mathop{\ast}_{n=1}^{\infty} \frac{1}{2}[E_{-(\frac{1}{3})^n} + E_{(\frac{1}{3})^n}].$$

Unsolved problems

1. Prove or disprove that any Gaussian distribution on the group $\mathbb{T}^\infty$ is either absolutely continuous or singular (with respect to $m_{\mathbb{T}^\infty}$). (See §§5.13–5.15.)

2. Prove or disprove that any two Gaussian distributions on the group X are either mutually absolutely continuous or mutually singular. (V. V. Sazonov and V. N. Tutubalin [ST]. (See Propositions 5.16 and 5.19.)

3. Prove or disprove that any symmetric Gaussian distribution of the class I_0 on a connected infinite-dimension group X is a continuous monomorphic image of a Gaussian distribution, concentrated on some linear subspace of $\mathbb{R}^\infty$. (See Proposition 5.29.)

4. Let $\gamma \in \Gamma^S(X)$ and $\sigma(\gamma)$ be a connected infinite dimension subgroup of X and let π be a Poisson distribution on X. If $\gamma, \pi \in I_0$, then $\gamma \ast \pi \in I_0$. (See Proposition 7.5.)

5. Let the group X be such that $\dim C_X = \infty$ and let γ be an arbitrary symmetric Gaussian distribution of the class I_0 on X. Prove or disprove that there exists a continuous measure $\phi \in M_+(X)$ such that $\gamma \ast e(\phi) \in I_0$. (See Theorem 7.13.)

6. Let $X = \mathbb{R} + D$, where D is a discrete group, $\gamma \in \Gamma(X)$, $\phi \in M_+(X)$ and $\gamma \ast e(\phi) \in I_0$. Must the measure ϕ be discrete? (See the Linnik theorem 7.6.)

7. For a given group X describe these sets of integers $\{a_j\}_1^m$ and $\{b_j\}_1^m$ admissible for X that have the following property: if ξ_j are X-valued independent random variables with distributions μ_j such that the linear forms $L_1 = a_1\xi_1 + \cdots + a_m\xi_m$ and $L_2 = b_1\xi_1 + \cdots + b_m\xi_m$ are independent, then $\mu_j \in I(X) \ast \Gamma(X)$, $j = 1, \ldots, m$. (See Theorem 10.5, 10.9.)

8. For a given group X describe those sets of integers $\{a_j\}_1^m$ and $\{b_j\}_1^m$ admissible for X that possess the following property: if ξ_j are X-valued independent random variables with distributions μ_j, satisfying

$$\prod_{j=1}^{m} \hat{\mu}_j(y) \neq 0$$

for any $y \in Y$, and if the linear forms $L_1 = a_1\xi_1 + \cdots + a_s\xi_s$ and $L_2 = b_1\xi_1 + \cdots + b_s\xi_s$ are independent, then $\mu_j \in \Gamma(X)$. (See Theorem 10.13 and 10.15–10.17.)

9. Describe those groups X for which the relation

$$\{\mu \in \Gamma_A(X) : \hat{\mu}(y) \neq 0 \text{ for any } y \in Y\} \subset \Gamma(X) \tag{1}$$

holds for any $A \in \mathscr{A}(X)$.

10. For a given group X describe those sets $A \in \mathscr{A}(X)$ for which equality 11.4(3) holds. (See Theorem 11.9.)

11. For a given group X describe those sets $A \in \mathscr{A}(X)$ for which relation (1), Problem 9, holds.

12. Extend the main results of §11 to the case of two linear forms having the same distributions instead of a monomial and a linear form with the same distributions.

13. Develop a technique for obtaining estimates of distribution decomposition stability for groups. (See Theorem A.2.3.) In particular, let γ be a Gaussian distribution of the class I_0. Estimate the value $\beta_{d_1 d_2}(\gamma, \varepsilon)$ for pairs of metrics d_1, d_2 on $\mathscr{M}^1(X)$.

14. Investigate problems of stability for characterization problems on groups. (See Chapter III.)

15. Investigate the structure of an infinitely divisible distribution μ, whose characteristic function has representation (x_0, ϕ, φ) (see §2.22), subject to the properties of the Lévy measure ϕ and the function φ. In particular, determine the structure of the distribution $\Pi(\phi)$ in the case $V = \infty$, $\phi \in \mathscr{L}_{1/n}$, $n \geq 2$, $\phi \in \mathscr{L}_0$ (see Appendix 3) (V. M. Zolotarev, V. M. Kruglov [ZK]).

Comments

Section 4

Theorem 4.1 and the group analogs of the Khinchin theorems (Theorems 4.3 and 4.4) were proved by Parthasarathy, Rao, and Varadhan in [PRV1]. Our exposition follows this article.

There are many sufficient conditions for a distribution on the groups $X = \mathbb{R}$ and $X = \mathbb{R}^n$ to be indecomposable [LinO, Chapter III, §3]. For a survey of results obtained up to 1974 see [LivOCh].

The main results on indecomposable distributions on topological groups satisfying the second axiom of countability were obtained by Parthasarathy, Rao, and Varadhan in [PRV2]. We present here the most important ones:

THEOREM C1. *Let* X *be an infinite complete group satisfying the second axiom of countability, perhaps nonabelian. Then the set of all indecomposable distributions is a dense* G_δ *set in* $\mathscr{M}^1(X)$.

Under the additional assumption that X is nondiscrete, the set of all nondiscrete indecomposable distributions taking positive values on open sets is also a dense G_δ set in $\mathscr{M}^1(X)$.

THEOREM C2. *Let* X *be a locally compact noncompact abelian group satisfying the second axiom of countability. Then the set of all absolutely continuous (with respect to* m_X*) indecomposable distributions is a dense* G_δ *set in the set of all absolutely continuous distributions on* X *endowed with the* $L^1(X, m_X)$ *topology.*

In [UU1], [UU2] V. G. Ushakov and N. G. Ushakov proved that for a wide class of groups X the set of all indecomposable distributions is dense in $\mathscr{M}^1(X)$ with respect to the variation distance, i.e., with respect to the norm $\| \cdot \|$ (see 2.26).

THEOREM C3 ([UU1], [UU2]). *Let* X *be an uncountable metric group, satisfying at least one of the following conditions:*

(I) X *is noncompact and any distribution is dense on* X.

(II) X *is nonbounded.*

(III) X *is separable.*

Then the set of all indecomposable distributions on X is dense in $\mathscr{M}^1(X)$
with respect to the variation distance.

The article [UU2] also contains conditions that ensure a mixture of discrete and continuous distributions to be indecomposable. Let X be an arbitrary metric group. Following [UU2] denote by $W(X)$ the class of all functions φ on $\mathscr{M}^1(X)$ that satisfy the following conditions:

1. Every function $\varphi \in W(X)$ takes only two values 0 and 1.
2. If $\varphi(\mu) = \varphi(\nu) = 0$, then $\varphi(\mu * \nu) = 0$.
3. If $\varphi(\mu) = 1$, then $\varphi(\mu * \nu) = \varphi(\nu * \mu) = 1$ for any $\nu \in \mathscr{M}^1(X)$.
4. If $\varphi(\mu) = 1$, then $\varphi(\alpha\mu + (1 - \alpha)\nu) = 1$ for any $\nu \in \mathscr{M}^1(X)$, $0 < \alpha < 1$.

THEOREM C4 [UU2]. *Let ν_1 be a discrete and ν_2 a continuous distribution such that for some $\varphi \in W(X)$ $\varphi(\nu_1) = 1$ and $\varphi(\nu_2) = 0$. For the mixture*

$$\mu = \alpha\nu_1 + (1 - \alpha)\nu_2, \qquad 0 < \alpha < 1,$$

to be indecomposable it is necessary and sufficient that ν_1 and ν_2 do not have common right or left nondegenerate divisors.

Among examples of application of Theorem C4 that are considered in [UU2], we note that for the group $X = \mathbb{R}$ the mixture of a Gaussian distribution with any discrete infinitely divisible distribution is indecomposable.

Two theorems by L. S. Kudina related to indecomposable distributions should also be noted. Their group analogs are still unknown.

1. For any closed subset $A \subset \mathbb{R}^n$ that contains at least two points there exists an indecomposable distribution μ such that $\sigma(\mu) = A$ [Kud1].
2. The weak closure of the set of all indecomposable distributions supported on A can differ from the set of all distributions μ such that $\sigma(\mu) = A$ [Kud2].

Proposition 4.6 was proved by G. M. Fel′dman. Decomposition 4.16(ii) was mentioned by Heyer in [He1]. Theorem 4.17 was proved by G. M. Fel′dman in [F2].

Section 5

Parthasarathy, Rao, and Varadhan defined a Gaussian distribution as a distribution satisfying conditions 5.30(i), (ii) and then proved that this class coincides with the class defined in 5.1. This equivalence is an important step in their proof of formula 2.22(i).

It was noted in [PRV1] that Gaussian distributions appear naturally from the point of view of limit theorems. Let $\{\Phi_n\}$ be a sequence of finite measures on a group X satisfying the following conditions:

1. $\lim_{n \to \infty} \Phi_n(X \backslash U) = 0$ for any neighborhood U of zero in X.
2. The distributions $e(\Phi_n)$ converge to a limit after being appropriately shifted.

If the sequence $\{\Phi_n\}$ is unbounded and the limit of $e(\Phi_n)$ is nondegenerate, then this limit is a nontrivial Gaussian distribution.

Consider the class of distributions μ that may be included into a one-parameter distribution semigroup (μ_t), $t \geq 0$, $\mu_0 = E_0$, i.e., there exists such a semigroup for which $\mu_1 = \mu$. Forst [Fo] (see also [BeF]) proved that Gaussian distributions of this class admit the following description.

THEOREM C5. *Suppose a distribution* $\mu \in \mathscr{M}^1(X)$ *may be included in a one-parameter distribution semigroup* (μ_t), $t \geq 0$, $\mu_0 = E_0$. *Then the following statements are equivalent*:

(i) $\mu \in \Gamma(X)$;

(ii) $\lim_{t \to 0}[\mu_t(X \backslash U)/0] = 0$ *for any neighborhood of zero in* X.

Remark 5.2 and Propositions 5.4 and 5.5 are due to Parthasarathy, Rao, and Varadhan [PRV1]. We follow [He2] in the proof of Proposition 5.4 and [ST] in that of Proposition 5.5. Remark 5.3 is due to G. M. Fel'dman.

Propositions 5.6 and 5.9 were proved by G. M. Fel'dman in [F3]. Concerning the linear spaces $\mathbb{R}^\infty$ and $\mathbb{R}_0^\infty$ see [RR]. It should be noted that the group $\mathbb{R}_0^\infty$ is not locally compact and does not satisfy the second axiom of countability. Therefore $\mathbb{R}_0^\infty$ is nonmetrizable. Nevertheless in [Se] it was proved that if s is a limit point of a set $B \subset \mathbb{R}_0^\infty$, then there exists a sequence of elements $\{s^{(k)}\} \subset B$ such that $s^{(k)} \to s$.

Remarks 5.11 and 5.12 belong to Parthasarathy, Rao, and Varadhan and were proved in [PRV1] by a different method.

Proposition 5.14 was proved in [F3]. Earlier Siebert had proved in [Si] that absolutely continuous Gaussian distribution exist only on connected locally connected groups.

Let μ be an infinitely divisible distribution without nondegenerate divisors, and let its characteristic function have representation $(0, \Phi, \varphi)$. V. M. Zolotarev and V. M. Kruglov [ZK] posed the problem of studying the structure of μ subject to properties of Φ and φ and solved this problem in the case $\varphi \equiv 0$ (see Appendix 3). Proposition 5.14 may be viewed as a partial solution of this problem for the case $\Phi = 0$.

Gaussian distributions on a finite-dimensional torus $\mathbb{T}^n$ were studied by Siebert [Si] (see also [He2, §5.5]). The detailed investigation of Gaussian distribution μ on an infinite-dimensional torus $\mathbb{T}^\infty$ corresponding to a diagonal matrix A in Proposition 5.9 was carried out by Berg [Be].

The comments to Chapter 5 in [He2] contain a complete bibliography related to those aspects of Gaussian distributions that are not considered in the present book.

Propositions 5.16 and 5.19 were proved in [F3]. They give a partial answer to the question (posed by V. V. Sazonov and V. V. Tutubalin in [ST]): are two Gaussian distributions on the group X either mutually singular or mutually absolutely continuous? Lemma 5.17 was proved by G. M. Fel'dman in [F1], [F3].

In 1936, answering a question of Lévy, Cramér proved that any divisor of a Gaussian distribution on $\mathbb{R}^n$ is Gaussian. This was the first result on distribution arithmetic. Somewhat later Marcinkiewicz noted [Mark] that any Gaussian distribution on $\mathbb{T}$ has non-Gaussian divisors (Theorem 5.20). This theorem together with the Lévy theorem on decomposition of a Poisson distribution on the group $\mathbb{T}$ (see the comment on §6 below) was the first result concerning distribution arithmetic for groups differing from $\mathbb{R}^n$. The Marcinkiewicz theorem was reproved by Martin-Löf (see [Gr, Remark 4.5.1]) and Carnal [C].

Theorems 5.22 and 5.23 were proved by G. M. Fel'dman in [F1]. It should be noted that the fact that a Gaussian distribution on the group $X = \mathbb{R} + \mathbb{Z}(2)$ belongs to the class I_0 was proved by V. M. Zolotarev [Z1]. Propositions 5.28 and 5.29 were proved by G. M. Fel'dman and A. E. Fryntov [FFr2].

It was Urbanik [Ur] who studied distributions, satisfying conditions 5.31(i), (ii). To be more precise, instead of condition (i) he considered the condition:

(i') μ can be included into a continuous one-parameter distribution semigroup (μ_t), $t \geq 0$, $\mu_0 = E_0$.

It was mentioned in Proposition 5.31, that the class of distributions, satisfying conditions (i), (ii), is the same as $\Gamma(X)$. At the same time an example was constructed in [Ur] of a non-Gaussian distribution on the group $X = \mathbb{T}^2$ satisfying condition 5.31(i). This work stimulated investigation of distributions Gaussian in the Urbanik sense (see Definition 5.32), that proved useful in the study of characterization problems on groups (see Chapter III). The results of §§5.32–5.37 are due to G. M. Fel'dman.

Section 6

The problem of whether the generalized Poisson distribution $\mu = e(\Phi)$ belongs to the class I_0 is important for distribution arithmetic on the groups $X = \mathbb{R}$ and $X = \mathbb{R}^n$. The first result in this direction was obtained by D. A. Raĭkov (1937) who got the affirmative answer for $X = \mathbb{R}$. For further results on this subject for the groups $X = \mathbb{R}$ and $X = \mathbb{R}^n$ see the monograph [LinO] and also the surveys [O4], [O2] by I. V. Ostrovskiĭ.

The problem of decomposition of a Poisson distribution on a group different from $\mathbb{R}^n$ was first considered by Lévy [Lévy2], who proved that the Poisson distribution $\mu = e(\psi(E_x))$, $\psi > 0$ on the group $X = \mathbb{T}$ belongs to the class I_0 if x is either of infinite order or of order two. Proposition 6.6 that generalizes this result was proved by A. L. Rukhin [Ruh2], [Ruh3]. Theorems 6.5 and 6.6 were proved by G. M. Fel'dman [F2]. It should also be noted that using Proposition 4.18 one may reduce the Rukhin theorem in the case of an element x of infinite order to Raĭkov's theorem. It suffices to consider the monomorphism $p: \mathbb{Z} \to X$ defined by the relation $p(nx) = nx$, $n \in \mathbb{Z}$.

The statement that the generalized Poisson distribution $\mu = e(\Phi)$, where $\Phi = \psi_1 E_{x_1} + \psi_2 E_{x_2}$, $x_i \in \mathbb{R}$, $\psi_i > 0$, $i = 1, 2$, belongs to the class I_0 on the group $\mathbb{R}$ was obtained by D. A. Raĭkov [Ra] in the case when x_1 and x_2 belong to the same semiaxis and are independent, and by Lévy [Lévy2] in the case when x_1 and x_2 belong to the same semiaxis and are dependent. Therefore, Theorem 6.7 in the case of dependent elements x_1, x_2 is a generalization of the Lévy theorem.

If the elements x_1 and x_2 belong to different semiaxes, then the inclusion $e(\Phi) \in I_0$ is a consequence of the Linnik theorem [LinO, Chapter I, §1]. Theorem 6.10 was proved by G. M. Fel'dman and A. E. Fryntov in [FFr1].

Theorem 6.13 was proved by G. M. Fel'dman in [F4], [F6]. The proof in the book belongs to G. P. Chistyakov [Ch2]. This theorem is new even for the case $X = \mathbb{R}$. It implies, for example, that if $\Phi_0 = \ast_{n=1}^{\infty} \frac{1}{2}(E_0 + E_{(2/3)^n})$ is the Lebesgue-Stieltjes distribution on the standard Cantor set, then $e(\Phi_0) \in I_0$ (the pairwise singularity of powers of the distribution Φ_0 is a consequence of a result of A. M. Vershik [V]; see also [BrM1]) and contains the Ostrovskiĭ-Cuppens theorem [LinO, Chapter VI, §4] stating that $e(\Phi) \in I_0$ if $\Phi \in \mathscr{M}_+(\mathbb{R}^n)$ and that the measure Φ is concentrated on an independent set of points. The Ostrovskiĭ-Cuppens theorem in its turn completed a series of results that involved some additional assumptions (see [LinO, Comment to Chapter VI]). For more details on distribution with mutually singular powers on a group X, see Appendix 4.

Theorem 6.18, Lemma 6.21, and Proposition 6.23 were proved in [F4]. Theorem 6.18 is the group analog of Ostrovskiĭ's theorem that was proved in connection with a problem posed by Yu. V. Linnik on the existence of distributions of the class I_0 on the group $X = \mathbb{R}$ with nondiscrete Lévy measure.

Section 7

The fact that the convolution of a Gaussian and a Poisson distribution on the group $X = \mathbb{R}$ belongs to the class I_0 was proved by Yu. V. Linnik [LinO, Chapter VI, §1]. The corresponding result for the group $X = \mathbb{R}^n$ was obtained by I. V. Ostrovskiĭ and Cuppens [LinO, Chapter VI, §3]. Theorem 7.2 was proved by G. M. Fel'dman and A. E. Fryntov in [FFr1].

Theorem 7.13 as well as Lemmas 7.7–7.12 were proved by G. M. Fel'dman in [F7]. It should be noted that Lemma 7.7 follows from a result by I. V. Ostrovskiĭ [O3] on the Cartesian product of one-dimensional distributions of the class I_0. The construction in Lemma 7.12 is a generalization of the construction used in [LinO, Chapter VI, §6] to prove that, for the case $X = \mathbb{R}^n$, both images and preimages of Borel sets in the isomorphism H of semigroups $M^+(A)$ and $M^+(A')$ are Borel sets.

Section 8

The fact that the class I_0 is dense in the class of all infinitely divisible distributions on the group $X = \mathbb{R}$ was proved by I. V. Ostrovskiĭ and on the group $X = \mathbb{R}^n$ by L. Z. Livshits and I. V. Ostrovskiĭ [LinO, Chapter VI, §4].

Lemma 8.2 is a consequence of Rudin's theorem [Rud] on the existence of a perfect independent set on a group any neighborhood of zero in which contains an element of infinite order. Propositions 8.3 and 8.6 were proved by G. M. Fel'dman in [F2], [F6]. Lemma 8.4 is due to Parthasarathy, Rao, and Varadhan [PRV1] and Lemma 8.5 to Dugué [LinO], [Fr].

The fact that any infinitely divisible distribution on the group $X = \mathbb{R}^n$ may be represented as a finite or infinite convolution of distributions of the class I_0 was proved by I. V. Ostrovskiĭ [O1]. Theorem 8.8 was proved by G. M. Fel'dman [F6].

Section 9

The problem of extending the Bernstein characterization of Gaussian distribution to groups was considered by A. L. Rukhin [Ruh1], [Ruh3] and independently by Heyer and Rall [HR] (see also [He2, Chapter 5, §3]). Proposition 9.5 was proved by these authors. Equality (9.5.2) was obtained by A. L. Rukhin [Ruh1], [Ruh3] under the assumption that Y is a Corwin group and by Heyer and Rall under the assumption that X is a Corwin group and condition 9.11(i) holds.

Theorem 9.10 and Lemmas 9.6–9.9 are due to G. M. Fel'dman [F8] as well as the results of 9.13–9.16. Some sufficient conditions for a group X to satisfy 9.13(i), 9.15(i) and 9.16(i) are mentioned in [HR].

Theorems 9.19 and 9.21 and Proposition 9.18 were proved by G. M. Fel'dman [F8]. It should also be noted that A. L. Rukhin proved [Ruh1], [Ruh3] that the assertion of Theorem 9.19 is true under the assumption that X and Y are Corwin groups.

Let $\tau: X^2 \to X^2$ be the mapping defined by the equality $\tau(x_1, x_2) = (x_1 + x_2, x_1 - x_2)$. Corwin [Co1] considered finite complex-valued measures on a group X satisfying the condition

$$\tau(\mu \otimes \mu) = (\mu^{*2}) \otimes (\mu * \overline{\mu}) \tag{1}$$

and studied their properties [Co2]–[Co4]. In the case $\mu \in \mathscr{M}^1(X)$ condition (1) is the same as $\mu \in \Gamma_B(X)$ (see [He2]).

Section 10

The results of this section are due to G. M. Fel'dman. It should be noted that in the proof of Theorem 10.3 the group analog of the following Marcinkiewicz theorem was used: if $\mu \in \mathscr{M}^1(\mathbb{R})$ and the characteristic function $\hat{\mu}(s)$ has the form $\hat{\mu}(s) = \exp\{P(s)\}$ where $P(s)$ is a polynomial, then $\mu \in \Gamma(\mathbb{R})$ [LinO, Chapter II, §5]. The complete description of those groups X for which this analog is valid is given in [F13].

Section 11

The results of this section are due to G. M. Fel'dman [F11]. Theorem 11.9 and the lemmas it uses were proved in [F14]. Theorem 11.16 for the case $m = 2$ was proved in [F10]. This proof is based on the results of 11.18–11.20. The results of 11.25–11.32 were obtained in [F12].

It should be noted that the problem of constructing a theory of equidistribution of forms on algebraic structures was posed by A. M. Kagan, Yu. V. Linnik, and Rao in [KLR].

Appendix 1

The results of Appendix 1 were obtained by G. M. Fel'dman [F13].

Appendix 2

Theorem A.2.2 was proved by A. P. Ushakova [Ush]. The first estimates of the decomposition stability were obtained by N. A. Sapogov (for Gaussian distributions on the group $X = \mathbb{R}$), see [LinO, Chapter VIII]. On the further development of this subject for the groups $X = \mathbb{R}$ and $X = \mathbb{R}^n$ see the survey article [53] by G. P. Chistyakov.

The results of A.2.3–A.2.5 were obtained by G. P. Chistyakov in [Ch2]. This is the only article known to the author that gives estimates of stability for distributions on general locally compact abelian groups.

Appendix 3

The results of this Appendix are due to V. M. Zolotarev and V. M. Kruglov [ZK]. It should be noted that Lemma A.3.3 is a group analog of the well-known result of Jessen and Wintner [JW]. Examples constructed in [ZK] for the group $X = \mathbb{R}$ demonstrate that the classes $\mathscr{L}_{1/n}$ for any $n \geq 2$ and $\mathscr{L}_0$ are nonempty.

Appendix 4

Infinite convolutions of discrete distributions appear naturally in many parts of analysis, number theory, probability, often as a source of various examples and counterexamples. We should single out the paper by Brown and Moran [BrM2] among a number of investigations devoted to this subject. They studied the Bernoulli convolution on the circle group $\mathbb{T}$. Lin and Saeki [LiSa] proved some analogs of the main result in [BrM2] for nondiscrete metric locally compact Abelian group. The results of A.4.2–A.4.7 belong to them. We give an account of their results following [LiSa]. It was Theorem A.4.7 that enabled us to prove the density of the class I_0 in the class of infinitely divisible distributions on a nondiscrete group X (Proposition 8.3).

In [BrM1] Brown and Moran considered more a general class of distributions, which they called ergodic distributions and used to prove that some symmetric Bernoulli convolutions and antisymmetric Bernoulli convolutions have relatively singular powers.

Let D be a countable subgroup of X and let $\lambda \in \mathscr{M}^1(K)$ be quasi-invariant under the action of D on X, i.e., if $N \in \mathscr{B}(X)$ and $\lambda(N) = 0$, then $\lambda(d + N) = 0$ for every $d \in D$. Such a distribution λ is called ergodic with respect to the action of D if whenever N is a D-invariant Borel set of X (i.e., $d + N = N$ for all $d \in D$), then either $\lambda(N) = 0$ or $\lambda(N) = 1$.

A group G is called a refinement of X if G is algebraically isomorphic to X but has a finer locally compact topology.

The main result of the paper [BrM1] is the following.

THEOREM C.6. *Let λ be ergodic with respect to the action of a countable subgroup D of X. Then either*

(i) *λ^{*m} is singular to λ^{*n} unless $m = n$, or*

(ii) *there are a refinement G of X and a positive integer p such that $\lambda \in \mathscr{M}^1(G)$ and $\lambda^{*p} \in L^1(G)$.*

Theorem A.4.9 follows from this theorem. Theorem A.4.9 and Corollary A.4.11 are due to Brown and Moran [BrM1]. Corollary A.4.10 is due to Kaufman [Kau]. We note that the relative singularity of powers of the Lebesgue-Stielties distribution μ_0 on the standard Cantor set follows from the Vershik's results [V].

Detailed references concerning these problems could be found in the papers [BrM1], [LiSa], and [BrM2].

References

[Be]   Christian Berg, *Potential theory on the infinite dimensional torus*, Invent. Math. **32** (1976), 49–100.

[BeF]   Christian Berg and Gunnar Forst, *Potential theory on locally compact Abelian groups*, Springer-Verlag, Heidelberg and New York, 1975.

[B1]   N. Bourbaki, *Topologie générale*, Chapitres 4–8, 2nd ed., Actualités Sci. Indust., nos. 1143, 1235, Hermann, Paris, 1951, 1955.

[B2]   _____, *Theories spectrales*, Chapitres 1, 2, Actualités Sci. Indust., no. 1332, Hermann, Paris, 1967.

[BrM]   Gavin Brown and William A. Moran, *A dichotomy for infinite convolutions of discrete measures*, Proc. Cambridge Philos. Soc. **73** (1973), 307–316.

[BrM2]   _____, *In general, Bernoulli convolutions have independent powers*, Studia Math. **47** (1973), 141–152.

[C]   Henri Carnal, *Non-validité du théorème de Lévy-Cramer sur le cercle*, Publ. Inst. Statist. Univ. Paris **13** (1964), 55–56.

[Ch1]   G. P. Chistyakov, *Stability of decompositions of laws of distributions*, Teor. Veroyatnost. i Primenen. **31** (1986), no. 3, 433–450; English transl. in Theory Probab. Appl. **31** (1986).

[Ch2]   _____, *Stability for a theorem of I. V. Ostrovskiĭ and R. Cuppens on groups*, Teor. Funktsii Funktsional. Anal. i Prilozhen. No. 50 (1988), 103–108; English transl. in J. Soviet Math. **49** (1990), no. 6.

[Ch3]   _____, *The sharpness of the estimates in theorems on the stability of decompositions of normal distribution and a Poisson distribution*, Teor. Funktsii Funktsional. Anal. i Prilozhen. No. 26 (1976), 119–128; English transl., Selected Transl. Math. Statist. and Probab., vol. 15, Amer. Math. Soc., Providence, RI, 1981, pp. 111–118.

[Co1]   Lawrence A. Corwin, *A "functional equation" for measures and a generalization of Gaussian measures*, Bull. Amer. Math. Soc. **75** (1969), 829–832.

[Co2]   _____, *Generalized Gaussian measures and a "functional equation". I*, J. Funct. Anal. **5** (1970), 412–427.

[Co3]   _____, *Generalized Gaussian measures and a "functional equation". II*, J. Funct. Anal. **6** (1970), 481–505.

[Co4]   _____, *Generalized Gaussian measures and a "functional equation". III. Measures on $\mathbf{R}^n$*, Adv. Math. **6** (1971), 239–251.

[F1]   G. M. Fel'dman, *On the decomposition of a Gaussian distribution on groups*, Teor. Veroyatnost. i Primenen. **22** (1977), no. 1, 136–143; English transl. in Theory Probab. Appl. **22** (1977).

[F2]   _____, *The generalized Poisson distribution on groups*, Zap. Nauchn. Sem. Leningrad. Otdel. Mat. Inst. Steklov. (LOMI) **72** (1977), 161–185; English transl. in J. Soviet Math. **23** (1984), no. 3.

[F3]   _____, *On Gaussian distributions on locally compact abelian groups*, Teor. Veroyatnost. i Primenen. **23** (1978), no. 3, 548–563; English transl. in Theory Probab. Appl. **23** (1978).

[F4]   _____, *On a decomposition of the generalized Poisson distribution on groups*, Teor. Veroyatnost. i Primenen. **27** (1982), no. 4, 725–738; English transl. in Theory Probab. Appl. **27** (1982).

[F5]   _____, *On Urbanik's characterization of Gaussian measures on locally compact abelian groups*, Studia Math. **73** (1982), 81–86.

[F6]   _____, *The generalized Poisson distribution of the class I_0 on groups*, Teor. Veroyatnost. i Primenen. **29** (1984), no. 2, 222–233; English transl. in Theory Probab. Appl. **29** (1984).

[F7]   _____, *Infinitely divisible distributions of the class I_0 on groups*, Teor. Veroyatnost. i Primenen. **30** (1985), no. 3, 449–461; English transl. in Theory Probab. Appl. **30** (1985).

[F8]   _____, *Gaussian distributions in the sense of Bernstein on groups*, Teor. Veroyatnost. i Primenen. **31** (1986), no. 1, 47–58; English transl. in Theory Probab. Appl. **31** (1986).

[F9]   _____, *A characterization of a Gaussian distribution on abelian groups*, Teor. Veroyatnost. i Primenen. **32** (1987), no. 3, 623–626; English transl. in Theory Probab. Appl. **32** (1987).

[F10]   _____, *The Polya characterization of a Gaussian measure on groups*, Studia Math. **87** (1987), 9–21.

[F11]   _____, *Characterization of Gaussian distribution on groups by the independence of linear statistics*, Dokl. Akad. Nauk SSSR **301** (1988), 558–560; English transl. in Soviet Math. Dokl. **38** (1989).

[F12]   _____, *Gaussian measures in Urbanik's sense and a characterization theorem for abelian groups*, Studia Math. **90** (1988), 165–174.

[F13]   _____, *Marcinkiewicz and Lukacs theorems on abelian groups*, Teor. Veroyatnost. i Primenen. **34** (1989), no. 2, 330–339; English transl. in Theory Probab. Appl. **34** (1989).

[F14]   _____, *On a characterization of the Gaussian distribution on groups by the equidistribution of the monomial and linear statistics*, Ukrain. Mat. Zh. **41** (1989), no. 8, 1112–1118; English transl. in Ukrainian Math. J. **41** (1989).

[F15]   _____, *On the groups analogue of the Khinchin theorem*, Analytic Methods in Probability and Operator Theory, Naukova Dumka, Kiev, 1990. (Russian)

[F16]   _____, *On groups that admit characterization of the Gaussian distribution by the identical distribution of a term and linear statistics*, Ukrain. Mat. Zh. **42** (1990), no. 1, 139–142; English transl. in Ukrainian Math. J. **42** (1990).

[FFr1]   G. M. Fel'dman and A. E. Fryntov, *On the decomposition of a convolution of two Poisson distributions on locally compact abelian groups*, Teor. Veroyatnost. i Primenen. **27** (1981), no. 3, 612–618; English transl. in Theory Probab. Appl. **27** (1981).

[FFr2]   _____, *On the decomposition of the convolution of a Gaussian and Poisson distribution on locally compact abelian groups*, J. Multivariate Anal. **13** (1983), 148–166.

[Fo]   Gunnar Forst, *Convolution semigroups of local type*, Math. Scand. **34** (1974), 211–218.

[Fr1]   A. E. Fryntov, *Factorization of compositions of a countable number of Poisson laws*, Mat. Sb. **99** (1976), no. 2, 176–191; English transl. in Math. USSR-Sb. **28** (1976).

[Fuks]   B. A. Fuks, *Introduction to the theory of analytic functions of several complex variables*, Fizmatgiz, Moscow, 1962; English transl., Amer. Math. Soc., Providence, RI, 1963.

[Fuchs]   Laszlo Fuchs, *Infinite abelian groups*, vol. 1, Academic Press, New York and London, 1970.

[Ga]   J. Gajek, *On a property of normal distributions of an arbitrary stochastic process*, Czechoslovak Math. J. **8** (1958), 610–618.

[Ge]   A. O. Gel'fond, *The calculus of finite differences*, Fizmatgiz, Moscow, 1959, 2nd rev. ed.; English transl. of 4th ed., Hindustan Publishing Corp., Delhi, 1971.

[Gr]   Ulf Grenander, *Probabilities on algebraic structures*, Wiley, New York, 1963.

[HR1]   Edwin Hewitt and Kenneth A. Ross, *Abstract harmonic analysis*, vol. 1, Springer-Verlag, Berlin and Heidelberg, 1963.

[HR2] _____, *Abstract harmonic analysis*, vol. 2, Springer-Verlag, Berlin and New York, 1970.

[He1] Herbert Heyer, *Factorization of probability measures on locally compact groups*, Z. Wahrscheinlichkeitstheorie und Verw. Gebiete **8** (1967), 231–258.

[He2] _____, *Probability measures on locally compact groups*, Springer-Verlag, Berlin and New York, 1977.

[HeRa] Herbert Heyer and Christian Rall, *Gaußsche Wahrscheinlichkeitsmaße auf Corwinschen Gruppen*, Math. Z. **128** (1972), 343–361.

[JW] B. Jessen and A. Wintner, *Distribution function and the Riemann zeta-function*, Trans. Amer. Math. Soc. **38** (1935), 48–88.

[KLR] A. M. Kagan, Yu. V. Linnik, and C. R. Rao, *Characterization problems in mathematical statistics*, Wiley, New York, 1973.

[Ka] Shizuo Kakutani, *On equivalence of infinite product measures*, Ann. of Math. (2) **49** (1948), 214–224.

[Kau] Robert P. Kaufman, *Some measures determined by mappings of the Cantor set*, Colloq. Math. **19** (1968), 77–83.

[Kud1] L. S. Kudina, *Indecomposable laws with a preassigned spectrum*, Teor. Funktsii Funktsional. Anal. i Prilozhen. No. 16 (1972), 206–212; English transl., Selected Transl. Math. Statist. and Probab., vol. 14, Amer. Math. Soc., Providence, RI, 1977.

[Kud2] _____, *The closure of the set of indecomposable distributions with a fixed spectrum*, Teor. Funktsii Funktsional. Anal. i Prilozhen. No. 17 (1973), 51–56; English transl., Selected Transl. Math. Statist. and Probab., vol. 15, Amer. Math. Soc., Providence, RI, 1981.

[Kur] K. Kuratovskiĭ, *Topology*, vol. 1, Academic Press, New York, 1968.

[Le] B. Ya. Levin, *Distribution of zeros of entire functions*, GITTL, Moscow, 1956; English transl., Amer. Math. Soc., Providence, RI, 1980, 2nd rev. ed..

[Lévy1] Paul Lévy, *Sur l'arithmétique des lois de probabilités enroulées*, C.R. Séances et Conf. Soc. Math. France **1938** (1939), 32–34.

[Lévy2] _____, *Sur les exponentielles de polynômes et sur l'arithmétique des produits finis de lois de Poisson*, Ann. Sci. École Norm. Sup. **54** (1937), 231–292.

[LiSa] Chung Lin and Sadahiro Saeki, *Bernoulli convolutions in LCA groups*, Studia Math. **58** (1976), 165–177.

[LivOCh] L. Z. Livshits, I. V. Ostrovskiĭ, and G. P. Chistyakov, *The arithmetic of probability laws*, Itogi Nauki i Tekhniki: Teor. Veroyatnost., Mat. Statist., Teoret. Kibernet., vol. 12, VINITI, Moscow, 1975, pp. 5–42; English transl. in J. Soviet Math. **6** (1976), no. 2.

[LinO] Yu. V. Linnik and I. V. Ostrovskiĭ, *Decomposition of random variables and vectors*, "Nauka", Moscow, 1972; English transl., Amer. Math. Soc., Providence, RI, 1977.

[Lu] Eugene Lukacs, *Characteristic functions*, Hafner, New York; Griffin, London, 1970.

[Mac] George W. Mackey, *Induced representations of locally compact groups. I*, Ann. of Math. (2) **55** (1952), 101–139.

[Marc] J. Marcinkiewicz, *Sur les variables aléatoires enroulées*, C.R. Séances et Conf. Soc. Math. France **1938** (1939), 34–36.

[Mark] A. I. Markushevich, *Theory of analytic functions*, vol. 2, "Nauka", Moscow, 1968, 2nd ed.; English transl. of 1st ed., Prentice-Hall, Englewood Cliffs, NJ, 1965.

[O1] I. V. Ostrovskiĭ, *Decompositions of infinitely divisible laws without a Gaussian component*, Dokl. Akad. Nauk SSSR **161** (1965), 48–51; English transl. in Soviet Math. Dokl. **6** (1965).

[O2] _____, *The arithmetic of probability distributions*, J. Multivariate Anal. **7** (1977), 475–490.

[O3] _____, *On the divisors of infinitely divisible distributions admitting a Cartesian product representation*, Teor. Veroyatnost. i Primenen. **27** (1982), no. 4, 772–777; English transl. in Theory Probab. Appl. **27** (1982).

[O4] _____, *The arithmetic of probability distributions*, Teor. Veroyatnost. i Primenen. **31** (1986), no. 1, 3–30; English transl. in Theory Probab. Appl. **31** (1986).

[Par]   R. P. Pakshirajan, *On analogue of Kolmogorov's three-series theorem for abstract random variables*, Pacific J. Math. **13** (1963), 639–646.

[P]   K. R. Parthasarathy, *Probability measures on metric spaces*, Academic Press, New York and London, 1967.

[PRV1]   K. R. Parthasarathy, R. R. Rao, and S. R. S. Varadhan, *Probability distributions on locally compact abelian groups*, Illinois J. Math. **7** (1963), 337–369.

[PRV2]   _____, *On the category of indecomposable distribution on topological groups*, Trans. Amer. Math. Soc. **102** (1962), 200–217.

[PS]   K. R. Parthasarathy and V. V. Sazonov, *On the representation of infinitely divisble distributions on a locally compact abelian group*, Teor. Veroyatnost. i Primenen. **9** (1964), no. 1, 118–122; English transl. in Theory Probab. Appl. **9** (1964).

[Ra]   D. A. Raĭkov, *On decomposition of Gauss and Poisson distributions*, Izv. Akad. Nauk SSSR Ser. Mat. **2** (1938), 91–124. (Russian)

[Ram]   B. Ramachandran, *Advanced theory of characteristic functions*, Statistical Publishing Society, Calcutta, 1967.

[RR]   A. P. Robertson and W. J. Robertson, *Topological vector spaces*, Cambridge Univ. Press, New York, 1964.

[Ro]   L. I. Ronkin, *Introduction to the theory of entire functions of several variables*,"Nauka", Moscow, 1971; English transl., Amer. Math. Soc., Providence, RI, 1974.

[Rud]   Walter Rudin, *Independent perfect sets in groups*, Michigan Math. J. **5** (1958), 159–161.

[Ruh1]   A. L. Rukhin, *A certain theorem of S. N. Bernstein*, Mat. Zametki **6** (1969), 301–307; English transl. in Math. Notes **6** (1969).

[Ruh2]   _____, *The Poisson law on groups*, Litovsk. Mat. Sb. **10** (1970), no. 3, 537–543. (Russian)

[Ruh3]   _____, *Certain statistical and probability problems on groups*, Trudy Mat. Inst. Steklov. **111** (1970), 52–109; English transl. in Proc. Steklov Inst. Math. **111** (1972).

[RS]   Imre Z. Ruzsa and Gabor J. Szekeley, *Algebraic probability theory*, Wiley, Chichester, 1988.

[Sal]   Raphael Salem, *Algebraic numbers and Fourier analysis*, Heath, Boston, 1963.

[Se]   José Sebastião e Silva, *Su certe classi di spazi localmente convessi importanti per le applicazioni*, Rend. Mat. e Appl. (5) **14** (1955), 388–410.

[Si]   Eberhard Siebert, *Einige Bemerkungen zu den Gauss-Verteilungen auf lokalkompakten abelschen Gruppen*, Manuscripta Math. **14** (1974), 41–55.

[ST]   V. V. Sazonov and V. N. Tutubalin, *Probability distributions on topological groups*, Teor. Veroyatnost. i Primenen. **11** (1966), no. 1, 3–55; English transl. in Theory Probab. Appl. **11** (1966).

[Sr]   Yu. A. Šreider, *The structure of maximal ideals in rings of measures with convolution*, Mat. Sb. **27** (1950), 297–318; English transl. in Amer. Math. Soc. Transl. (1) **8** (1962).

[Ur]   K. Urbanik, *Gaussian measures on locally compact abelian topological groups*, Studia Math. **19** (1960), 77–88.

[UU1]   V. G. Ushakov and N. G. Ushakov, *Indecomposable probability distributions on groups*, Teor. Veroyatnost. i Primenen. **29** (1984), no. 2, 348–351; English transl. in Theory Probab. Appl. **29** (1984).

[UU2]   _____, *Decomposition of mixtures of probability distributions*, Teor. Veroyatnost. i Primenen. **31** (1986), no. 2, 369–372; English transl. in Theory Probab. Appl. **31** (1986).

[Ush]   A. P. Ushakova, *Stability of decomposition of probability laws*, Teor. Veroyatnost. i Primenen. **28** (1983), no. 3, 572–574; English transl. in Theory Probab. Appl. **28** (1983).

[V]   A. M. Vershik, *A spectral and metrical isomorphism of normal dynamical systems*, Dokl. Akad. Nauk SSSR **144** (1962), 255–257; English transl. in Soviet Math. Dokl. **3** (1962).

[VS]   A. M. Vershik and V. N. Sudakov, *Probability measures in infinite-dimensional spaces*, Zap. Nauchn. Sem. Leningrad. Otdel. Mat. Inst. Steklov. (LOMI) **12** (1969), 7–67. (Russian)

[Z1]   V. M. Zolotarev, *General theory of the multiplication of random variables*, Dokl. Akad. Nauk SSSR **142** (1962), 788–791; English transl. in Soviet Math. Dokl. **3** (1962).

[Z2]   _____, *On the problem of the stability of the decomposition of the normal law*, Teor. Veroyatnost. i Primenen. **13** (1968), no. 3, 738–742; English transl. in Theory Probab. Appl. **13** (1968).

[ZK]   V. M. Zolotarev and V. M. Kruglov, *Structure of infinitely divisible distributions on a locally bicompact abelian group*, Teor. Veroyatnost. i Primenen. **20** (1975), no. 4, 712–724; English transl. in Theory Probab. Appl. **20** (1975).

Notation

$A(Y, G)$ - annihilator of a subgroup, 7

$A_f(r)$, 19, 20

$A_1 + A_2$ - arithmetic sum, 5

$\beta_d(\mu, \varepsilon)$, 181

$\beta_{d_1, d_2}(\mu, \varepsilon)$, 182

$\mathscr{A}(X)$, 152

B_r, 20

$\mathscr{B}(X)$ - σ-algebra of Borel subset of X, 10

C_X - component of zero of the group X, 7

$\mathbb{C}$ - complex plane, 2

$\Gamma(X)$ - Gaussian distributions on a group X, 34

$\Gamma^s(X)$ - symmetric Gaussian distributions on a group X, 34

$\Gamma_A(X)$, 151

$\Gamma_B(X)$ - Gaussian distributions in the Bernstein sense on a
           group X, 121

$\Gamma_B^s(X)$ - symmetric Gaussian distributions in the Bernstein sense on a
           group X, 121

$\Gamma_U(X)$ - Gaussian distributions in the Urbanik sense on a group X, 52

$\Gamma_\infty(X)$, 165

$D(X)$ - degenerate distributions on a group X, 11

$D(\mu)$, 58

$\dim X$ - dimension of a group X, 8

Δ_p - additive group of all p-adic integers, 6

Δ_h - finite difference operator, 145

E_x - degenerate distribution, 11

E - (mathematical) expectation, 121

$e(\Phi)$, 15

$F(N)$ - set of all divisors of elements of N, 11

f_n, 5

$f_1 * f_2$, 82

f^{*n}, 82

$I(X)$ - idempotent distributions on a group X, 14

$I_A(X)$, ·152

$I_B(X)$, 122

$I_B^s(X)$, 129

$I_\infty(X)$, 165

I_0, 24

$\| k \|$, 20

L_h, 53

$M(A)$, 58

$M^+(A)$, 58

M_x, 7

$M_f(r)$, 18

m_K - the Haar distribution on a group K, 14

$\mathscr{M}_+(X)$ - finite measures on a group X, 10

$\mathscr{M}^1(X)$ - distributions on a group X, 10

$\hat{\mu}(y)$ - the characteristic function of the measure μ, 12

$\mu * \nu$ - convolution of measures, 10

μ^{*n}, 10

$\bar{\mu}$, 10

$(n)A$, 5

$\tilde{p}$, 9

$p(\mu)$, 12

$\mathbb{Q}$ - additive group of rational numbers, 6

$\mathbb{R}$ - additive group of real numbers, 6

$\mathbb{R}^\infty$ - space of all real sequences, 37

$\mathbb{R}_0^\infty$ - space of all finite real sequences, 37

$r(G)$ - rank of the group G, 8

$\Sigma_{\mathbf{a}}$ - $\mathbf{a}$-adic solenoid, 6

$\sigma(\mu)$ - the support of the measure μ, 10

$\mathbb{T}$ - the group of rotations of the unit circle, 6

$\mathbb{T}^\infty$, 40

X^* - the group of characters of the group X, 6

X^∞, 5

X_0 - the set of all compact elements of a group X, 7

X^n, 5

X^{n}, 5

$X^{(n)}$, 5

$\mathsf{P}_{i \in I} X_i$, 5

$[x]$, 5

(x, y), 6

(x, Φ, φ), 17

$\chi_A(x)$, 31

$\mathbb{Z}$ - additive group of integers, 6

$\mathbb{Z}(n)$ - multiplicative group of primitive roots of degree n, 6

$\mathbb{Z}(p^{\infty})$, 6

$\mathbb{Z}_0^{\infty}$, 40

$\approx$ - topological isomorphism of groups, 6

Subject Index

Annihilator of a subgroup, 7
Charge, 18
Adjoint homomorphism, 9
Convolution of measures, 10
Distribution, 10
- degenerate, 11
- Gaussian, 33
- - in the Bernstein sense, 121
- - in the Urbanik sense, 52
- - symmetric, 33
- generalized Poisson, 57
- idempotent, 14
- indecomposable, 11
- infinitely divisible, 15
- probability, 10
Divisor, 11
Element
- compact, 7
- dependent on elements, 53
- infinitely divisible, 52
Function
- characteristic, 12
- entire, 18
- entire characteristic, 21
- entire of exponential type, 19
- positive definite, 12
Group
- Corwin, 122
- periodic, 5
- torsion-free, 5
Matrix symmetric positive semidefinite, 36

Measure, 10
- concentrated on a set, 11
- finite, 10
- Haar, 14
- Lévy, 17
One-parameter subgroup, 9
Rank of group, 8
Set
- admissible, 135
- independent, 8
- shift-compact, 11
Structure of a measure, 40
Theorem
- Bochner-Khinchin, 12
- Cramér, 43
- Lévy, 21
- Lévy-Raïkov, 21
- Linnik, 21
- Marcinkiewicz, 43
- Ostrovskiĭ, 78
- Pólya, 151
- Skitovich-Darmois, 135

Author Index

Berg, Ch., 1, 205
Bernstein, S. N., 2, 3, 121
Bochner, S., 12
Brown, G., 209

Carathéodory, C., 19
Carnal, H., 206
Chistyakov, G. P., 2, 207
Corwin, L., 122, 123, 208
Cramér, H., 1, 2, 34, 43, 147, 206
Cuppens, R., 21, 88, 207

Darmois, G., 2, 135, 136, 145, 147
Dugué, D., 208

Fel'dman, G. M., 204–209
Forst, G., 1, 205
Fryntov, A. E., 206, 207

Hartogs, F., 20, 68
Heyer, H., 1, 3, 205, 208

Jessen, B., 209

Kagan, A. M., 2, 209
Kakutani, S., 41
Khinchin, A. I., 2, 12, 23, 24, 203
Kolmogorov, A. N., 38
Kronecker, S., 58
Kruglov, V. M., 202, 205, 209
Kudina, L. S., 204

Lévy, P., 2, 21, 69, 70, 79, 88, 181, 206
Lin, Ch., 78
Lindelöf, E., 19
Linnik, Yu. V., 2, 21, 70, 80, 86–88, 97, 151, 207–209
Livshits, L. Z., 208
Lukacs, E., 177, 178

Mackey, G. W., 23
Marcinkiewicz, J., 2, 43, 173, 206
Martin-Löf, P., 206
Moran, W., 209

Ostrovskiĭ, I. V., 2, 21, 78, 88, 206–208

Parthasarathy, K. R., 1, 2, 11, 14, 17, 203, 205, 208
Phragmén, E., 19
Pólya, G., 151, 163
Pontryagin, L. S., 6
Prokhorov, Yu. V., 114, 181

Raĭkov, D. A., 2, 21, 70, 206, 207
Rall, Ch., 3, 208
Rao Ranga, R., 2, 11, 14, 17, 203, 205, 208
Rao, C. R., 2, 209
Rudin, W., 78, 106, 208
Rukhin, A. L., 3, 206, 208
Ruzsa, I. Z., 1

Saeki, S., 78
Salem, R., 200
Sapogov, N. A., 209
Sazonov, V. V., 17, 205
Schwarz, H. A., 20
Siebert, E., 205
Skitovich, V. P., 2, 135, 136, 145, 147
Stein, K., 49
Székely, G. J., 1

Tutubalin, V. N., 205

Urbanik, K., 52, 206
Ushakov, N. G., 203
Ushakov, V. G., 203
Ushakova, A. P., 209

Varadhan, S. R. S., 2, 11, 14, 17, 203, 205, 208
Vershik, A. M., 207

Wintner, A., 209

Zolotarev, V. M., 2, 17, 202, 205, 206, 209

Recent Titles in This Series

(Continued from the front of this publication)

81 I. M. Gelfand and S. G. Gindikin, Editors, Mathematical problems of tomography, 1990

80 Junjiro Noguchi and Takushiro Ochiai, Geometric function theory in several complex variables, 1990

79 N. I. Akhiezer, Elements of the theory of elliptic functions, 1990

78 A. V. Skorokhod, Asymptotic methods of the theory of stochastic differential equations, 1989

77 V. M. Filippov, Variational principles for nonpotential operators, 1989

76 Phillip A. Griffiths, Introduction to algebraic curves, 1989

75 B. S. Kashin and A. A. Saakyan, Orthogonal series, 1989

74 V. I. Yudovich, The linearization method in hydrodynamical stability theory, 1989

73 Yu. G. Reshetnyak, Space mappings with bounded distortion, 1989

72 A. V. Pogorelev, Bendings of surfaces and stability of shells, 1988

71 A. S. Markus, Introduction to the spectral theory of polynomial operator pencils, 1988

70 N. I. Akhiezer, Lectures on integral transforms, 1988

69 V. N. Salii, Lattices with unique complements, 1988

68 A. G. Postnikov, Introduction to analytic number theory, 1988

67 A. G. Dragalin, Mathematical intuitionism: Introduction to proof theory, 1988

66 Ye Yan-Qian, Theory of limit cycles, 1986

65 V. M. Zolotarev, One-dimensional stable distributions, 1986

64 M. M. Lavrent'ev, V. G. Romanov, and S. P. Shishat·skii, Ill-posed problems of mathematical physics and analysis, 1986

63 Yu. M. Berezanskii, Selfadjoint operators in spaces of functions of infinitely many variables, 1986

62 S. L. Krushkal', B. N. Apanasov, and N. A. Gusevskii, Kleinian groups and uniformization in examples and problems, 1986

61 B. V. Shabat, Distribution of values of holomorphic mappings, 1985

60 B. A. Kushner, Lectures on constructive mathematical analysis, 1984

59 G. P. Egorychev, Integral representation and the computation of combinatorial sums, 1984

58 L. A. Aizenberg and A. P. Yuzhakov, Integral representations and residues in multidimensional complex analysis, 1983

57 V. N. Monakhov, Boundary-value problems with free boundaries for elliptic systems of equations, 1983

56 L. A. Aizenberg and Sh. A. Dautov, Differential forms orthogonal to holomorphic functions or forms, and their properties, 1983

55 B. L. Roždestvenskii and N. N. Janenko, Systems of quasilinear equations and their applications to gas dynamics, 1983

54 S. G. Krein, Ju. I. Petunin, and E. M. Semenov, Interpolation of linear operators, 1982

53 N. N. Čencov, Statistical decision rules and optimal inference, 1981

52 G. I. Èskin, Boundary value problems for elliptic pseudodifferential equations, 1981

51 M. M. Smirnov, Equations of mixed type, 1978

50 M. G. Krein and A. A. Nudel'man, The Markov moment problem and extremal problems, 1977

49 I. M. Milin, Univalent functions and orthonormal systems, 1977

48 Ju. V. Linnik and I. V. Ostrovskii, Decomposition of random variables and vectors, 1977

(See the AMS catalog for earlier titles)